Introductory Electrical
Engineering with Math
Explained in Accessible
Language

Introductory Electrical Engineering with Math Explained in Accessible Language

Magno Urbano
Independent Author
Lisbon, Portugal

Registered Office
John Wiley & Sons, Inc., 111 River Street, Hoboken, NJ 07030, USA

Editorial Office
111 River Street, Hoboken, NJ 07030, USA

For details of our global editorial offices, customer services, and more information about Wiley products visit us at www.wiley.com.

Wiley also publishes its books in a variety of electronic formats and by print-on-demand. Some content that appears in standard print versions of this book may not be available in other formats.

Library of Congress Cataloging-in-Publication Data

Names: Urbano, Magno, author.
Title: Introductory electrical engineering with math explained in
  accessible language / Magno Urbano.
Description: Hoboken, NJ : Wiley, 2020. | Includes index.
Identifiers: LCCN 2019030445 (print) | LCCN 2019030446 (ebook) | ISBN 9781119580188
  (paperback) | ISBN 9781119580225 (adobe pdf) | ISBN 9781119580201 (epub)
Subjects: LCSH: Electrical engineering–Mathematics.
Classification: LCC TK153 .U73 2020 (print) | LCC TK153 (ebook) | DDC
  621.3–dc23
LC record available at https://lccn.loc.gov/2019030445
LC ebook record available at https://lccn.loc.gov/2019030446

Cover design by: Wiley
Cover image: © Andrew Brookes/Getty Images

Set in 10/12pt Warnock by SPi Global, Pondicherry, India

Printed in the United States of America

V10014728_101519

Contents

About the Author

 With more than 25 years of experience in the field, Magno Urbano started his professional life by graduating as an Electrical Engineer from USU (Brazil) and worked in fields like computer graphics, visual effects, and programming for two of the largest broadcast companies in two continents: Globo TV Network in Brazil and Radio and Television of Portugal (RTP).

Prior to this volume, Magno has written 16 books on multimedia themes and authored nearly 100 articles and 50 multimedia courses for the most important magazines in Europe and South America.

Magno has been developing apps for Apple devices since 2008. During this period, he has developed and published about 120 applications for iPhone, iPad, macOS, and Apple TV, some of them hitting 1 or placing in the top 10 in multiple countries for several weeks.

Some of his apps can be seen at www.katkay.com.

Right now, he is excited to deliver his first book published by Wiley & Sons! Magno can be reached at apps@katkay.com.

Preface

Magno Urbano, a professional with an amazing educational background in electrical engineering and an autodidact in programming, is always trying to learn the most he can.

He worked for the Visual Effects Post Production Center of Globo Television in Brazil, revealing himself as a great professional.

He decided to move to Portugal in 2000 to expand his possibilities.

Working as an entrepreneur in Portugal, he created a lot of multimedia courses, published a dozen books, and created hundreds of apps for Apple devices.

In everything he does, he always shows competence, know-how, qualification, and knowledge.

The good persons are always those who make the difference.

Taulio Mello
Visual Effects Artist, former VFX specialist for TV Globo

I know Magno Urbano for two decades.

In the areas that he has worked, like in multimedia, visual effects, Internet, and software development and in the publishing world, I have always recognized in Magno Urbano a huge effort and a notable dedication.

The enthusiasm and quality he applies to his works always lead to something that, without doubt, will be of great interest to those who like this branch of universal culture.

Fernando Salgado
Former Responsible for the Post-Production Center of RTP
(Radio and Television of Portugal)

Acknowledgement

Thanks to my mother and sister and to people inside Wiley & Sons, specially Brett Kurzman, Victoria Bradshaw, Blesy Regulas, and Grace Paulin who made this project possible.

Introduction

1 Knowledge is Everything

We live in a world where technology is advancing at an astonishing speed.

We have arrived at this point in knowledge, because man always had this curiosity about the things around.

At first, it was a matter of survival, but then curiosity turned into discovery, and we have started to unveil the secrets of the universe.

The human race produced geniuses like Isaac Newton, Copernicus, Galileo Galilei, Alessandro Volta, Michael Faraday, Albert Einstein, Pierre-Simon Laplace, Carl Friedrich Gauss, Richard Feynman, and so many others, who created the pillars of knowledge and contributed to the various branches of science, including physics and mathematics, and in fields that are pertinent to this book, such as electromagnetism and electricity.

In this book, we try to present knowledge in a simple and straightforward way and in an accessible language, using as many illustrations and examples as possible. Unlike other similar books, we provide the mathematical theory necessary to the reader, so every theme can be fully understood.

Many electrical engineering teachers complain that students arrive at their courses without the previous mathematical skills required to understand the theories. This book tries to fill this gap, by providing such mathematical skills every time they are necessary.

We invite the reader to this journey.

Conventions

Used by this Book

1 Introduction

In this chapter, we will examine the several conventions used in this book.

2 Equations

In mathematics, several conventions are used to show the product or multiplications of two elements like

$A \times B$

$A \cdot B$

AB

$(A)(B)$

Every one of these conventions represents value A multiplied by value B.

In this book we will use all these forms, the one that makes more sense for the context being explained.

3 Electric Schematics

3.1 Connection

Every time a small black circuit is drawn in the intersection between two elements, it means that these elements are electrically connected.

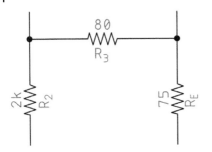

Figure 0.1 Connected components.

In Figure 0.1, resistor R_3 is electrically connected to resistor R_2 in one side and to RE on the other.

3.2 Unspecified Value

If a component does not have its value specified in a schematic in this book, the following rules must be followed: every resistor is always expressed in Ohms, every capacitor in Farads, every inductor in Henries, and every power supply in Volts.

All resistors in Figure 0.1, for example, have their values in Ohms.

4 Mathematical Concepts

Unlike other works, this book provides the reader with the mathematical concepts necessary for understanding a given segment or the development of an equation.

In the following example, we see an equation followed by a mathematical concept. The equation following the concept is a result of the concept applied:

$$du = d[V_1] - d\left[\frac{Q}{C}\right]$$

MATHEMATICAL CONCEPT   The derivative of a constant is 0.

$$du = d[\cancel{V_1}]^0 - d\left[\frac{Q}{C}\right]$$

5 Examples and Exercises

This book contains several examples and exercises.

The examples are used to illustrate the theory and are resolved promptly. The exercises should be done by the readers. The endings of the chapters contain the answers to the exercises.

1

Scientific Method

General Concepts

1.1 Introduction

In this chapter, we will examine the scientific method and its concepts. These concepts must be well understood and are fundamental to achieve a solid foundation in electrical engineering.

1.2 Powers of 10

In science it is common to represent numbers like 100, 1000, and 10000, as 10^2, 10^3, and 10^4, respectively, in what is called powers of ten.

Powers of 10 show how many times 10 has to be multiplied by itself to obtain the final value.

In other words,

$$10^4 = 10 \times 10 \times 10 \times 10$$

is equal to 10 multiplied by itself 4 times.

However, if we want to represent very small numbers like 0.1, 0.01, and 0.001, we have to use negative exponents, as 10^{-1}, 10^{-2}, 10^{-3}, respectively.

This negative exponent represents how many times the number must be divided by 10.

Numbers like
10^{-1}, 10^{-2}, 10^{-3} represent
1/10, 1/100, 1/100, respectively.

Any number raised to 0 is equal to 1. Thus, $10^0 = 1$.

Introductory Electrical Engineering with Math Explained in Accessible Language,
First Edition. Magno Urbano.
© 2020 John Wiley & Sons, Inc. Published 2020 by John Wiley & Sons, Inc.

1.3 Roots

Roots are the inverse operation of powers.
Roots are expressed in scientific notation as a fractional exponent:

$$\sqrt[2]{5} = 5^{\frac{1}{2}}$$

$$\sqrt[3]{5} = 5^{\frac{1}{3}}$$

$$\sqrt[4]{5} = 5^{\frac{1}{4}}$$

$$\sqrt[4]{5^2} = 5^{\frac{2}{4}}$$

Any number raised to a fractional exponent in the form $x^{\frac{m}{n}}$ is equal to $\sqrt[n]{x^m}$.

1.4 Scientific Notation as a Tool

In science, we constantly have to deal with numbers that are very hard to write in natural form, for example, 0.00000000003434033323. Imagine how hard it would be to compare this number with 0.00000000000000212817716 and know which one is bigger.

To solve this problem and make it easy to compare and write huge or minuscule numbers, a methodology called scientific notation was born.

1.4.1 Very Large Numbers

To convert a large number like 324484738, for example, to scientific notation, we divide this number by 10, several times, until we get a nonzero number with just one digit in the integer part.
Example:

324484738	÷	10	=	32448473.8
32448473.8	÷	10	=	3244847.4
3244847.38	÷	10	=	324484.74
324484.738	÷	10	=	32448.474
32448.4738	÷	10	=	3244.8474
3244.84738	÷	10	=	324.48474
324.484738	÷	10	=	32.448474
32.4484738	÷	10	=	3.2448474

These numbers show that we have to divide the initial number 8 times by 10 to get a single digit alone in the integer part, like 3.24484738.

For that reason, the given number in scientific notation is represented as 3.24484738×10^8.

1.4.2 Very Small Numbers

To convert a very small number to scientific notation, we must multiply the number by 10 several times, until we get a single-digit number in the integer part. The number of times the multiplication happens will be the exponent.

In order to convert, for example, 0.00156 to scientific notation, it would be necessary to multiply it 3 times by 10 to obtain a single nonzero-digit number:

0.00156	×	10	=	0.0156
0.0156	×	10	=	0.156
0.156	×	10	=	1.56

For that reason, 0.00156 is represented in scientific notation as 1.56×10^{-3}.

Like we said before, negative exponents mean division. So, 1.56×10^{-3} can be also written as

$$\frac{1.56}{10^3} = \frac{1.56}{1000}$$

That is exactly 0.00156.

1.4.3 Operations with Powers of 10

Operations like multiplication, division, etc. can be done with numbers represented in powers of 10.

This is how they are done.

1.4.3.1 Multiplication

To multiply two numbers, for example, 1.56×10^{-3} and 3.33×10^8, we must first multiply the numbers without the powers

$$1.56 \times 3.33 = 5.19$$

and then sum the exponents

$$-3 + 8 = 5$$

The result would be 5.19×10^5.

If the final result is not a number with a single nonzero digit in the integer part, we must make sure it is.

A final result of 24×10^{11} should be corrected to 2.4×10^{12}.

1.4.3.2 Division

As explained before, any number raised to a negative exponent is equal to that number divided by the same power:

$$1.56 \times 10^{-3} = \frac{1.56}{10^3} = \frac{1.56}{1000} = 0.00156$$

If we want to divide 3×10^9 by 8×10^2, we can represent that division as

$$\frac{3 \times 10^9}{8 \times 10^2}$$

The denominator can be converted to a nominator if we invert the exponent

$$\frac{3 \times 10^9}{8 \times 10^2} = \left(3 \times 10^9\right) \times \left(8 \times 10^{-2}\right)$$

and we just converted a division into a multiplication.

We can now multiply the numbers without the powers obtaining

$$8 \times 3 = 24$$

and add the powers

$$9 + \left(-2\right) = 7$$

and the final result will be 24×10^7, which should be converted to a single-digit number equal to 2.4×10^8.

1.4.3.3 Adding and Subtracting

To add or subtract numbers represented in powers of 10, we must first make the numbers have the same exponent.

To add, for example, 2.4×10^8 and 3×10^9, we must first make both numbers have the same base. We chose 10^8. That will make the second number change to 30×10^8. We can now add both numbers:

$$2.4 + 30 = 32.4$$

The final result will be 32.4×10^8, corrected to 3.24×10^9.

1.4.3.4 Raising to a Number

To raise a number expressed in powers of 10 to a given exponent, we must raise the number without the power and then multiply the power by the exponent.

The cube of 2.4×10^8 will be equal to the number alone raised to the cube

$$2.4^3 = 13.824$$

and the base multiplied by the exponent

$$8 \times 3 = 24$$

The final result will be 13.824×10^{24}, corrected to 1.3824×10^{25}.

1.4.3.5   Expressing Roots as Exponents

All roots in the form $\sqrt[n]{a^m}$ can be expressed as an exponent in the form $a^{\frac{m}{n}}$. For example,

$$\sqrt{3} = 3^{\frac{1}{2}}$$

$$\sqrt[3]{7} = 7^{\frac{1}{3}}$$

$$\sqrt{8^5} = 8^{\frac{5}{2}}$$

1.4.3.6   Extracting the Root

To extract a root of a number expressed in powers of 10, we have to raise the number alone, without the power, to the correspondent exponent and multiply the power by that exponent.

To calculate the square root of

$$\sqrt{2.4 \times 10^8}$$

we first have to write the number in the exponent form

$$\left(2.4 \times 10^8\right)^{\frac{1}{2}}$$

We then extract the root of the number alone

$$\sqrt{2.4} = 1.549$$

and then multiply the base by the exponent

$$8 \times \frac{1}{2} = 4$$

The result is 1.549×10^4.

If the root is cubical, like

$$\sqrt[3]{2.4 \times 10^9}$$

we first extract the cubical root of the number alone

$$\sqrt[3]{2.4} = 1.338$$

and multiply the base by the equivalent exponent

$$9 \times \frac{1}{3} = 3$$

The result is 1.338×10^3.

A problem arises if we need to extract a root of a number expressed in powers of 10 that is raised to another number, like

$$\sqrt[3]{2.4 \times 10^8}$$

This root can be written as

$$\left(2.4 \times 10^8\right)^{\frac{1}{3}}$$

As usual, we could start by extracting the root of the number alone and then multiplying the base by the exponent.

But when we do that, we have an exponent that is 8 divided by 3, resulting in $10^{8/3}$ or $10^{2.666}$, which is not a very pretty result.

To improve the result, we have to convert the exponent first to something that can be divided by 3. The closest candidate is 9.

Hence, we convert 2.4×10^8 into 0.24×10^9, that is exactly the same number written differently.

Now we can proceed. We first extract the cubical root of the number alone

$$\sqrt[3]{0.24} = 0.6214465$$

and multiply the base by the exponent

$$9 \times \frac{1}{3} = 3$$

The result is 0.6214465×10^3, which must be corrected to 6.214465×10^2.

1.4.4 Computers and Programming

Numbers written in scientific notation inside computer programs and other computer contexts use a specific syntax.

A number like 1.38×10^{25}, for example, is normally written as 1.38E25 or 1.38e25.

Numbers with negative exponents have the sign before the number, for example, 1.38e – 25 or –3.43e – 12.

In almost all the cases, computer languages use decimal numbers as fraction number separators.

1.4.5 Engineering Notation

In engineering, scientific notation numbers are expressed with exponents of 10 that are multiples of 3, so they match one of the multiples of the International System (see Appendix A).

Following that rule, a number like 2×10^{-7} must be converted to 0.2×10^{-6} or 200×10^{-9}. Normally it is always preferred to have numbers with first digits nonzero. For that reason, the later form will be generally preferred. In this case, the term 10^{-9} is equivalent to the prefix nano. So, the given number will be written as 200 nano or 200 n.

2

Infinitesimal Calculus

A Brief Introduction

2.1 Introduction

In this chapter we will do a brief introduction to infinitesimal calculus or differential and integral, known as simply, calculus.

Wikipedia has as good definition about calculus:

> *Calculus is the mathematical study of continuous change, in the same way that geometry is the study of shape and algebra is the study of generalizations of arithmetic operations.*

> *It has two major branches, differential calculus (concerning instantaneous rates of change and slopes of curves) and integral calculus (concerning accumulation of quantities and the areas under and between curves). These two branches are related to each other by the fundamental theorem of Calculus. Both branches make use of the fundamental notions of convergence of infinite sequences and infinite series to a well-defined limit.*

> *Generally, modern calculus is considered to have been developed, independently, in the 17th century by Isaac Newton and Gottfried Wilhelm Leibniz.*

> *Today, Calculus has widespread uses in science, engineering, and economics.*

> *Calculus is a part of modern mathematics education.*

2.2 The Concept Behind Calculus

Suppose we have a car traveling in a straight line for 1 h, at a constant speed of 100 km/h, and later reducing the speed in half and traveling for another hour. What is the average speed of that car?

Introductory Electrical Engineering with Math Explained in Accessible Language,
First Edition. Magno Urbano.
© 2020 John Wiley & Sons, Inc. Published 2020 by John Wiley & Sons, Inc.

The answer is

$$\frac{100\,km/h + 50\,km/h}{2\,h} = 75\,km/h$$

Now let us see a more complex problem.

The car travels 15 min at a nonconstant speed of 15 km/h, stops for 5 min, travels for 5 km at a nonconstant speed of 8 km/h, stops again, and then travels a distance of 2 km at 15 km/h. What is the average speed now?

The answer is not evident because the method we have cannot deal with variable entities. Another method is required to perform this calculation.

This is exactly the kind of calculation that is possible using infinitesimal calculus.

2.2.1 Limits

Limits are one tool of calculus that helps to calculate values that would be impossible by other methods.

Consider the function

$$f(x) = \frac{4x - 4}{4x - 4}$$

Apparently, we can have the temptation to say that this function would be equivalent to 1, because both nominator and denominator are the same.

The problem arises if we make x = 1, making the denominator equal to 0 and producing an indefinite result.

If we plot the graph for this function, we get a horizontal line and an indefinite result when x = 1 (see Figure 2.1).

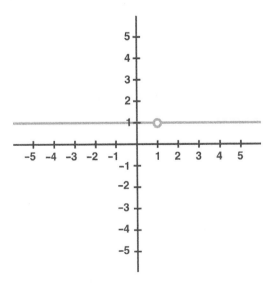

Figure 2.1 Function plot.

We can say that the function is indefinite for x = 1, but how will the function behave for values very close to 1, like 0.9?

If this is the case, we get

$$f(x) = \frac{4x - 4}{4x - 4}$$

$$f(x) = \frac{4(0.9) - 4}{4(0.9) - 4}$$

$$f(x) = \frac{-0.4}{-0.4}$$

$$f(x) = 1$$

And what happens if x is even closer to 1, like 0.999999999999? The result will be the same, 1.

The same will be true if for values of x are bigger but closer to 1, like 1.0000000000000001, then the result would continue to be 1.

So, we conclude that the function is equal to 1, for any value closer to 1.

It is like getting closer and closer to a dangerous point but never reaching it. This is the exact definition of limit.

For the given function, we can say the same thing using the following mathematical notation:

$$\lim_{x \to 1} f(x) = \frac{4x - 4}{4x - 4} = 1$$

the limit of f(x) when x approaches 1.

2.2.2 Derivatives

Derivatives are the second tool of calculus.

Suppose we want to calculate the slope of the curve seen in Figure 2.2.

Figure 2.2 A random curve.

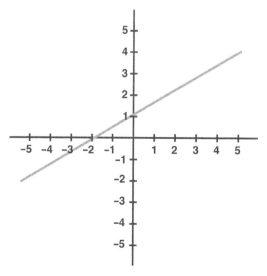

To calculate the slope, we can simply select two points at random, A and B, like shown in Figure 2.3, measure how much these points vary horizontally in X and vertically in Y, and find the slope by dividing the variation in Y by the variation in X.

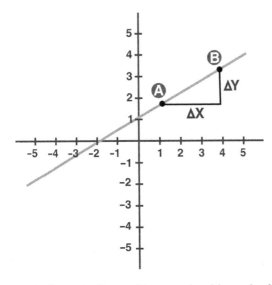

Figure 2.3 The slope of a curve.

Mathematically speaking, we should use the following formula for the slope:

$$s = \frac{\Delta Y}{\Delta X}$$

The slope of a line is easy, but imagine a function like $f(x) = \sin(x)$. How do we calculate the slope of this curve if the slope varies constantly in time? See Figure 2.4.

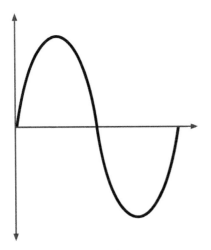

Figure 2.4 Sinusoidal curve.

Obviously, it is clear that such curve has multiples slopes, depending on the point we select (see Figure 2.5). It is also clear that the curve has no slope per se. Only points belonging to that curve will have slopes.

Figure 2.5 Slope for a particular point on a curve.

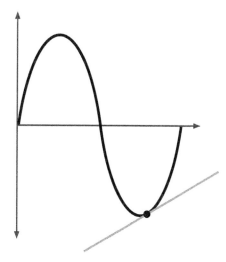

Because the function varies in time, points on the curve will represent instantaneous values for a time in particular.

To calculate the slope for a particular point on the curve, we can use the same method as before. We first define a second point on the curve, B, near the first one, as seen in Figure 2.6.

Figure 2.6 A second point on the curve.

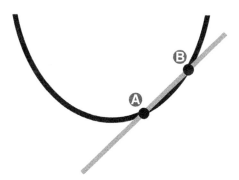

Now we can calculate the difference between both points, in terms of X and Y, as seen in Figure 2.7, and calculate the slope using the same formula as before.

The problem with this kind of approach is that the result is just an approximation, because the formula we have calculates the slope of a line between A and B and the line is not the same as the curve, in terms of curvature.

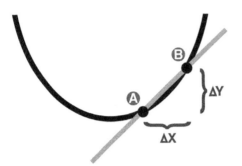

Figure 2.7 The variation of X and Y.

We could improve the result by choosing a second point closer to the first one, as shown in Figure 2.8. By reducing the distance between the points, we make the line connecting these points closer, in terms of curvature, to the curve itself. The slope calculation now will give us a result closer to the real value. But we can improve that!

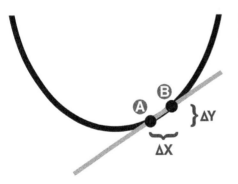

Figure 2.8 A line between point A and a second point closer.

If bringing the points together improves the result, how close can we get to have the correct result? The answer is: we can make the points infinitely close, and that is the exact definition of derivatives.

Mathematically, the slope of point A can be defined as the difference in Y divided by the difference in X, which in other words is equal to

$$s = \frac{f(A + \Delta Y) - f(A)}{\Delta X}$$

f(A) is the function value at point A.

If we make $\Delta X = \Delta Y$, we can use a generic Greek letter β to represent this variation.
Thus,

$$s = \frac{f(A + \beta) - f(A)}{\beta}$$

To get maximum precision, we have to make sure that the distance β is infinitely closer to 0, which will bring points A and B closer.
We can, indeed, use the definition of limits to help us:

$$s = \lim_{\beta \to 0} \frac{f(A + \beta) - f(A)}{\beta}$$

That is the slope of function f(x) as β gets closer to 0.

2.2.3 Integral

Integral is the third and last tool of calculus.
To understand what Integral is, consider the following curve shown in Figure 2.9.

Figure 2.9 A random curve represented on the X and Y axes.

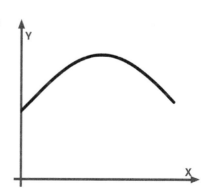

Imagine this curve represents some kind of variation, like the temperatures of a city during a year, and we want to calculate the area below the curve, shaded area in Figure 2.10.

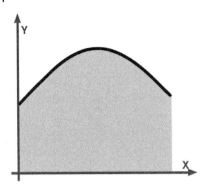

Figure 2.10 Area below the curve.

To perform such calculation, we can use a technique similar to the one we used with derivatives. We can divide the area into rectangles, calculate the area of each rectangle, and sum them all, like shown in Figure 2.11.

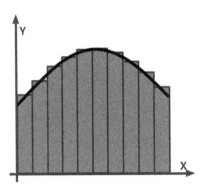

Figure 2.11 Rectangles being used to calculate the area under a curve.

Mathematically, the total area A equivalent to the sum of all small areas is equal to

$$A = \sum_{i=0}^{n} Ai$$

where A_i represents the areas of each rectangle.

The area under the curve can be also written in terms of f(x):

$$A = \sum_{i=0}^{n} f(x) \Delta x$$

where f(x) is the function represented by the curve and Δx is the width of each individual rectangle.

Looking at Figure 2.11, it is clear that if we could create an infinite number of rectangles and sum all their areas, we would obtain the exact number representing the area below the curve.

Mathematically, this area (A) can be expressed as a limit:

$$A = \lim_{n \to \infty} \left(\sum_{i=0}^{n} f(x)\,\Delta x \right)$$

which is the exact definition of an integral, normally written in the following notation:

$$A = \int f(x)\,dx$$

3

Atom

Quarks, Protons, and Electrons

3.1 Introduction

In this chapter, we will make a brief introduction about the physics behind electricity and electromagnetism, starting at the atomic level.

3.2 Atoms and Quarks

The word atom comes from ancient Greek word meaning "indivisible," because it was believed at the time that the atom was the smallest part of matter and could not be divided.

At the atomic level, atoms are composed of protons, neutrons, and electrons.

Today we know that these elements are composed of more elementary elements called quarks, and known to exist in six variations called flavors: top, bottom, charm, strange, up, and down.

As far as physicists know today, quarks are themselves elements that cannot be divided. Every time you try to divide them, a new quark is created.

Protons for example, are composed of two up quarks and one down quark, and neutrons are composed of two down quarks and one up quark (Figure 3.1).

Protons have a positive electric charge and neutrons have no electric charge.

Electric charge is normally represented, in physics, by the letter q or e.

In terms of reference charge, a proton is said to have an electric charge equal to 1. This charge, generally referred as e^+, is equal to $1.602176634 \times 10^{-19}$ C.[1]

---

[1] This is the new value revised by the SI in 2018. The old value was $1.6021766208 \times 10^{-19}$ C.

Introductory Electrical Engineering with Math Explained in Accessible Language,
First Edition. Magno Urbano.

Figure 3.1 A proton and its quarks.

3.3 Electrons

Electrons are stable subatomic particles with a negative charge that can be found in all atoms and act like charge carriers.

In terms of reference charge, an electron is said to have a charge equal to -1 or, in other words, a charge opposite to the charge of a proton. For that reason, its charge, generally referred as e^-, is equal to $-1.602176634 \times 10^{-19}$C.

In the early days, science pictured electrons spinning around atomic nucleus like planets spinning around the sun. Today we know that this image is false. Electrons oscillate frenetically around the atomic nucleus, and their position cannot be determined,[2] like they were in all positions at the same time, a kind of "energetic" cloud.

The idea of an electron orbiting an atomic nucleus like a planet around a sun is wrong because it makes us think about electrons as a solid and discrete element, like a particle. In real life, electrons behave both as particles and waves.[3]

Physicists knows that electrons occupy orbits around the nucleus at a certain distance, like they were shells. Each orbit is at a discrete level of energy and can only be occupied by electrons having that particular level of energy. If an electron gains energy, it may pass to a superior orbit. If it loses energy, it will decay to a lower orbit and release the extra energy by emitting a photon.[4]

---

2 Heisenberg's uncertainty principle.
3 The famous wave–particle duality from quantum mechanics, demonstrated brilliantly in 1802 by Thomas Young, for the first time, by the double-slit experiment.
4 The photon is a type of elementary particle, the quantum of the electromagnetic field including electromagnetic radiation such as light, and the force carrier for the electromagnetic force. The photon has zero rest mass and always moves at the speed of light within a vacuum (source Wikipedia).

Figure 3.2 Atom representation.

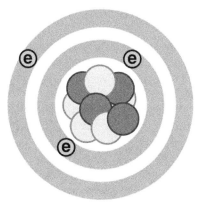

A representation of an atom is shown in Figure 3.2. The spheres at the center represent the nucleus, nothing more than an agglomeration of protons and neutrons. The circles represent the energetic orbits. Electrons are shown in the diagram as "e" inside the orbits but in real life they can be at any point and in all points at the same time inside a particular orbit.

3.4   Strong Force and Weak Force

Atoms and their protons and neutrons are kept together by powerful attraction forces.

The laws of physics say that elements with the same electric charge repel each other and that elements with opposite electric charge attract each other.

Atomic nuclei may be formed by several protons and neutrons. These protons could never be kept together with other protons, at the required close proximity, to form these atomic nuclei, simply because they would repel each other. A bigger force must exist to overpower the electric repulsion force. This force is called the strong nuclear force or strong interaction.

At a closer range, the strong force is hundreds of times as strong as electromagnetism, a million times as strong as the weak force, and a thousand times as strong as gravitation. The strong force keeps most ordinary matter together and binds neutrons and protons to create atomic nuclei.

The weak force, weak nuclear force or weak interaction, takes place only at very small, subatomic distances, less than the diameter of a proton. It is one of the four known fundamental interactions of nature, alongside the strong interaction, electromagnetism, and gravitation.

This weak force is responsible for radioactive decay and thus plays an essential role in nuclear fission.

3.5 Conductors and Electricity

As mentioned before, atoms are composed of protons, holding a positive electric charge and neutrons that have no electric charge. Electrons are found frenetically oscillating in a cloud of energy around this agglomerate of protons and neutrons.

We know that electrons are charge carriers and that as they move inside a material, they transport electric charges around. This movement of charges is called electric current.

In order for an electron to be able to move around a material, it is necessary, for the material itself, to have a certain kind of atomic structure.

Hydrogen (H), for example, is the simplest atomic element and for that reason the most abundant element in the universe. Hydrogen's nucleus is formed of a single proton and one electron is found orbiting around (see Figure 3.3).

Figure 3.3 Hydrogen atom.

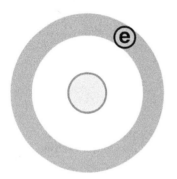

Helium (He), on the other hand, is composed of two protons, two neutrons, and two electrons, like shown in Figure 3.4.

Figure 3.4 Helium atom.

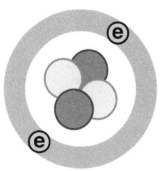

Hydrogen and helium are not good conductors of electricity. They have very few electrons, and these electrons are closer to the nucleus, subjected to the strong force, and, for that reason, are very difficult to move out of the atomic structure.

Copper (Cu), for example, an excellent conductor of electricity, is much more complex than hydrogen and helium. Its nucleus contains 29 protons, 35 neutrons, and 29 electrons. The electrons are distributed in several energetic shells, as represented in Figure 3.5.

Figure 3.5 Copper atom.

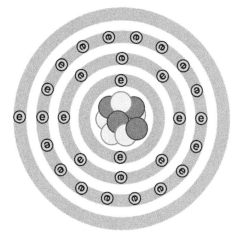

Copper contains 29 electrons disposed in four shells. From inside out, these shells have 2, 8, 18, and 1 electrons, respectively.

3.6   The Shells

Atoms that have a lot of electrons have these electrons distributed through several orbits, or shells, around the nucleus, like shown in Figure 3.5.

An electron needs a certain level of energy to occupy a particular orbit. Orbits that require more energy are farthest from the nucleus.

Each orbit, or shell, allows a certain limited number of electrons. The first six shells, for example, known by the letters K, L, M, N, O, and P, can have 2, 2, 18, 32, 50, and 72 electrons, respectively. The outermost shell of an atom is known as the valence shell.

The number of electrons allowed by the atomic shells follows a mathematical pattern equal to $2n^2$, where n is the shell number, starting with 0.

Electrons are not allowed to exist between shells. If an electron is on a particular shell and absorbs the required energy level to jump to an outer shell, it does by disappearing from the current shell and appearing on the outer shell.

If it is the other way around, that is, an electron is on an outer shell and loses energy to be on a vacant position in an inner shell, it disappears from the current shell and appears on the inner shell. In the process, it releases the extra energy by emitting a photon.

Electrons that are on the outermost shells are less subject to the strong force, exerted by the nucleus. Hence, these electrons can be more freely moved, if energy is applied.

Imagine a block of copper, containing millions and millions of atoms of this element. When these atoms are packed together, a magic happens and something known as a metallic bonding is formed, leaving a huge number of electrons detached from the nucleus. These electrons are now completely free to move if energy is applied.

3.7 Electric Potential

Like explained before, protons have positive electric charge and electrons have negative electric charge.

A large number of electrons on one side of a given material compared with another represent a difference in electric potential between the two points. This difference is electric potential energy stored in the form of electric charge.

A difference in electric potential between two points will create an electric field, and this field will force electrons to move into a specific direction.

The bigger the difference in potential, the bigger is the electric field, and, consequently, the bigger the number of electrons moving, the bigger is the current. A large difference in potential will also cause electrons to move more aggressively through the field.

The difference in electric potential is a kind of potential energy and as an energy it can create work.

The electric potential energy can be defined as the work, in Joules, to move electrons on an electric field.

The difference in electric potential, also known as electromotive force, is measured in Volts[5] and is represented by the uppercase letter V in the SI.[6]

---

5 In honor of Alessandro Giuseppe Antonio Anastasio Volta, simply known as Alessandro Volta (1745–1827), an Italian physicist, a chemist, and a pioneer of electricity and power who is also credited as the inventor of the electric battery.

6 French's Système Internationale d'Unités (International System of Unities), abbreviated as SI in all languages.

3.8 Current

Like mentioned previously, an electric field applied to a material will create a kind of pressure that will force the electric field to be transmitted from electron to electron at a speed about 63% of the light speed.

The electrons themselves move very slowly, from the lower potential point to the higher potential point, at a speed, called "drift velocity" that is something like 1.5 cm/min. The movement is irregular, like in a random zigzag motion, as they collide with atoms in the conductor.

The effect of having the electric field transmitted at a speed higher than the electron speed can be illustrated, roughly, to what happens on the Newton cradle toy (see Figure 3.6).

 **Figure 3.6** Newton cradle.

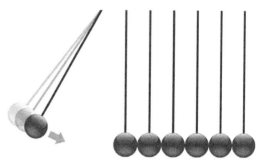

All balls are at initially at rest until we apply a speed to the first one in the left and make it collide with the other ones. As soon as the collision happens, the last ball in the right jumps. In other words, the balls practically do not move, but the force is transmitted from one to the other through the entire chain.

Current is measured in Ampere[7] and is represented by the uppercase letter A in the SI.

3.9 Electric Resistance

Free electrons inside a material can move forced by an electric field, but not all electrons are completely free to move, due to restrictions imposed by the material's structure. These restrictions prevent the free movement of electrons and resist their passage, generating friction and heat.

This restriction is called electric resistance, measured in Ohm,[8] and is represented by the Greek letter omega (Ω) in the SI.

---

7 In honor of André-Marie Ampère (1775–1836), French mathematician and physicist, credited as the father of electrodynamics.
8 In honor of Georg Simon Ohm (1787–1854), German physicist and mathematician.

4

Voltage and Current

Direct and Alternating Current and Voltage

4.1 Introduction

In the last chapter we have described the characteristics behind difference in electric potential, electric fields, and current.

In this chapter, we will describe direct current (DC) and alternating current (AC) and voltage.

4.2 Terminology

The term "direct current" is used indiscriminately to refer to direct voltage and to direct current. In this book we will differentiate the terms to prevent confusion. The same thing is valid for "alternating current," which in the real world is used to refer to both alternating voltage and alternating current.

Other terms also used are VDC to refer to "DC voltage" or direct (continuous) voltage and VAC to refer to "AC voltage" or alternating voltage.

4.3 Batteries

Around 1800, Alessandro Volta reported the results of an experiment where he was able to create electricity by piling discs of zinc and copper separated by felt spacers soaked in salt water. With this invention, Volta proved that electricity could be generated chemically and debunked the prevalent theory that electricity was generated solely by living beings.

A new invention was born: the battery.

A device that is able to generate electricity is generally referred as power supply.

Introductory Electrical Engineering with Math Explained in Accessible Language,
First Edition. Magno Urbano.
© 2020 John Wiley & Sons, Inc. Published 2020 by John Wiley & Sons, Inc.

4.3.1 Battery Life

Batteries are electrochemical devices created by embedding two different metallic plates into a third one. This third one, the electrolyte, is responsible, broadly speaking, for creating a chemical reaction in such a way that one of the elements ends with excess of electrons (negative charges) and the other one with an excess of positive charges – in other words, to create an imbalance of charges between the plates.

To summarize how chemical energy is converted into electrical energy, let us take an example of a typical car battery: one plate of pure lead and another one made of lead dioxide both immersed in a solution of sulfuric acid diluted with water. Chemical reaction causes the lead to become negatively charged and the lead dioxide to become positively charged.

The element with excess of electrons will be the negative pole (cathode) and the other element will be the positive pole (anode).

This difference in the number of electrons between the poles is called "difference in electric potential" or popularly, "voltage."

Any element, like a light bulb, for example, connected between the poles, will provide a path for the excess of electrons of the lead element to migrate to the lead dioxide, which lacks electrons. As the reaction happens, most of the sulfuric acid in the electrolyte is consumed and used to convert both plates into lead sulfate, and the electrolyte turns into, mostly, regular water.

At some point, all three elements are electrically balanced, current stops flowing, and the battery dies.

In electricity's early days, voltage was known as electromotive force (EMF), because voltage is considered to be a kind of "pressure" that pushes the electric field through a conducting loop.

It is known that electrons flow from the negative to the positive pole, at a certain slow speed; when compared with the speed, the electric field travels in the same direction. Current is said to flow in the opposite direction, from positive to negative, because its direction was established before the electron discovery.

The process of charging the battery will apply energy to the plates and electrolyte to reverse the reaction, making the plates to be lead dioxide and lead and the electrolyte sulfuric acid and water, again.

All power supplies have a limited voltage and current capacity. Current has a direct relation to the number of electrons that a power supply is able to move over time.

4.3.2 Batteries in Series

Consider two identical batteries connected in series, like shown in Figure 4.1. The positive pole of one battery is connected to the negative of the other and two wires are connected to points A and B.

Figure 4.1 Batteries in series.

If each battery has a voltage of 1.5 V, two batteries connected in series will provide a combined voltage of 3 V. This will be the value measured across A and B.

Batteries can only provide a limited amount of current. If each battery on the given example can provide 0.2 A of current, connecting any number of batteries in series will not change that.

Conclusions:

- Two power supplies in series will have their voltages added.
- Two power supplies in series will not have their currents added.

4.3.3 Batteries in Parallel

Figure 4.2 shows two identical batteries connected in parallel. The positive poles and negative poles are connected together.

Figure 4.2 Batteries in parallel.

Supposing each battery can provide 1.5 V and 0.2 A,[1] the voltage across AB will be 1.5 V, and together the batteries will be able to provide 0.4 A of current, which is the sum of their individual currents.

---

1 This can be written as 1.5 V @ 0.2 A.

Conclusions:

- Two power supplies in parallel have their currents added.
- Two power supplies in parallel will not have their voltages added.

4.4 Danger Will Robison, Danger!

When connecting batteries in series or parallel, a few rules must be followed.

4.4.1 Never Invert Polarities

Inverting a battery in a block of batteries in series, like shown in Figure 4.3, or in parallel may represent a danger situation.

Figure 4.3 Incorrect connection.

The other batteries will force current circulation across the inverted one, generating excessive heat that may lead to fire hazard and explosions.

4.4.2 Never Use Different Batteries

A battery is only capable of providing a limited amount of current and voltage.

If you connect batteries with different current capabilities in series, for example, 2, 4, and 6 A, the largest battery will force 6 A of current through batteries that cannot handle that amount of current. Excessive heat may be generated, again creating a danger situation.

The same is true for batteries with different voltages in parallel.

4.4.3 Short-Circuiting Batteries

Never short-circuit a battery like shown in Figure 4.4. Current will flow between them, at maximum level, making both batteries explode and catch fire.

Figure 4.4 Batteries in short circuit.

4.5 Direct Current

Direct current or DC is a term used to refer to the kind of voltage or difference in potential that keeps its value constant over time and does not vary in amplitude or frequency.

Figure 4.5 shows the symbol used to refer to a source of DC. In that case, the symbol is used to refer to batteries.

Figure 4.5 Battery symbol.

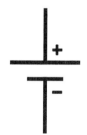

Batteries are sources of DC, meaning that once created they will provide a constant voltage and current during its lifetime.

4.5.1 DC Characteristics

- Current and voltage are constant over time.
- They are generated by chemistry and are very difficult to deploy in large scale.
- They never change polarity over time.
- Current only flows in one direction during all time.
- A DC voltage is not able to cross large distances. They lose power, considerably, if transmitted over long distances.

4.6 Relative Voltages

Consider two different electric potentials, like two batteries in series. How relative is the potential of one battery in respect with the other?

If each battery provides 1.5 V and two batteries in series provide 3 V, what are the references to which we are measuring these numbers? What potential is positive and what potential is negative? Is a potential really positive or negative? In relation to what?

4.6.1 Mountains

To understand why, consider two mountains like shown in Figure 4.6 and a path that one can climb from A to C, including a point B in the middle.

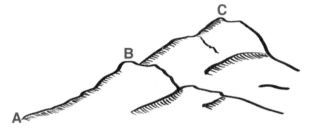

Figure 4.6 Two mountains.

A person standing on point A can climb to C in two phases: from A to B, resting at B a little bit, and then from B to C.

In terms of physics, the climber is going against gravity and, by doing so, accumulates gravitational potential energy.

Gravitation potential energy is calculated by the following formula.

POTENTIAL ENERGY

$$E_{gp} = mgh$$

E_{gp} is the gravitational potential energy, in Joules.[2]
m is the object's mass, in kilograms.
g is the gravity acceleration on Earth, equal to 9.834 m/s^2.
h is the vertical distance traveled by the object, in meters.

If point A is at sea level, B is at 100 m, C is at 200 m, and a person weights 70 kg, we can use the given formula and calculate that the person will gain a potential energy of 68.6 kJ by climbing from A to B and the same amount, again, by climbing from B to C.

If we choose point A as zero reference, we conclude that the person will have a potential energy of zero at point A, 68.6 kJ at B, and 137.2 kJ at C.

---

2 In honor of James Prescott Joule (1818–1889), English physicist. Joule is equivalent to kg.m^2/s^2 and represented by the uppercase letter J in SI.

What happens if instead of point A we select point B as zero reference?
If we do, we conclude that point C is 68.6 kJ over B and A is 68.6 kJ under B. In other words, in relation to B, the energy at C is 68.6 kJ and the energy at A is −68.6 kJ.

The negative sign is used to indicate that the potential level of A is below the reference point.

> Climbing down from B to A means losing potential energy, because we walk from a point with a higher potential energy to another with a lower potential energy.

Now consider two batteries in series, like shown before, but, this time, we take them apart to include a third point B, like shown in Figure 4.7.

Figure 4.7 Batteries in series with an intermediary point.

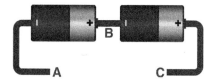

Figure 4.7 shows three points: A, B, and C. No need to remember the mountain analogy.

If each battery can provide a voltage of 1.5 V and we use point A as zero reference, we see that point B will be 1.5 V and point C will be 3 V above A.

On the other hand, if we use point B as a zero reference, C will be at 1.5 V above B and A will be −1.5 V below B. Again, the negative sign is used to show that A is below B.

This example shows the concept of negative electric potentials that are nothing more than a way to tell if a voltage is above or below a zero reference. This zero-reference point is called ground point, or simply ground, or in some cases, earth point.

4.7 Ground

In an electric circuit, the zero reference is called ground. The most used abbreviation in schematics is GND and its symbols are shown in Figure 4.8.

Figure 4.8 Ground symbols.

4.8 Alternating Current

AC is a term used to refer to an alternating voltage or to an alternating current source.

An alternating voltage changes its amplitude values constantly over time in a repeating pattern.

Like the alternating voltage, an alternating current also changes its amplitude values over time as well but additionally inverts polarity in a repeating pattern.

When current reverses polarity, it means a change in direction.

Figure 4.9 shows the symbol used for AC voltage sources in schematics.

Figure 4.9 Alternating current voltage source symbol.

4.8.1 AC Characteristics

- Can be easily created in large or small scale with electric generators.
- The levels of voltage and current vary cyclically with time.
- Voltage polarity may change with time.
- Current direction changes with time.
- Can travel large distances over wires with minimum losses.

4.8.2 AC Cycles

Figure 4.10 shows a graph representing an alternating voltage.

In Figure 4.11, we can see that alternating waves have a positive cycle and a negative cycle and vary over time in intensity.

The wave starts at 0 V, increases until it reaches V^+, and then decreases to a negative level V^-, repeating the whole pattern after that, *ad infinitum.*

Because the alternating voltage follows a cyclic pattern going up and down constantly, over time, it oscillates at a specific constant frequency. This

Figure 4.10 Alternating voltage.

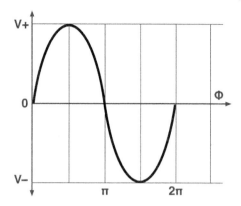

Figure 4.11 Alternating current cycles.

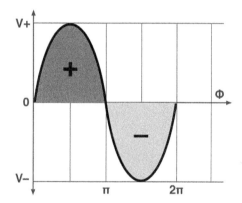

frequency is measured in Hertz,[3] meaning cycles per second,[4] and is represented in the SI by the letters Hz.

4.8.3 Period and Frequency

Figure 4.12 shows that the alternating wave starts at 0 and goes up and down and after a period equal to T starts repeating it.

In the given example, T is equal to 4 s and is known as the wave's period of oscillation.

The frequency and the period of a wave are related by the following formula.

---

3 In honor of Heinrich Rudolf Hertz, the first to prove, conclusively, the existence of electromagnetic waves.
4 A cycle is the period of time a wave needs to complete itself. Old schematics and books used the term cycles per second (cps) as a way to refer to wave frequency. It is not rare to find old texts referring to frequency as kcps (kilo cycles per second), instead of Hz.

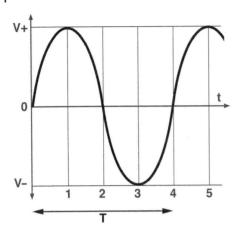

Figure 4.12 Period of a sinusoidal wave.

WAVE FREQUENCY

$$f = \frac{1}{T}$$

f is the wave frequency, in Hertz.
T is the wave period, in seconds.

4.8.4 Peak-to-Peak Voltage

Consider the wave shown in Figure 4.13, where voltages vary from 20 to −20 V. These values are measured against the 0 and are called peaks.

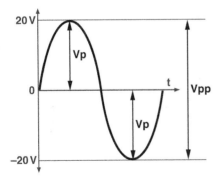

Figure 4.13 Alternating current peaks.

Remembering the mountain analogy, we can now use the negative peak as zero reference. If so, we can say that the positive peak is 40 V above that. In other words, 40 V is the difference between both peaks, also known as peak-to-peak voltage.

Peak-to-peak voltage is generally represented by the abbreviation V_{pp}.

4.8.5   DC Offset

We have described DC and AC as two separate entities, but we never said that they cannot exist together.

Observe the alternating voltage shown in Figure 4.14. The wave reference rests, vertically, at 0 V. The wave varies from 20 to –20 V.

Figure 4.14 40 Vpp Alternating voltage.

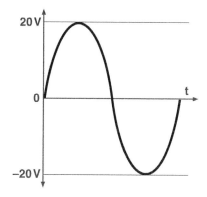

Now suppose we add 10 V continuous to that wave. The result is seen in Figure 4.15.

Figure 4.15 Alternating current plus 10 V DC.

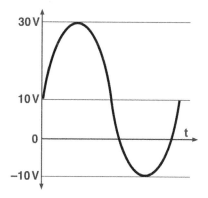

By observing Figure 4.15, it is clear that the wave shifted up by 10 V. Now, its positive peak is at 30 V and its negative peak is at −10 V. The wave now rests, vertically, at 10 V and not at 0 V as before. We say, in that case, that an offset of 10 V was applied to the alternating voltage.

If the offset is negative, −10 V, for example, the wave of Figure 4.14 will shift down and be like the one shown in Figure 4.16. The wave now rests at −10 V and variates up and down to 10 V and −30 V, respectively. In other words, we pushed the wave down by 10 V.

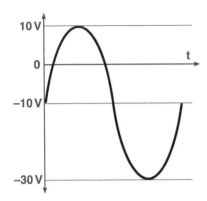

Figure 4.16 Alternating current minus 10 V DC.

Exercises

1   Consider the circuit shown in Figure 4.17, composed of several power supplies. Find the voltage between points A and B.

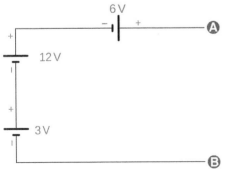

Figure 4.17 Batteries in series (Exercise 1).

2 Consider the same circuit shown on Figure 4.17. The three batteries of 3 V, 12 V, and 6 V are capable of providing 2 A, 4 A, and 3 A, respectively. Find the maximum current this circuit can provide.

3 Circuit shown on Figure 4.18 is composed of three power supplies. Find the voltage between points A and B.

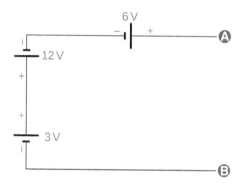

Figure 4.18 Batteries in series (Exercise 3).

4 Consider a 12 V alternating voltage power supply in series with an 8 V direct voltage power supply. What are the resulting peak voltages of this combination?

5 Find the period of oscillation of a 60 Hz AC power supply.

Solutions

1 If we follow the circuit from point B to A, we cross the 3 V power supply from the negative to the positive pole. It means that we are going from a point of low electric potential to a point of high electric potential and consequently we went up 3 V.

If we continue to follow the circuit, we cross the 12 V battery, again, from the negative to the positive pole, increasing the potential an additional 12 V.

The same will happen to the third battery, raising the potential an additional 6 V.

So, the voltage across AB is the sum of these three potential increases.

FINAL RESULT

$$V_{AB} = 3 + 12 + 6 = 21\,V$$

2 Batteries with different current capabilities should never be connected in series and may lead to fire hazard or explosion.

3 If we follow the circuit from B to A, we will cross the first voltage source (3 V) from its negative to its positive pole, a 3 V increase in potential.

  As we continue to follow the circuit, we cross the second battery (12 V), from the positive to the negative pole, meaning a potential decrease of 12 V.

  As we cross the third battery (6 V), we see an increase of 6 V.

  So, the voltage across AB is equal to

FINAL RESULT

$$V_{AB} = 3 + 6 - 12 = -3\,V$$

The negative sign means that B's potential is lower than A's.

4 If a source of alternating voltage has an amplitude of 12 V, it means it can reach a positive peak of 12 V and a negative peak of –12 V.

  An 8 V direct voltage source added to this alternating source will shift the wave 8 V up. So, the wave will now reach a positive peak of 20 V(12 + 8) and a negative peak of –4 V(–12 + 8).

5 The period of a wave is given by

$$T = \frac{1}{f}$$

$$T = \frac{1}{60}$$

FINAL RESULT

$$T = 0.0167\,s$$

5

Resistors

The Most Fundamental Component

5.1 Introduction

This chapter is about resistors, the most basic electric component.

5.2 Resistor

Resistors are the most fundamental and commonly used of all the electronic components.

Having two terminals, this component opposes current flow. This opposition is called electric resistance, measured in Ohms, and is represented by the Greek letter omega (Ω) in the SI.

5.3 Electric Resistance

All materials offer, to a certain degree, resistance to the flow of electrons. Metals, for example, offer very little electric resistance, but the resistance is there and it is not 0.

Factors like surface area, thickness, and chemical properties can alter electric resistance, and by manipulating these characteristics correctly, several types of resistors can be created.

5.4 Symbols

There are two different symbols to represent a resistor in schematics, one in the American standard and the other in the international standard. Figure 5.1 shows both symbols.

Introductory Electrical Engineering with Math Explained in Accessible Language,
First Edition. Magno Urbano.
© 2020 John Wiley & Sons, Inc. Published 2020 by John Wiley & Sons, Inc.

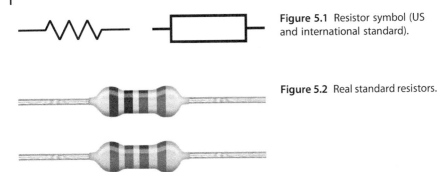

Figure 5.1 Resistor symbol (US and international standard).

Figure 5.2 Real standard resistors.

Figure 5.2 shows a picture of real resistors where color stripes describing their resistance values and other characteristics can be seen around their bodies.

5.5 Types of Resistor

There are several types of resistors produced using a variety of technologies: wire, carbon, metallic oxide film, etc.

Every type of resistor is useful for a specific application. There are resistors suitable for high stability, high voltage, high current, and high precision, among other uses.

Resistors have different degrees of precision called "tolerance." The more precise or more suitable for a specific kind of application, the higher the cost.

General-purpose cheaper resistors may have a tolerance of 20%, for example, meaning that a resistor of 1000 Ω, for instance, is only guaranteed to have a real resistance between 800 and 1200 Ω.

5.6 Power

Resistors are produced to handle a certain amount of current and, consequently, a certain amount of heat and power in Watts.[1]

Typical values for power are 1/8 W, 1/4 W, 1/2 W, 1 W, etc.

5.7 Color Code

Standard regular resistors have their characteristics encoded in four color stripes around their bodies.

---

1 We will talk about power in a future chapter.

The first two stripes define the first two digits of their resistance, the third represents the multiplier, and the fourth represents the tolerance. See Table 5.1.

Table 5.1 Resistor color code.

Color	First stripe	Second stripe	Multiplier	Tolerance (%)
Black		0	× 1	
Brown	1	1	× 10	± 1
Red	2	2	× 100	± 2
Orange	3	3	× 1000	
Yellow	4	4	× 10k	
Green	5	5	× 100k	± 0.5
Blue	6	6	× 1M	± 0.25
Violet	7	7	× 10M	± 0.1
Gray	8	8	× 100M	± 0.05
White	9	9	× 1G	
Gold			× 0.1	± 5
Silver			× 0.01	± 10

A resistor with the stripes

Brown	1
Black	0
Yellow	×10k
Gold	±5%

will be equal to $10 \times 10000 = 100$ kΩ with a tolerance of ±5%.

Some types of resistors may have a fifth stripe. If so, the first three stripes represent the value, the fourth is the multiplier, and the last is the tolerance.

A resistor of this kind, with the stripes shown next, will have the following characteristics:

Orange	3
Orange	3
Black	0
Brown	×10
Violet	±0.1%

The resistor will be 330 × 10 = 3.3 kΩ with a tolerance of ±0.1%.

Resistors in the range of kΩ and MΩ are also identified by using uppercase letters K and M, like 1K2 and 2M7 to identify 1.2 kΩ and 2.7 MΩ, respectively.

On the other end, the uppercase letter R is used to represent very small values of resistance. A resistor of 1.2 Ω may be referred to as 1R2.

We have compiled a table shown in Appendix B, which makes easy to find resistance values for color-coded resistors.

Appendix G shows several tables containing the values of resistances available in the market and their tolerances.

5.8 Potentiometer

Regular resistors have two terminals and a fixed value of resistance.

Many situations in real life require resistors that can vary their resistance. For this reason, a new kind of resistor was born: the potentiometer.

The potentiometer is basically a three-terminal device with a central axis that can be rotated or moved to make between its terminals. The volume rotary device used in amplifiers is one example.

Symbols for the potentiometer in the American and international standards can be seen in Figure 5.3.

5.9 Trimpots

Trimpots or trimmer potentiometers are miniature potentiometers used for adjustment, tuning, and calibration in circuits.

The resistance of trimpots and potentiometers can vary linearly or logarithmically.

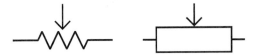

Figure 5.3 Potentiometer symbols (US and international standard).

5.10   Practical Usage

Incandescent light bulbs are examples of resistors with a very small value of resistance.

When used in circuits, resistors are used to control the flow of current. This property can be used to protect sensitive components from excessive current.

Other uses involve dividing voltages and creating timers in conjunction with capacitors.

5.11   Electric Characteristics

Resistors are "passive components," because they have no active role in a circuit. They simply resist current flow.

As they operate, they reduce current, and by doing so, they create an electric potential difference between their terminals. A resistor may have 10 V on one terminal and just 8 V on the other. For that reason, it is said that resistors create voltage drops across their terminals.

5.11.1   Important Facts

Obviously, voltage drops will just happen if current flows across a resistor. If current is 0, both terminals will have the same voltage.

Even without any current, resistors have what is called thermal noise, related to the internal chaotic movement of electrons inside the material's structure. When current flows across a resistor, it creates friction against its internal structure, generating heat and increasing the thermal noise. This noise leaks to the circuit and may interfere with other components, distort signals, and create other undesired problems. Different types of resistors were created to minimize problems like this.

5.12   Resistors in Series

Resistors connected in series, like shown in Figure 5.4, will have a total resistance equal to the sum of each individual resistance, or

Figure 5.4 Resistors in series.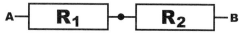

$$R = R_1 + R_2$$

For an infinite number of resistors, the following formula applies.

EQUIVALENT RESISTANCE: SERIES

$$R = R_1 + R_2 + \cdots + R_\infty$$

R is the equivalent resistance, in Ohms.
$R_1, R_2, \ldots$ are the values of the individual resistors, in Ohms.

5.13 Resistors in Parallel

Resistors connected in parallel, like shown in Figure 5.5, will have a total resistance that can be found by the following formula:

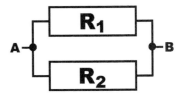

Figure 5.5 Resistors in parallel.

$$\frac{1}{R} = \frac{1}{R_1} + \frac{1}{R_2}$$

If we have only two resistors, we can use this other friendly formula:

$$R = \frac{R_1 \times R_2}{R_1 + R_2}$$

For an infinite number of resistors, we get the following formula.

EQUIVALENT RESISTANCE: PARALLEL

$$\frac{1}{R} = \frac{1}{R_1} + \frac{1}{R_2} + \cdots + \frac{1}{R_\infty}$$

R is the equivalent resistance, in Ohms.
$R_1, R_2, \ldots$ are the values of the individual resistors, in Ohms.

5.14 DC and AC Analysis

Differently from other components we will still examine in this book, resistors are the only components that behave the same way when connected to direct current or to alternating current.

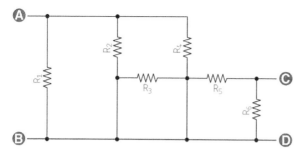

Figure 5.6 A simple resistive circuit.

Suppose we connect points A and B of the circuit shown in Figure 5.6, to a battery.

Current flows across the several components in the circuit and a certain voltage would be measured across points C and D. This voltage would be less than the one applied between A and B.

If the voltage across AB varies, voltage across CD will vary instantly.

If we replace the battery with a source of alternating voltage, the result will be the same in essence: we will measure a voltage between C and D that is less than the one applied to the input.

Again, if the voltage across AB varies, voltage across CD will vary instantly.

5.15   Input and Output Synchronism

Imagine the circuit in Figure 5.6, initially at rest. No input or output voltage.

At time equal to 0, we apply an alternating voltage at the circuit's input and start monitoring what happens at the output. We will call t_0 this instant in time.

At the time the voltage is applied across the input, a smaller voltage appears across the output.

As the input voltage varies, the output voltage also varies, at the same time, without any delay. We say that both input and output are in synchronism or in phase or have a phase equal to 0°.

In Figure 5.7 we show an input signal (larger wave) and the respective output signal (smaller wave) in synchronism.

If we want to make an analogy to the real world, we could say that resistors have zero inertia and react instantly to any variations in voltage or current.

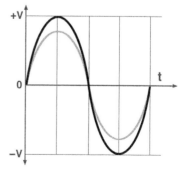

Figure 5.7 Input and output signals in phase.

Exercises

1 Consider three resistors in series: 220 Ω, 180 Ω, and 1 kΩ. What is the equivalent resistance?

2 Consider three resistors in parallel: 1.2 kΩ, 100 kΩ, and 1 Ω. What is the equivalent resistance?

3 Figure 5.8 shows a purely resistive circuit. What is the resistance measured between points A and B?

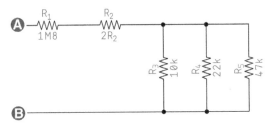

Figure 5.8 Resistive circuit (exercise).

4 A resistive circuit is connected to an alternated voltage source. What is the phase angle between output and input?

Solutions

1 The equivalent resistance of resistors in series is the sum of them all. Hence,

$$R_{EQ} = R_1 + R_2 + R_3$$

$$R_{EQ} = 220 + 180 + 1000$$

FINAL RESULT

$$R_{EQ} = 1400\ \Omega$$

2 The equivalent resistance of resistors in parallel is found by the formula

$$\frac{1}{R_{EQ}} = \frac{1}{R_1} + \frac{1}{R_2} + \frac{1}{R_3}$$

$$\frac{1}{R_{EQ}} = \frac{1}{1200} + \frac{1}{100000} + \frac{1}{1}$$

FINAL RESULT

$R_{EQ} = 0.9992 \ \Omega$

3 Resistors R_1 and R_2 are in series. Their equivalent resistance (R_{EQ1}) will be the sum of their individual resistances:

$R_{EQ1} = R_1 + R_2$

$R_{EQ1} = 1800000 + 2.2$

$R_{EQ1} = 1800002.2 \ \Omega$

The remaining resistors, R_3, R_4, and R_5, are in parallel. Their equivalent resistance follows the formula

$$\frac{1}{R_{EQ2}} = \frac{1}{R_3} + \frac{1}{R_4} + \frac{1}{R_5}$$

$$\frac{1}{R_{EQ2}} = \frac{1}{10000} + \frac{1}{22000} + \frac{1}{47000}$$

$R_{EQ2} = 5997.6769 \ \Omega$

R_{EQ1} and R_{EQ2} are in series. We must sum them to obtain the total resistance between points A and B:

$R_{AB} = R_{EQ1} + R_{EQ2}$

$R_{AB} = 1800002.2 + 5997.6768$

FINAL RESULT

$R_{AB} = 1805999.88 \ \Omega$

4 Pure resistors do not affect the phase of a signal.

FINAL RESULT

Phase $= 0°$

6

Ohm's Laws

Circuit Analysis

6.1 Introduction

In this chapter we will examine Ohm's law, postulated by German physicist Georg Simon Ohm in 1827, that relates resistance, current, and voltage in conductors.

6.2 Basic Rules of Electricity

A few basic rules of electricity are always true, independent of which components we connect in a circuit:

- Components in parallel will always have the same voltage and divide the current.
- Components in series will always have the same current and divide the voltage.
- Components with the same voltage in both terminals do not have any current flowing across.

Consider the circuit shown in Figure 6.1.

Figure 6.1 shows three resistors in parallel. These resistors have the same voltage across. Current, on the other hand, will divide between them.

Figure 6.1 Resistors in parallel with a battery.

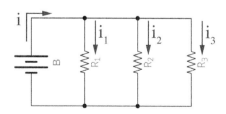

Introductory Electrical Engineering with Math Explained in Accessible Language,
First Edition. Magno Urbano.
© 2020 John Wiley & Sons, Inc. Published 2020 by John Wiley & Sons, Inc.

If all resistors have the same resistance, each one will receive one-third of the total current. On the other hand, if they have different resistances, each one will have a current inversely proportional to their resistance.

Anyway, the total current will be always equal to the sum of the currents of all parallel branches.

In Figure 6.2 we show another circuit. This time, three resistors are in series with a battery.

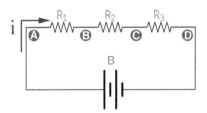

Figure 6.2 Resistors in series with a battery.

In the circuit shown in Figure 6.2, all resistors have the same current across, because they are in series. On the other hand, they will divide the battery's voltage among them.

If all resistors have the same resistance, each one will have one-third of the voltage provided by the battery. On the other hand, if they have different resistances, their voltages will depend on a relation between their resistances and the sum of all three resistances.

Anyway, the sum of their voltages will always be equal to the voltage provided by the battery.

6.3   First Ohm's Law

The first Ohm's law postulates that the current across two points of a conductor is directly proportional to the voltage across these same two points, according to the following relation.

FIRST OHM'S LAW

$$V = RI$$

V is the voltage, in Volts.
R is the resistance, in Ohms.
I is the current, in Amperes.

6.4   Second Ohm's Law

The second Ohm's law states that the resistance of a homogeneous conductor of constant cross section is directly proportional to its length and is inversely proportional to the area of its cross section.

$$R = \frac{\rho L}{A}$$

V is the voltage, in Volts.
ρ is the conductor resistivity, in Ohm meters (depends on the material and its temperature).
L is the conductor length, in meters.
A is the conductor cross section, in squared meters.

6.5   Examples

6.5.1   Example 1

Figure 6.3 shows a circuit where a 1000 Ω resistor is connected to a 12 V battery. We want to find the current flowing in the circuit.

6.5.1.1   Solution
The firs Ohm's law postulates that

$$V = RI$$

Hence, current will be

$$I = \frac{V}{R} = \frac{12}{1000}$$

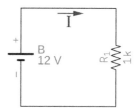

Figure 6.3 One resistor in series with a battery.

$$I = 0.012\,A = 12\,mA$$

6.5.2   Example 2

Figure 6.4 shows a circuit where two resistors are in parallel with a battery. The battery provides 12 V and the total current drained by the circuit is 3 A. We want to find the current flowing across each resistor.

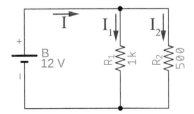

Figure 6.4 Two resistors in parallel with a battery.

6.5.2.1 Solution

We know that the total current I is equal to the sum of the branch currents, I_1 and I_2, that are the currents flowing across R_1 and R_2, respectively:

$$I = I_1 + I_2$$

R_2 is 500 Ω, exactly half the resistance of R_1, which is 1000 Ω. Thus, R_2 will receive twice the current of R_1:

$$I_2 = 2I_1$$

We know that

$$I = I_1 + I_2$$

Upon substituting the first condition,

$$I = I_1 + 2I_1$$
$$I = 3I_1$$

The total current I is equal to 3 A; thus

$$3I_1 = 3$$

Hence

FINAL RESULT

$$I_1 = \frac{3}{3} = 1\,A$$

To find I_2, we know that

$$I_2 = 2I_1$$

Hence,

FINAL RESULT

$I_2 = 2\,A$

6.5.2.2 Another Method to Obtain the Same Result

R_1 and R_2 are in parallel; thus they have the same voltage. We know that the voltage across a resistor is

$$V = RI$$

Thus,

$$R_1 I_1 = R_2 I_2$$

This formula tells us that the voltage across R_1 ($R_1 I_1$) is equal to the voltage across R_2 ($R_2 I_2$).

Hence,

$$I_1 = \frac{R_2 I_2}{R_1}$$

Upon substituting the values, we get

$$I_1 = \left(\frac{500}{1000}\right) \times I_2$$

$$I_1 = 0.5\,I_2$$

This is the same conclusion we have found before, that is, I_2 is twice I_1.

If I_2 is twice I_1, we can divide the total current into three equal parts and designate two parts for I_2 and one for I_1.

FINAL RESULT

$I_1 = 1\,A$

$I_2 = 2\,A$

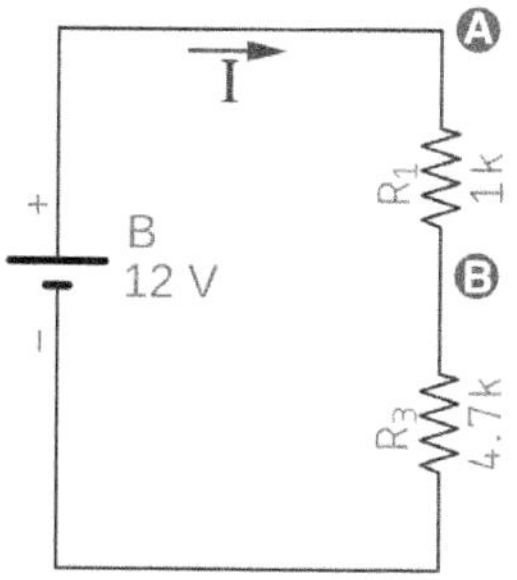

Figure 6.5 Two resistors in series with a battery.

6.5.3 Example 3

Consider the circuit shown in Figure 6.5, containing two resistors in series with a battery.

The battery provides 12 V, and the resistors, R_1 and R_2, have resistances of 1000 Ω and 4700 Ω, respectively. We want to find the total current flowing in the circuit.

6.5.3.1 Solution

Like we have learned before, the equivalent resistance of resistors in series is the sum of their individual resistances. Thus, the total resistance is 5700 Ω.

By Ohm's law,

$$V = RI$$

Upon substituting the values, the current flowing in the circuit is

$$12 = 5700\, I$$

$$I = \frac{12}{5700}$$

FINAL RESULT

$$I = 0.0021052 = 2.1052\, \text{mA}$$

6.5.4 Example 4

Using the same circuit shown in Figure 6.5, find the voltage drop across R_1.

6.5.4.1 Solution

A current flowing in a resistor produces a voltage drop, which can be calculated by using

$$V = RI$$

We know that the total current, flowing across both R_1 and R_2, is 2.1 mA. Therefore, the voltage drop across R_1 will be its resistance multiplied by the current flowing across itself:

$$V_{R_1} = R_1 I$$

$$V_{R_1} = 1000 \times 0.0021052$$

FINAL RESULT

$$V_{R_1} = 2.1052\, \text{V}$$

6.5.4.2 Another Method to Obtain the Same Result

We know that R_2's resistance is 4700 Ω and that current flowing across that resistor is 2.1 mA.

By Ohm's law,

$$V = RI$$

Thus, voltage drop across R_2 is

$$V_{R_2} = R_2 I$$

$$V_{R_2} = 4700 \times 0.0021052$$

$$\boxed{V_{R_2} = 9.8947\,\text{V}}$$

R_1 and R_2 are in series. The voltage across both resistors is 12 V.

So, the sum of the voltage drops across each resistor must be also equal to 12 V:

$$V_{R_1} + V_{R_2} = 12\,\text{V}$$

Hence,

$$V_{R_1} = 12 - V_{R_2}$$

$$V_{R_1} = 12 - 9.8947$$

And we obtain the same value as before.

FINAL RESULT

$$V_{R_1} = 2.1052\,\text{V}$$

6.5.5 Example 5

Suppose we have 100 m of wire with a cross section equal to 5 mm^2. We apply 200 V to this wire and measure a current of 10 A. What is this conductor's resistivity?

6.5.5.1 Solution

We use the first Ohm's law to find the wire's resistance:

$$V = RI$$

Thus,

$$R = \frac{V}{I} = \frac{200}{10} = 20\,\Omega$$

By the second Ohm's law,

$$R = \frac{\rho L}{A}$$

or,

$$\rho = \frac{RA}{L}$$

Substituting the values, we get

$$\begin{cases} R = 20\,\Omega \\ A = 5\,\text{mm}^2 = 5 \times 10^{-6}\,\text{m}^2 \\ L = 100\,\text{m} \end{cases}$$

$$\rho = \frac{20 \times (5 \times 10^{-6})}{100} = 1 \times 10^{-6}\,\Omega.\text{m}$$

FINAL RESULT

$$\rho = 1\,\mu\Omega.\text{m}$$

Exercises

1   Consider the circuit shown in Figure 6.6, composed of three resistors and a battery.

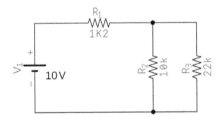

Figure 6.6 Resistive circuit and a battery.

Find the current flowing across R_1.

2   Find the currents flowing across R_2 and R_3 in the circuit shown in Figure 6.6.

3   Find the voltage drop caused by the total current across R_1.

4   1000 m of wire with a cross section equal to 8 mm^2 is subjected to 100 V. A current of 4 A is measured across this wire. What is the wire resistivity?

Solutions

1 First, we need to calculate the circuit's total resistance.
Resistors R_2 and R_3 are in parallel. Their equivalent resistance is

$$\frac{1}{R_{EQ}} = \frac{1}{R_2} + \frac{1}{R_3}$$

$$\frac{1}{R_{EQ}} = \frac{1}{10000} + \frac{1}{22000}$$

$$R_{EQ} = 6875\,\Omega$$

This equivalent resistance is in series with R_1. We must sum both to obtain the total resistance:

$$R_T = R_1 + R_{EQ}$$

$$R_T = 1200 + 6875$$

$$R_T = 8075\,\Omega$$

Now that we have the total resistance, we can use Ohm's law and find the current across R_1:

$$V = RI$$

$$10 = 8075\,I$$

$$I = \frac{10}{8075}$$

FINAL RESULT

$$I = 1.2384\,\text{mA}$$

2 Current flowing across R_1 will split between R_2 and R_3.
Thus, the sum of currents across R_2 and R_3 will be equal to the total current flowing across R_1:

$$I_1 = I_2 + I_3$$

If R_2 is 10 kΩ and R_3 is 22 kΩ, we conclude that R_3 is 2.2 times bigger than R_2, or

$$R_3 = 2.2R_2$$

Hence, current in R_2 will be 2.2 times bigger than the current flowing across R_3:

$$I_2 = 2.2\,I_3$$

Upon substituting the values, we get

$$I_1 = 2.2\,I_3 + I_3$$

$$I_1 = 3.3\,I_3$$

Thus,

$$0.0012384 = 3.3\,I_3$$

FINAL RESULT

$$I_3 = 375.2697\,\mu A$$

Now that we know I_3, we can find I_2:

$$I_2 = 2.2\,I_3$$

FINAL RESULT

$$I_2 = 825.5933\,\mu A$$

3 Current across R_1 causes a voltage drop of

$$V_{R_1} = R_1 I_1$$

$$V_{R_1} = (1200)(0.0012384)$$

FINAL RESULT

$$V_{R_1} = 1.4860\,V$$

4 We first apply the first Ohm's law to find the conductor's resistance:

$$V = RI$$

Thus,

$$R = \frac{V}{I}$$

$$R = \frac{100}{4}$$

$$R = 25\,\Omega$$

The second Ohm's law states that the conductor's resistance is

$$R = \frac{\rho L}{A}$$

or,

$$\rho = \frac{RA}{L}$$

Upon substituting the values, we obtain

$$\begin{cases} R = 25\,\Omega \\ A = 8\,\text{mm}^2 = 8 \times 10^{-6}\,\text{m}^2 \\ L = 1000\,\text{m} \end{cases}$$

Thus,

$$\rho = \frac{25 \times (8 \times 10^{-6})}{1000} = 2 \times 10^{-7}$$

FINAL RESULT

$$\rho = 200\,\text{n}\Omega.\text{m}$$

7

Delta–Wye Conversions

Circuit Analysis

7.1 Introduction

In this chapter, we will examine two kinds of circuit configurations, Delta and Wye (Y), and how to convert from one to the other.[1]

Converting a circuit between Delta and Y configurations and vice versa can make a circuit easier to understand and analyze.

7.2 Delta Circuit

Figures 7.1 and 7.2 show examples of Delta and Y circuit configurations.

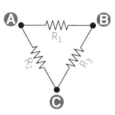

Figure 7.1 Delta or triangle circuit.

7.3 Delta–Wye Conversion

Consider that we want to find the resistance between points A and B in the circuit shown in Figure 7.1 and that point C is not connected to anything.

If point C is not connected to anything, we can redraw the circuit into a friendlier version shown in Figure 7.3.

Thus, the resistance between A and B will be the resistance of R_1 in parallel with the sum of R_2 and R_3, because these later resistors are in series.

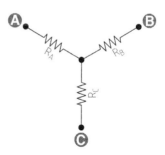

Figure 7.2 Y or Wye circuit.

---

1 Delta configuration is also known as triangular or π and Y configuration is also known as T or Wye.

Introductory Electrical Engineering with Math Explained in Accessible Language, First Edition. Magno Urbano.

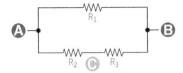

Figure 7.3 Delta circuit modified.

Mathematically speaking,

$$R_{AB} = R_1 // (R_2 + R_3)$$

By applying the equations for the equivalent resistance of parallel resistors,

$$R_{AB} = \frac{R_1 \times (R_2 + R_3)}{R_1 + (R_2 + R_3)}$$

If we want to convert the original Delta circuit into a Wye circuit, we must look at the Wye circuit in Figure 7.2 and observe what resistance would be seen between A and B if point C was not connected to anything. The answer is obvious: the sum of RA and RB, or

$$R_{AB} = R_A + R_B$$

We have now two equations that calculate the resistance between points A and B when C is not connected to anything: one for the Delta circuit and one for the Wye circuit.

We can equate both equations for nodes A and B:

$$\frac{R_1 \times (R_2 + R_3)}{R_1 + (R_2 + R_3)} = R_A + R_B$$

If we do the same for the other Delta nodes, we obtain two more equations. For nodes A and C,

$$\frac{R_2 \times (R_1 + R_3)}{R_2 + (R_1 + R_3)} = R_A + R_C$$

and for nodes B and C,

$$\frac{R_3 \times (R_1 + R_2)}{R_3 + (R_1 + R_3)} = R_B + R_C$$

If we solve these three equations, we obtain the equations that can be used to convert any Delta circuit into a Wye circuit.

DELTA–WYE CONVERSION EQUATIONS

$$R_A = \frac{R_1 R_2}{R_1 + R_2 + R_3}$$

$$R_B = \frac{R_1 R_3}{R_1 + R_2 + R_3}$$

$$R_C = \frac{R_2 R_3}{R_1 + R_2 + R_3}$$

R_A, R_B, R_C are the resistors in the Wye configuration, in Ohms.
R_1, R_2, R_3 are the resistors in the Delta configuration, in Ohms.

7.4 Wye–Delta Conversion

If we want to convert Wye circuits to Delta circuits, we must follow the same kind of principles we have used to convert from Delta to Wye.

If we do, we find the following equations that can be used to convert any Wye circuit into a Delta one.

Y-DELTA CONVERSION EQUATIONS

$$R_1 = \frac{R_A R_B + R_B R_C + R_A R_C}{R_C}$$

$$R_2 = \frac{R_A R_B + R_B R_C + R_A R_C}{R_B}$$

$$R_3 = \frac{R_A R_B + R_B R_C + R_A R_C}{R_A}$$

R_A, R_B, R_C are the resistors in the Wye configuration, in Ohms.
R_1, R_2, R_3 are the resistors in the Delta configuration, in Ohms.

If we superimpose the Delta and the Wye circuits, we notice that a new node was created by the conversion from Delta to Wye, in the middle of the diagram, as shown in Figure 7.4.

In the next example, it will become clear how useful is the ability to convert a circuit from Delta to Wye or vice versa and how this can help to simplify circuits in real life.

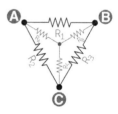

Figure 7.4 A new node is created after a Delta–Wye conversion.

7.5 Examples

7.5.1 Example 1

Find the resistance measured between points A and B in the circuit shown in Figure 7.5.

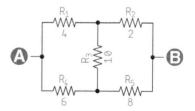

Figure 7.5 Delta–Wye conversion.

7.5.1.1 Solution

Finding the resistance between A and B is not easy as it appears. The circuit has several resistors that are in series or parallel, and there is nothing obvious that can be replaced by an equivalent resistance.

We could apply Ohm's law, for example, to discover the voltages and currents across the components, but we would end with a lot of equations to solve.

Things will get easy if we use the Delta–Wye conversion equations.

To see how this circuit can be converted into a Delta configuration, we stretch the diagram and redraw it like shown in Figure 7.6.

It is now clear that the circuit in Figure 7.5 is, in fact, a double Delta configuration, sharing resistor R_3.

To simplify the circuit, let us replace R_1, R_3, and R_4 by the Y configuration shown superimposed in Figure 7.7.

According to the Delta–Wye conversion rules, R_1, R_3, and R_4 must be replaced by R_a, R_b, and R_c, and their values can be calculated by the following equations:

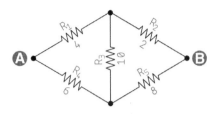

Figure 7.6 Stretching the circuit to see the Delta.

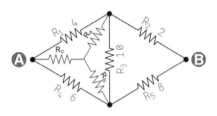

Figure 7.7 Wye superimposed into de Delta.

$$R_a = \frac{R_1 R_3}{R_1 + R_3 + R_4}$$

$$R_a = \frac{4 \times 10}{4 + 10 + 6}$$

$$R_a = 2\ \Omega$$

$$R_b = \frac{R_3 R_4}{R_1 + R_3 + R_4}$$

$$R_b = \frac{10 \times 6}{4 + 10 + 6}$$

$$R_b = 3\ \Omega$$

$$R_c = \frac{R_1 R_4}{R_1 + R_3 + R_4}$$

$$R_c = \frac{4 \times 6}{10 + 4 + 6}$$

$$R_c = 1.2 \, \Omega$$

Now that we have the values, let us rewrite the final circuit as shown in Figure 7.8.

Circuit in Figure 7.8 can be redrawn into the one in Figure 7.9.

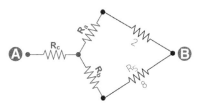

Figure 7.8 Simplified circuit.

Now we can clearly see that R_c is in series with two branches of resistors in parallel and that each branch has two resistors in series, R_a/R_2 and R_b/R_5.

Mathematically speaking, the resistance between A and B is

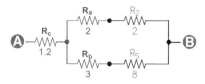

Figure 7.9 Final simplified circuit.

$$R_{AB} = R_c + (R_a + R_2)//(R_b + R_5)$$

$$R_{AB} = R_c + \frac{(R_a + R_2) \times (R_b + R_5)}{(R_a + R_2) + (R_b + R_5)}$$

$$R_{AB} = 1.2 + \frac{(2+2) \times (3+8)}{(2+2) + (3+8)}$$

$$R_{AB} = 1.2 + \frac{4 \times 11}{4 + 11}$$

$$R_{AB} = 1.2 + \frac{44}{15}$$

FINAL RESULT

$$R_{AB} = 4.133 \, \Omega$$

7.5.2 Example 2

Figure 7.10 shows what appears to be a Delta circuit. We want to find the resistance between points A and B.

7.5.2.1 Solution

The circuit in Figure 7.10 appears to show that R_3, R_4, and R_5 are in a Delta configuration. This is not true. This circuit is what we call a fake Delta circuit, an impostor, because these resistors are not in Delta.

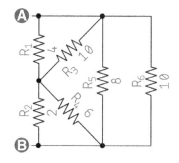

Figure 7.10 Peculiar Delta circuit.

The circuit in Figure 7.10 can be redrawn as the one shown in Figure 7.11.

We can improve the circuit more by redrawing it like it is shown in Figure 7.12.

Figure 7.12 shows that R_1/R_3 and R_2/R_6 are in parallel and both groups are in series. There is no resistor in Delta.

The resistance between A and B can be then calculated as

$$R_{AB} = R_{EQ}//R_5//R_6$$

where R_{EQ} is

$$R_{EQ} = (R_1//R_3) + (R_2//R_4)$$

$$R_{EQ} = \frac{R_1 \times R_3}{R_1 + R_3} + \frac{R_2 \times R_4}{R_2 + R_4}$$

$$R_{EQ} = \frac{4 \times 10}{4 + 10} + \frac{2 \times 6}{2 + 6}$$

$$R_{EQ} = \frac{40}{14} + \frac{12}{8}$$

$$R_{EQ} = \frac{61}{14}\Omega$$

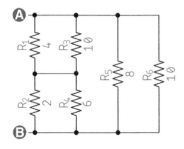

Figure 7.11 Fake Delta circuit

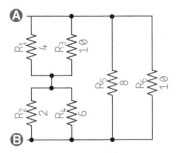

Figure 7.12 Peculiar Delta circuit decomposed.

Hence, the resistance between A and B is

$$\frac{1}{R_{AB}} = \frac{1}{R_{EQ}} + \frac{1}{R_5} + \frac{1}{R_6}$$

$$\frac{1}{R_{AB}} = \frac{1}{\frac{61}{14}} + \frac{1}{8} + \frac{1}{10}$$

$$\frac{1}{R_{AB}} = \frac{14}{61} + \frac{1}{8} + \frac{1}{10}$$

$$\frac{1}{R_{AB}} = \frac{1109}{2440}$$

FINAL RESULT

$$R_{AB} = 2.2\ \Omega$$

Exercises

1 Find the resistance between A and B in the circuit shown in Figure 7.13.

2 Find the resistance between A and B in the circuit shown in Figure 7.14.

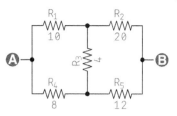

Figure 7.13 Delta circuit (Exercise 1).

Solutions

1 We can convert from Delta to Wye by replacing R_1, R_3 and R_4 with R_a, R_b and R_c, redrawing the circuit as a Wye configuration, and using the following equations:

$$R_a = \frac{R_1 R_3}{R_1 + R_3 + R_4}$$

$$R_a = \frac{10 \times 4}{10 + 4 + 8}$$

$$R_a = 1.8182 \ \Omega$$

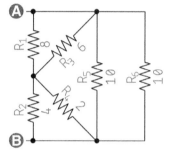

Figure 7.14 Delta circuit (Exercise 2).

$$R_b = \frac{R_3 R_4}{R_1 + R_3 + R_4}$$

$$R_b = \frac{4 \times 8}{10 + 4 + 8}$$

$$R_b = 1.4545 \ \Omega$$

$$R_c = \frac{R_1 R_4}{R_1 + R_3 + R_4}$$

$$R_C = \frac{10 \times 8}{10 + 4 + 8}$$

$$R_c = 3.6364 \ \Omega$$

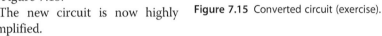

This operation will convert the original circuit into the one shown in Figure 7.15.

Figure 7.15 Converted circuit (exercise).

The new circuit is now highly simplified.

Resistors R_a/R_2 and R_b/R_5 are in series. Both groups are in parallel and the block is in series with R_c.

Mathematically speaking, the resistance between A and B can be described by the following equations:

$$R_{AB} = R_c + (R_a + R_2)//(R_b + R_5)$$

$$R_{AB} = R_c + \frac{(R_a + R_2) \times (R_b + R_5)}{(R_a + R_2) + (R_b + R_5)}$$

$$R_{AB} = 3.6364 + \frac{(1.8182 + 20) \times (1.4545 + 12)}{(1.8182 + 20) + (1.4545 + 12)}$$

FINAL RESULT

$$R_{AB} = 11.9588 \ \Omega$$

2 The circuit shown in Figure 7.14 is a Delta impostor and can be redrawn as shown in Figure 7.16.

Thus, the resistance between points A and B can be found by the following formula:

$$R_{AB} = \left(R_{EQ}//R_5//R_6\right)$$

where

$$R_{EQ} = (R_1//R_3) + (R_2//R_4)$$

Thus,

$$R_{EQ} = \left(\frac{R_1 \times R_3}{R_1 + R_3}\right) + \left(\frac{R_2 \times R_4}{R_2 + R_4}\right)$$

$$R_{EQ} = \left(\frac{8 \times 6}{8 + 6}\right) + \left(\frac{4 \times 2}{4 + 2}\right)$$

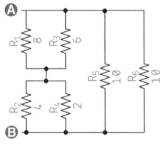

Figure 7.16 Fake Delta circuit (exercise).

$$R_{EQ} = 4.7619 \ \Omega$$

Hence, the resistance between A and B is

$$R_{AB} = \left(R_{EQ}//R_5//R_6\right)$$

$$\frac{1}{R_{AB}} = \frac{1}{R_{EQ}} + \frac{1}{R_5} + \frac{1}{R_6}$$

$$\frac{1}{R_{AB}} = \frac{1}{4.7619} + \frac{1}{10} + \frac{1}{10}$$

FINAL RESULT

$$R_{AB} = 2.4390 \ \Omega$$

8

Capacitors

And Electric Charges

8.1 Introduction

In this chapter, we will examine capacitors, basic electronic components very popular in all kinds of circuits.

8.2 History

The capacitor or originally known as condenser[1] is a two-terminal electric component that can store potential energy in the form of electric field.

Technically, a capacitor is composed of two metal plates, separated by a medium. This medium, called dielectric, can be any nonconductive element like air, oil, plastic, etc.

Historically the idea for the capacitor is based on the Leyden Jar.[2]

8.3 How It Works

Capacitors are basically two metal plates in parallel. The plates are put very close to each other, without touching. These plates are at rest and have roughly the same number of electrons (see Figure 8.1).

---

1 The term condenser used to refer to capacitors was first used around 1782. It is interesting to know that, in that year, James Watt had just patented a new revolutionary component for his steam engine, called the "separate condenser," making the term a success among scientists.
2 Leyden Jar (or Leiden Jar) is a kind of primitive capacitor that can store high voltage electric charges between two electric conductors. Invented by accident in 1746 by Pieter van Musschenbroek.

Introductory Electrical Engineering with Math Explained in Accessible Language,
First Edition. Magno Urbano.
© 2020 John Wiley & Sons, Inc. Published 2020 by John Wiley & Sons, Inc.

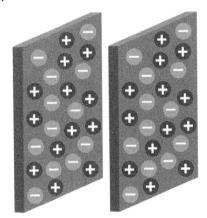

Figure 8.1 Two metallic plates in parallel, at rest.

Something magical happens when a battery is connected to the plates. Suddenly, an electric field (F) is formed between the plates, as shown in Figure 8.2.

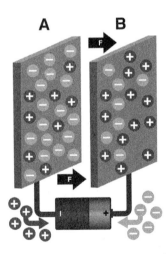

Figure 8.2 A battery is connected between the plates.

The electric field makes the plate connected to the battery's negative pole (A) to accumulate an excess of negative charges (electrons) and the other one to accumulate positive charges, or lack of electrons (B).

Because both plates are very close and charges with the same polarity repel each other, this excess of electrons in plate A will create a force that will repel electrons on plate B, expelling them from the plate and forcing them to migrate to the battery's positive pole.

At this time, one may have the illusion of current flow between the plates. The truth is that the electric field generated by plate A forced electrons from plate B to move off the plate to the battery's positive pole. No electron crossed through the dielectric from plate A to B.

A real current may flow across the dielectric in some cases, but this is an undesired effect that may destruct the capacitor.

As plate A receives more electrons, it saturates and an increasing repulsion force develops to prevent more electrons from arriving at plate A. The force between plates stops increasing and cannot remove electrons from the other plate anymore. At this point, current stops "flowing."

Charging a capacitor is like compressing a coil to its physical limit. At some point it is not possible to compress the coil anymore.

Capacitors are normally represented on schematics by the uppercase letter C. Figure 8.3 shows the symbols used to represent a capacitor in the American and in the international standards.

Figure 8.3 Capacitor symbols.

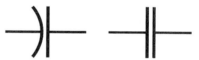

8.3.1 Dielectric

The kind of capacitor we are examining so far is composed of two metallic plates separated by air. Air is the dielectric.

We know that air is not a very good conductor. In fact, air is an insulator.

What makes a material a good insulator is having very few free electrons. Electrons are charge carriers. A small number of free electrons make it hard to conduct electricity, because all available electrons are held in place by the nuclear strong force, making very hard for them to move. A lot of energy would be required to strip these electrons from their positions and make them flow.

Therefore, we conclude that given enough energy an insulator can be transformed into a conductor.

Air, for example, is an insulator and a lightning is an example of how current can flow through an insulator if a colossal energy is provided. Air insulation is about 3 million volts per meter.

A lightning measures, in average, about an inch wide (2.5 cm) and 5 miles (8 km) long and can reach 200 million volts. The most impressive lightning ever recorded occurred in Dallas and crossed an impressive 118 miles (190 km).

A unity called permittivity, measured in Farads per meter (F/m) and referred by the symbol ε_r or κ, denotes how much the molecules of a material oppose an external electric field. A second unity called relative permittivity, also known as dielectric constant, shows how this permittivity relates to the permittivity of the vacuum.

No current will flow across the dielectric until the electric field is strong enough to break it. If so, little "lightnings" or sparks will fly across the plates and current will flow. This is not a desired effect and will cause damage to the capacitor.

The voltage at which the dielectric breaks is called the breakdown voltage.

In real life, capacitors that can work with very high voltages are needed and air is not enough insulator for these cases. Thus, new dielectrics and construction methods had to be developed.

8.3.2 Construction Methods

Capacitors with simply round or rectangle plates parallel to each other are not practical. They are generally huge in size and have low capacitances.

One method for building better capacitors is using long two thin strips of metal foil (A and C) with dielectric material sandwiched between them (B) – everything wounded into a tight roll and sealed in paper or metal tubes – as shown in Figure 8.4.

Figure 8.4 Rolled-up capacitors.

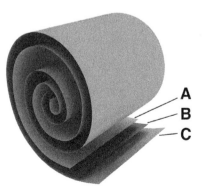

A
B
C

This method, in combination with diverse dielectrics, allows building high capacitance capacitors in a small package.

Different dielectrics also make possible to create capacitors with large capacitances, large breakdown voltages, and better performances under certain circumstances, as high frequencies, for example.

Table 8.1 shows several materials and their permittivities.

Table 8.1 Permittivity.

Material	Permittivity (F/m)
Vacuum	1
Glass	3.7–10
Teflon	2.1
Polyurethane	2.25
Polyamide	3.4
Polypropylene	2.2–2.36
Polystyrene	2.4–2.7
Titanium dioxide	86–173
Strontium titanate	310
Strontium barium titanate	500
Barium titanate	1250–10,000
Diverse polymers	1.8–10,000
Calcium copper titanate	> 250,000

8.4 Electric Characteristics

These are the main characteristics of capacitors:

- They store electric charges in the form of an electric field.
- An ideal capacitor can store an electric charge indefinitely. In real life, they discharge slowly through the dielectric (leakage), because dielectrics are not perfect insulators and let a little bit of current flow.
- Capacitors block direct current.
- Capacitors let alternating current (AC) pass.

Like resistors, capacitors are passive components, because they do not have an active role inside a circuit.

8.5 Electric Field

Considering that the plates of a capacitor have an area equal to A and are separated by a distance d, the electric field between the plates can be found by using the following equation.

ELECTRIC FIELD

$$E = \frac{qd}{\varepsilon A}$$

E is the electric field, in Newtons per Coulomb or volts per meter.
q is the amount of stored charge, in Coulombs.
ε is the dielectric permittivity, in Farads per meter (F/m).
A is the plate area, in squared meters.
d is the distance between the plates, in meters.

8.6 Capacitance

Capacitance is the ratio of the change in an electric charge in a system to the corresponding change in its electric potential.

CAPACITANCE

$$C = \frac{q}{V}$$

C is the capacitance, in Farads.
q is the amount of charge, in Coulombs.
V is the electric potential, in Volts.

Capacitance is measured in Farads[3] and represented by the uppercase letter F in the SI.

---

3 In honor of Michael Faraday (1791–1867), an English scientist that pioneered the first studies about electromagnetism and electrochemistry.

According to this formula, a capacitor of 1 F can store a charge of 1 C when subjected to 1 V.

There are two kinds of capacitance: one produced by the object itself and one produced by proximity to other objects.

Any object that can be electrically charged will have its own value of capacitance.

Also, any object in the proximity of others will generate mutual capacitance.

Every component in a circuit has capacitance. Most of the time this is a parasitic characteristic.

Capacitance can also be expressed by the following formula.

CAPACITANCE

$$C = \frac{\varepsilon A}{d}$$

C is the capacitance, in Farads.
ε is the dielectric permittivity, in Farads per meter (F/m).
A is the plate's area, in squared meters.
d is the distance between the plates, in meters.

8.7   Stored Energy

Capacitors can store a large amount of energy in the form of electric field in a relatively short amount of time, differently from batteries that take a long time to charge.

However, capacitors cannot supply energy or hold their electric charge for a long time because they slowly discharge through their own dielectrics or through the circuit they are connected.

However, their charge and discharge characteristics can be utilized to create very interesting circuits, like audio filters, blinkers, and timers.

Capacitors are frequently used to stabilize voltage levels on power supplies.

The energy stored inside a capacitor can be expressed in terms of work to move charges from a plate to another. In other words, the energy stored inside a capacitor is equivalent to the work necessary to keep positive and negative charges, +q and −q, in their respective plates.

Mathematically, this work and its equivalent charge can be related by the following equation.

STORED ENERGY OR WORK

$$dW = \frac{q}{C}dq$$

W is the work, in Joules.
q is the stored charge, in Coulombs.
C is the capacitance, in Farads.

To find the total energy, we can integrate the previous equation:

$$W = \frac{1}{C}\int_0^Q q.dq$$

MATH CONCEPT   The integral of

$$\int f(x)dx \longrightarrow \frac{x}{2} + C$$

The constant of integration (C) can be ignored when dealing with defined integrals. See Appendix F.

Hence,

$$W = \frac{1}{C} \times \left(\frac{q^2}{2}\right)\Bigg|_0^Q dq$$

$$W = \frac{1}{C} \times \left(\frac{Q^2}{2} - \frac{0^2}{2}\right)$$

TOTAL ENERGY

$$W = \frac{Q^2}{C}$$

W is the total energy, in Joules.
Q is the total charge, in Coulombs.
C is the capacitance, in Farads.

We know that Q = CV, and thus we can convert the previous formula into the following equation.

TOTAL ENERGY

$$W = \frac{CV^2}{2}$$

W is the total energy, in Joules.
V is the voltage, in Volts.
C is the capacitance, in Farads.

8.8   Voltage and Current

Like mentioned before, when a discharged capacitor is connected to a continuous voltage source, such as a battery, one of its plates will be filled with an excess of charges and the other will have a lack of charges. This excess of charges in one of the plates forms a strong electric field that expels electrons from the other plate and makes them move to the battery. Over time, the first negative plate will become saturated with charges and the process will stop.

For all intents and purposes, we can say that a current flowed through the capacitor.

8.8.1   Current on a Charging Capacitor

During the charge process of an initially discharged capacitor, the charging current will start at the maximum value possible and will exponentially decay to 0 as the capacitor charges.

Figure 8.5 shows how the current behaves during charge.

Figure 8.5 Current on a charging capacitor.

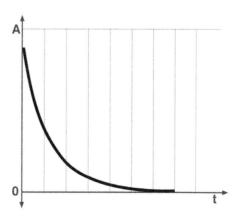

Current across a capacitor follows the following equation.

CURRENT ON A CHARGING CAPACITOR

$$i = C\frac{dv}{dt}$$

i is the current, in Amperes.
C is the capacitance, in Farads.
dv/dt is the ratio of voltage change over time.

The previous equation tells us that current in a capacitor is proportional to the ratio between voltage variations over time.
We know that charge is also equal to Q = CV.
Therefore, if we derive both sides,

$$\frac{dQ}{dt} = C\frac{dV}{dt}$$

Upon substituting this relation in the previous formula, we obtain the following equation.

CURRENT ACROSS A CAPACITOR

$$i = \frac{dQ}{dt}$$

i is the current, in Amperes.
dQ/dt is the ratio of charge change over time.

8.8.2 Voltage on a Charging Capacitor

During the charge process of an initially discharged capacitor, voltage will start at 0 and increase exponentially until it reaches the charging level. As the voltage across the capacitor approaches the charging one, the charging speed will decrease until it reaches 0 and the capacitor is fully charged.

Figure 8.6 shows how the voltage across a capacitor behaves over time in a charging capacitor.

Figure 8.6 Voltage on a charging capacitor.

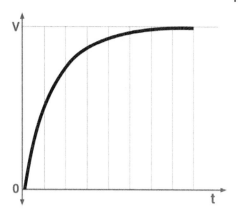

Rearranging the previous equation, we see that the voltage across a capacitor can be written as

$$i = C\frac{dv}{dt}$$

$$\frac{dv}{dt} = \frac{i}{C}$$

$$dv = \frac{i}{C}dt$$

We can integrate both sides and obtain the equation of voltage across a capacitor during charge.

VOLTAGE ACROSS A CAPACITOR

$$v(t) = \frac{1}{C}\int i(t)dt$$

v(t) is the equation of voltage across a capacitor as a function of time, in Volts.
C is the capacitance, in Farads.
i(t) is the equation of current as a function of time, in Amperes.

If we integrate for a specific interval, we get the following equation.

VOLTAGE ACROSS A CAPACITOR

$$v(t) = \frac{1}{C} \int_0^T i(t)dt + V_0 \qquad (8.1)$$

v(t) is the equation of voltage across a capacitor as a function of time, in Volts.
C is the capacitance, in Farads.
i(t) is the equation of current as a function of time, in Amperes.
T is the considered time, in seconds.
V_0 is the capacitor's initial voltage, in Volts.

8.9   Examples

8.9.1   Example 1

Suppose we apply a current of 5 mA for 4 ms across an ideal 10 µF capacitor. The capacitor is initially discharged. We will call this sudden current a "current pulse."

What is the voltage across the capacitor before and after the current pulse?

8.9.1.1   Solution

We know that integrals always calculate the area below the functions they represent.

If we want to find the voltage across a capacitor at a specific instant in time, we must consider the equation for the voltage across the capacitor during charge:

$$v(t) = \frac{1}{C} \int_0^T i(t)dt + V_0$$

This equation describes voltage in terms of current as a function of time. In other words, this equation describes the current's area for a specific interval.

For this example, the area is the one shown on Figure 8.7.

Figure 8.7 Current area.

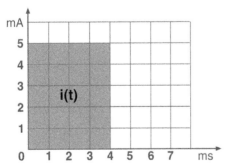

8.9.1.2 Before the Pulse

Before the pulse, the capacitor is completely discharged. Hence, its voltage is 0.

8.9.1.3 During the Pulse

During the pulse, the capacitor charges exponentially and therefore its voltage increases.

Voltage across a capacitor is ruled by the following equation:

$$v(t) = \frac{1}{C} \int_0^T i(t)dt + V_0$$

The current pulse we have applied is constant. Hence, we can substitute the function i(t) by a fixed value i and remove it from the integral. Also, V_0 is equal to 0, because the capacitor is initially discharged.

Upon substituting these values, we get

$$v(t) = \frac{1}{C} i \int_0^T dt$$

MATH CONCEPT The integral of

$$\int dx \longrightarrow x + C$$

The constant of integration (C) can be ignored for defined integrals. See Appendix F.

Solving the integral,

$$v(t) = \frac{i}{C}t \Big|_0^t$$

$$v = \frac{i}{C}(T-0)$$

$$v = \frac{i}{C}T$$

Upon substituting the values in the formula, we obtain the voltage across the capacitor after 4 ms:

$$\begin{cases} i = 5 \text{ mA} \\ C = 10 \text{ μF} \\ t = 4 \text{ ms} \end{cases}$$

$$v = \frac{i}{C} T$$

$$v(4\,\text{ms}) = \frac{5 \times 10^{-3}}{10 \times 10^{-6}} \left(4 \times 10^{-3}\right)$$

FINAL RESULT

$$v(4\,\text{ms}) = 2\,\text{V}$$

8.9.1.4 After the Pulse

After the pulse, current drops to 0 and the capacitor stops charging.
An ideal capacitor will never discharge and hold its charge indefinitely.
Consequently, the capacitor will keep its voltage equal to 2 V forever.

8.9.2 Example 2

We repeat the same example as before but using a 1000 µF capacitor.
What is the voltage across the capacitor after the same current pulse is applied?

8.9.2.1 Solution

Upon substituting the values in the formula, we obtain the voltage across the capacitor after 4 ms:

$$\begin{cases} i = 5\,\text{mA} \\ C = 1000\,\text{µF} \\ t = 4\,\text{ms} \end{cases}$$

$$v = \frac{i}{C} T$$

$$v(4\,\text{ms}) = \frac{5 \times 10^{-3}}{1000 \times 10^{-6}} \left(4 \times 10^{-3}\right)$$

FINAL RESULT

$$v(4\,\text{ms}) = 20\,\text{mV}$$

For the same pulse, the voltage across the capacitor is very small now. This is expected because the capacitance is bigger and, for that reason, the capacitor takes more time to charge.

8.10 AC Analysis

We will examine, now, what happens when a capacitor is connected to an AC source.

Capacitors have the property to block direct current and let pass AC under certain conditions.

Capacitors react when subjected to AC by offering a resistance to the current flow. This resistance, known as reactance, varies with the wave's frequency. This property can be used to create filters and manipulate, for example, audio[4] signals.

8.10.1   Pool Effect

Some of the properties of capacitors can be explained by an analogy with pools.

Imagine an empty pool with two pipes. One of the pipes feeds the pool with water and the other drains the water out. The input pipe is larger and is always closed. It opens, when necessary, to let water in. The output pipe is thin and is always open, letting water out, continuously.

If the pool is empty and we let water in, it will take a while to fill. The time it takes to fill depends on how much water enters the pool per second and the size of the pool.

If the pool is full and we close the input pipe, the pool will continue to hold its level of water for a while but the water level will decrease slowly, because the output pipe is draining the pool.

If the pool is half full and the water flow at the input pipe starts to fail or variate dramatically, the output flux will not be affected and continue to flow as before. We conclude that the pool creates a kind of buffer or cushion effect, keeping the output flow stabilized. Looking through another point of view, we can say that the pool resists drastic changes in the input flow of water, a kind of inertia that will "smooth" drastic changes.

Capacitors have the same behavior. As soon as a voltage is applied, they will start to charge. If the voltage increases beyond the capacitor current level, it will charge to match the input level. If the input voltage decreases, the capacitor will discharge to match the input level. If the input voltage is removed, the capacitor will keep its charge for a while but will slowly discharge.

---

4 Yes, audio is an alternating voltage/current.

Sudden and drastic variations of the input voltage will be softened and damped, and the voltage across the capacitor will remain relatively constant.

8.11   Capacitive Reactance

Capacitors resist to any sudden changes in voltage. This resistance is called capacitive reactance, represented, normally, by the symbol X_c and measured in Ohms (Ω).
Capacitive reactance is represented by the following formula.

CAPACITIVE REACTANCE

$$X_C = \frac{1}{2\pi f C}$$

X_C is the capacitive reactance, in Ohms.
f is input's frequency, in Hertz.
C is the capacitance, in Farads.

or by

$$X_C = \frac{1}{\omega C}$$

---

considering that $\omega = 2\pi f$.
Capacitive reactance and frequency are inversely proportional; if frequency increases, the reactance decreases.
If frequency reduces gradually toward 0,[5] capacitive reactance will skyrocket to infinity. This explains why capacitors block direct current.[6]

8.12   Phase

Like explained before, capacitors resist sudden changes in voltage. This resistance represents a kind of inertia that prevents capacitors from reacting instantly to voltage changes.
When a capacitor starts to charge, current will be maximum and voltage minimum. When the capacitor is charged, voltage will be maximum and current will be 0.

---

5 Remember limits.
6 Direct current is a kind of alternating current with a frequency equal to 0.

It is clear the existence of a delay between voltage and current and that voltage is 90° behind current.

Technically speaking, voltage and current are 90° out of phase.

8.12.1 Mathematical Proof

Like examined before, the voltage across a capacitor can be described in terms of current by this formula:

$$v(t) = \frac{1}{C} \int_0^T i(t) dt \qquad (8.2)$$

If we apply an alternated current across a capacitor, this current follows a sinusoidal form:

$$I = I_p \sin(\omega t) \qquad (8.3)$$

where I_p is the peak current and I is the instantaneous current.

Substituting (8.3) in (8.2), we get

$$v(t) = \frac{1}{C} I_p \int_0^T \sin(\omega t) dt \qquad (8.4)$$

To solve this integral we must use the substitution rule.

Let $u = \omega t$

Thus, $du = \omega dt$

Hence,

$$\frac{du}{\omega} = dt$$

Substituting this on the integral,

$$v(t) = \frac{I_p}{C} \int_0^T \sin(u) \frac{du}{\omega}$$

$$v(t) = \frac{I_p}{\omega C} \int_0^T \sin(u) du$$

MATH CONCEPT   The integral of
$\int \sin(x) dx \longrightarrow -\cos(x)$

By applying the previous concept, we get

$$v(t) = \left(\frac{I_P}{\omega C}\right)(-\cos(u))$$

MATH CONCEPT   The integral of
$\int \cos(x)dx \longrightarrow \sin\left(\theta - \frac{\pi}{2}\right)$

By applying this other concept, we obtain the final equation for the voltage across a capacitor.

VOLTAGE ACROSS A CAPACITOR

$$v(t) = \left(\frac{I_P}{\omega C}\right)\sin\left(\omega t - \frac{\pi}{2}\right) \tag{8.5}$$

v(t) is the equation for the voltage across the capacitor, in Volts.
I_P is the current's peak current, in Amperes.
ω is the angular frequency, in radians per second, equal to $2\pi f$.
C is the capacitance, in Farads.
t is the instant in time, in seconds.

What do we see when we compare equations (8.3) and (8.5)?

We see that the first equation describes current in terms of $\sin(\omega t)$ and the second describes the voltage in terms of $\sin(\omega t - \pi/2)$.

The voltage equation is telling us to subtract $\pi/2$, equivalent to 90°, to the angle, meaning that the voltage is delayed 90° in relation to current.

Figure 8.8 shows a diagram of voltage curve (second curve) that is 90° behind the current curve (first curve).

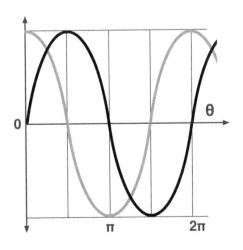

Figure 8.8 Voltage (in black) is behind current (in red).

In real life, the angle between current and voltage angles may not be exactly equal to 90° and may vary, due to imperfections and different capacitance values, but it will always be less than 90°.

8.13 Electrolytic Capacitor

To increase the capacitance of a capacitor, we can do a few things:

- Increase the plate areas, but this is not practical and will make capacitors huge.
- Create new dielectrics that can increase the capacitance by increasing the insulation between the plates.
- Reduce the distance between the plates, but this is a problem, because the distance between the plates depends on how thin we can make the dielectric.

To overcome these problems, a new kind of capacitor was invented, called electrolyte capacitor, where the plates are formed by an aluminum foil with an oxide layer, a tissue soaked in electrolyte liquid, and a regular aluminum foil.

The oxide layer on the first aluminum foil insulates it from the electrolyte liquid, becoming the dielectric and making the electrolyte liquid itself the other "plate," a kind of virtual plate.

For that reason, the electrolyte capacitor is not formed by two metal plates, but rather by an oxidized metal plate and an electrolyte element. The oxidized layer serves as a dielectric and the second regular aluminum foil serves just to supply voltage to the electrolyte.

Therefore, an electrolyte capacitor is composed of a metal foil (A) with an aluminum oxide layer (B), a tissue soaked in electrolyte liquid (C), and a regular metal foil (D), as shown in Figure 8.9.

Figure 8.9 Several layers of an electrolyte capacitor (cross section).

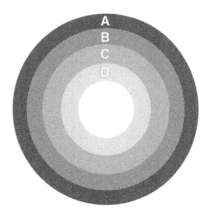

The electrolyte capacitor is far from being a traditional capacitor. In fact, the process that creates the oxide layer on one of the plates makes the electrolyte capacitor only able to conduct current in one direction. If an electrolyte capacitor is subjected to a reverse current flow, it will be destroyed and probably explode.

Therefore, these electrolyte capacitors, also known as "polarized capacitors," have a minus sign on their bodies to identify their polarity.

The advantages of electrolytic capacitors are as follows:

- Their dielectric layer is very thin, allowing the plates to be very close, thus increasing the capacitance.
- They are extremely tough, have a high degree of insulation, and can heal themselves up to a certain degree during their lifespan.
- High capacitance electrolyte capacitors have very small sizes.
- Very cheap to produce.

Figure 8.10 shows a representation of an electrolytic capacitor where a huge minus sign can be seen on the stripe and an X is seen on the metallic end.

Figure 8.10 A radial electrolyte capacitor.

This X at ending is, in fact, a groove to create a weakened part of the casing, which will bulge upward and help the capacitor release its contents in case of failure. If so, they will not explode violently hurting people or causing damage to the whole equipment. These grooves are designed to prevent catastrophic failures from happening.

Figure 8.11 shows the symbols for the electrolytic capacitor in the United States (first symbol) and in the world (last two symbols).

Figure 8.11 Electrolyte capacitor symbols in the United States and internationally.

8.14 Variable Capacitors

Some applications in the real world require capacitors that can have a variable capacitance. For that reason, a new capacitor was born, the variable capacitors or trimmers.

The first line in Figure 8.12 shows the symbols for the variable capacitors used in the United States, and the second line shows the symbols used internationally.

Figure 8.12 Variable capacitors.

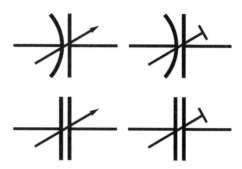

8.15 Capacitors in Series

If we connect two capacitors in series, like shown in Figure 8.13, the equivalent capacitance between points A and B can be found by using the following formula:

$$\frac{1}{C} = \frac{1}{C_1} + \frac{1}{C_2}$$

Figure 8.13 Capacitors in series.

For an infinite number of capacitors, a more generic formula must be used.

$$\frac{1}{C} = \frac{1}{C_1} + \frac{1}{C_2} + \cdots + \frac{1}{C_\infty}$$

C is the total equivalent capacitance, in Farads.
$C_1, C_2, \ldots$ is the capacitance of each individual capacitor, in Farads.

If there are just two capacitors in series, a friendlier formula can be used:

$$C = \frac{C_1 \times C_2}{C_1 + C_2}$$

8.16 Capacitors in Parallel

If we connect two capacitors in parallel, like shown in Figure 8.14, the equivalent capacitance between points A and B can be found by using the following formula:

$$C = C_1 + C_2$$

For an infinite number of capacitors, a more generic formula must be used.

$$C = C_1 + C_2 + \cdots + C_\infty$$

C is the total equivalent capacitance, in Farads.
$C_1, C_2, \ldots$ is the capacitance of each individual capacitor, in Farads.

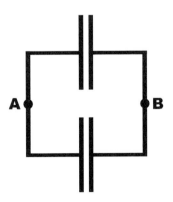

Figure 8.14 Capacitors in parallel.

8.17   Capacitor Color Code

In the past, capacitors, as well as resistors, were identified by a color code painted on their bodies. Today this is not the case anymore.

However, it is worth to know how this code works because it is not rare to find capacitors like that on old equipment.

8.17.1   The Code

The code was composed of four or five stripes painted on their bodies.

The capacitance is represented by the four-color code: the first two stripes represent the first digits, the third stripe is the multiplier, and the last stripe is the maximum operational voltage.

In the five-color code, the first two stripes represent the first two digits, the third stripe is the multiplier, the fourth stripe represents the tolerance, and the last one represents maximum operational voltage (Figure 8.15).

Figure 8.15 Four- and five-stripe capacitors.

Table 8.2 shows the values for each stripe in both codes.

The color code result is always expressed in picofarads.

A capacitor with the stripes shown next will have the following value:

Brown	1
Black	0
Orange	×1000
Brown	±1%

The first three colors mean 10×1000 pF or 10000 pF, which is equivalent to 10 nF.

Table 8.2 Capacitor color code.

Color	First Stripe	Second Stripe	Multiplier	Tolerance >10 pF (%)	Tolerance <10 pF
Black	0	0	×1	±20	±2 pF
Brown	1	1	×10	±1	±0.1 pF
Red	2	2	×100	±2	±0.25 pF
Orange	3	3	×1000	±3	
Yellow	4	4	×10k	±4	
Green	5	5	×100k	±5	±0.5 pF
Blue	6	6	×1M	+80, −20	
Violet	7	7		±10	±1 pF
Gray	8	8	×0.01	±5	
White	9	9	×0.1	±10	
Gold			×0.1		
Silver			×0.01		

The last strip is the tolerance, equal to ±1%, because the capacitor is bigger than 10 pF.

Therefore, the capacitor is 10 nF ± 1% of tolerance.

8.18 Capacitor Markings

Modern capacitors use a numeric code to identify themselves like 473 J or 104 K.

The first two digits represent the capacitance, the third is the multiplier, and the letter is the tolerance.

Tables 8.3 and 8.4 show the multipliers and the tolerances.

Therefore, a capacitor marked as 473 J will be equal to 47 × 1000 pf = 47000 pf or 47 nF, and the J letter would be a tolerance of ±5%.

Capacitor will be 47 nF ± 5% tolerance.

Some capacitors may contain additional numeric codes, written apart from the main code, to represent their maximum operational voltage. Table 8.5 shows these codes.

Appendix H lists several types of capacitors produced worldwide.

Table 8.3 Multipliers.

Third digit	Multiplier
0	×1
1	×10
2	×100
3	×1k
4	× 10 k
5	× 100 k
6	× 1 M
7	
8	×0.01
9	×0.1

Table 8.4 Tolerance.

Letter	Tolerance <10 pF
B	±0.1 pF
C	±0.25 pF
D	±0.5 pF
F	±1%
H	±2%
G	±3%
J	±5%
K	±10%
M	±20%
P	+100%, −0%
Z	+80%, −20%

Table 8.5 Additional numeric codes.

Code	Voltage (V)	Code	Voltage (V)
0G	4	2C	160
0L	5.5	2Z	180
0J	6.3	2D	200
1A	10	2P	220
1C	16	2E	250
1E	25	2F	315
1H	50	2V	350
1J	63	2G	400
1K	80	2W	450
2A	100	2H	500
2Q	110	2J	630
2B	125	3A	1000

Exercises

1   A capacitor with an area equal to 10 mm², a distance between plates equal to 2 mm, and a dielectric permittivity of 2.1 F/m has a charge of 1 C. Find the electric field intensity.

2   A voltage of 80 V is measured across a 100 µF capacitor. Find the stored charge.

3   A 11 µF capacitor has a charge of 22 C. What is the total stored energy?

4   An initially discharged 47 µF capacitor receives a current pulse of 10 mA for 8 ms. What voltage develops across its terminals after that time?

5   What is the total capacitance of three capacitors, 47 µF, 100 µF, and 330 µF, connected in series?

6   What is the total capacitance of three capacitors, 47 µF, 100 µF, and 330 µF, connected in parallel?

7   An alternating voltage with a frequency equal to 300 Hz is applied to a 200 µF capacitor. What is the capacitor's reactive capacitance?

8   A capacitor has the markings 104Z on its body. What capacitor is that?

Solutions

1   The electric field is calculated by the following formula:

$$E = \frac{qd}{\varepsilon A}$$

Substituting the values, we have

$$\begin{cases} q = 1\,C \\ d = 2\text{ mm} = 2 \times 10^{-3}\,m \\ \varepsilon = 2.1\,F/m \\ A = 10\,mm^2 = 10 \times 10^{-6}\,m^2 \end{cases}$$

on the formula

$$E = \frac{(1)(2 \times 10^{-3})}{(2.1)(10 \times 10^{-6})}$$

FINAL RESULT

E = 95.2381 N/C

---

2 The following formula relates charge and voltage for a given capacitance:

$$C = \frac{q}{V}$$

Substituting the values, we get

$$100 = \frac{q}{80}$$

FINAL RESULT

q = 8000 C

---

3 The energy formula that relates charge and capacitance is

$$W = \frac{Q^2}{C}$$

Upon substituting the values, we obtain the total stored energy:

$$W = \frac{22^2}{11}$$

FINAL RESULT

W = 44 J

---

4 The capacitor was initially discharged. During the current pulse the capacitor starts to charge and develops a voltage across its terminals. Capacitor's voltage equation, in terms of current, is

$$v(t) = \frac{1}{C} \int_0^T i(t)dt + V_0$$

The current pulse is constant. We can remove it from the integral. The capacitor is initially discharged; thus,

$$V_0 = 0.$$

Hence, our integral is now

$$v(t) = \frac{1}{C} i \int_0^T dt$$

MATH CONCEPT The integral of

$$\int dx \longrightarrow x + C$$

The constant of integration (C) can be ignored for defined integrals. See Appendix F.

Using this concept, we can solve the integral:

$$v(t) = \frac{i}{C} t \Big|_0^T$$

$$v = \frac{i}{C}(T - 0)$$

$$v = \frac{i}{C}T$$

If we use the values, we have

$$\begin{cases} i = 10 \text{ mA} = 0.01 \text{ A} \\ C = 47 \text{ μF} = 47 \times 10^{-6} \text{ F} \\ t = 8 \text{ ms} = 0.008 \text{ s} \end{cases}$$

We get the voltage across the capacitor after the pulse:

$$v = \frac{0.01}{47 \times 10^{-6}}(0.008)$$

FINAL RESULT

$$v = 1.7021 \text{ V}$$

---

5 Series capacitance is calculated by the following formula:

$$\frac{1}{C_{EQ}} = \sum_{n=1}^{\infty} \frac{1}{C_n}$$

For this current example, we get

$$\frac{1}{C_{EQ}} = \frac{1}{C_1} + \frac{1}{C_2} + \frac{1}{C_3}$$

$$\frac{1}{C_{EQ}} = \frac{1}{47 \times 10^{-6}} + \frac{1}{100 \times 10^{-6}} + \frac{1}{330 \times 10^{-6}}$$

FINAL RESULT

$C_{EQ} = 29.1487 \ \mu F$

6 Parallel capacitance is calculated by the following formula:

$$C_{EQ} = \sum_{n=1}^{\infty} C_n$$

For this current example, we get

$$C_{EQ} = C_1 + C_2 + C_3$$

$$C_{EQ} = 47 \times 10^{-6} + 100 \times 10^{-6} + 330 \times 10^{-6}$$

FINAL RESULT

$C_{EQ} = 477 \ \mu F$

7 Reactive capacitance is given by the formula

$$X_C = \frac{1}{2\pi f C}$$

Substituting the values, we have

$$X_C = \frac{1}{2\pi(300)(200 \times 10^{-6})}$$

FINAL RESULT

$X_C = 2.6525 \ \Omega$

8   Code 104Z means 10 followed by 4 zeros. The answer expressed in picofarads, meaning 100000 pF or 100 nF.

Letter Z means a tolerance between −20% and +80%.

Therefore, the result is shown below.

FINAL RESULT

100 nF with a tolerance between −20% and +80%

Due to the tolerance range, the capacitor can have any value from 80 to 180 nF.

9

Electromagnetism

And the World Revolution

9.1 Introduction

In this chapter, we will examine electromagnetism and the effects of electromagnetic induction, one of the greatest discoveries of physics of all times that changed the world forever.

9.2 The Theory

The effects of magnets over certain materials and their interference in compasses were well known for centuries.

The discovery that magnets have poles and that identical poles repel and opposite poles attract each other came later.

Electricity and magnetism were seen as separate entities, with no relation to each other and no practical usage in daily life.

9.3 Hans Christian Ørsted

In 1890, Hans Christian Ørsted, a Danish physicist, discovered that an electric current was able to produce a magnetic field.

The proof was simple: a battery connected to a wire suspended over compasses and a switch initially open.

Without current, the compasses all point to the Earth's magnetic north, as expected, as illustrated in Figure 9.1.

As soon as the switch is closed, current flows across the wire and gives rise to a magnetic field. This magnetic field interferes with the compasses that move to follow the field, as illustrated in Figure 9.2.

Placing the compasses over the wire would make their needles point to the opposite direction. We will know why in the following paragraphs.

Introductory Electrical Engineering with Math Explained in Accessible Language,
First Edition. Magno Urbano.
© 2020 John Wiley & Sons, Inc. Published 2020 by John Wiley & Sons, Inc.

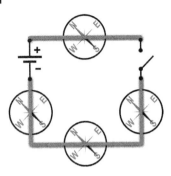

Figure 9.1 Ørsted experiment (open circuit).

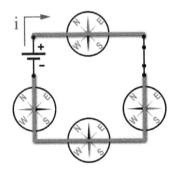

Figure 9.2 Ørsted experiment (switch closed).

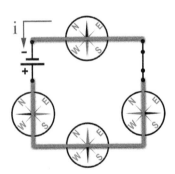

Figure 9.3 Ørsted experiment (battery is inverted).

Keeping the compasses below the wire and inverting the battery make current flow in the opposite direction, giving rise to a magnetic field with an opposite direction. This will force the compasses to rotate 180° to point to the new magnetic field direction, as shown in Figure 9.3.

Once more, placing the compasses over the wire will make their needles point to the opposite direction.

Magnetic fields have a specific orientation that depends on the current flow direction.

To find the magnetic field orientation, we can use the right-hand rule.

9.4 The Right-Hand Rule

Imagine a right hand grabbing a wire. The thumb points to the current flow direction as illustrated by the black arrow pointing up in Figure 9.4.

For a right hand grabbing a wire like this, the magnetic field circulates the wire and points in the direction of the other fingers, as illustrated in Figure 9.4.

Figure 9.4 Right-hand rule.

> The right-hand rule helps to find the magnetic field orientation in a wire.

Magnetic fields are normally designated in technical literature by the uppercase letter B and have the circular form, shown in Figure 9.5, for a current flowing in the direction of I.

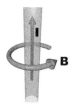

> The rise of a magnetic field by a current flow shows a connection between electricity and magnetism. One cannot exist without the other. The term "electromagnetism" was born.

Figure 9.5 Magnetic field created by a current flow.

9.5 Faraday First Experiment

After Ørsted discovery, Michael Faraday[1] suggested that if a current could generate a magnetic field, perhaps a magnetic field could produce current.

To prove this hypothesis, Faraday wounded a few turns of wire forming two coils and placed one inside the other, as shown in Figure 9.6.

> Coils formed by turns of wires are also known as solenoids, chokes, or inductors.

The first coil was connected to a circuit with a battery and a switch initially open and the second coil was connected to the outside coil and a device that could measure current flow,[2] as seen in Figure 9.7.

---

1 Michael Faraday was an English physicist and chemist. One of the most brilliant and influent scientists of all times, despite not having a formal education.
2 Known as ammeter.

Figure 9.6 A coil inside another one.

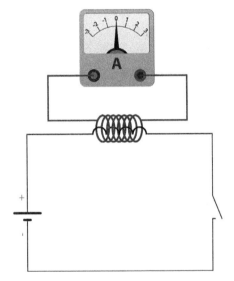

Figure 9.7 Faraday experiment: primary and secondary circuit.

After turning the switch on or off, Faraday noticed a brief current on the secondary circuit. The secondary current was only present the moment the switch was open or closed.

Faraday noticed that current would only appear on the secondary circuit when current was varying in the primary. Then, he concluded that a variable magnetic field could create a current on the secondary circuit.

9.6 Faraday Second Experiment

If a variable current produces magnetic field, then a variable magnetic field should produce current.

To confirm this hypothesis, Faraday imagined the following experiment: a fixed coil connected to a circuit and a magnet. The magnet was adjusted to move in and out of the coil, as shown in Figure 9.8.

Figure 9.8 An inductor and a magnet.

The hypothesis confirmed to be true. Current is created in the coil as soon as the magnet moves in or out. The faster the movement, the bigger the current. The bigger the number of wounds of wire or the strongest the magnet, the bigger the current.

Faraday also noticed that moving the magnet in the coil would produce current in one direction and moving the magnet off the coil would produce current in the other direction, producing fields in opposite directions.

Another thing later discovered is that a coil moving in a fixed magnetic field could also produce current.

9.7 Conclusion

The principles of electromagnetism discovered by Faraday are, possibly, one of the most important discoveries of all times.

These discoveries lead to future creations by Siemens, like the electric elevator, electric generator, and Tesla's alternating current generator, responsible for the first transmissions of electric power over long distances and the development of a new society.

The base for the world revolution was created.

Without electromagnetism, we would not have motors, airplanes, television, Internet, computers, or anything a modern society offers, and we would still be living with carriages and horses.

10

Inductors

Temperamental Devices

10.1 Introduction

In this chapter, we will examine an important kind of passive component called inductor, also known as solenoid, coil, or choke.

10.2 The Inductor

It is known that a current flowing in a wire creates a magnetic field, but what happens to the magnetic field when the wire is wounded to a coil, like the one shown in Figure 10.1? Is the field the same as if the wire was straight?

Figure 10.1 Wire wounded to a coil.

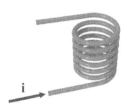

No, the field is not the same. The winding creates a single longitudinal magnetic field that is the combination of all magnetic fields of the coil sections.

In Figure 10.2, the arrow shows the magnetic field created by a coil for a current that enters by the base.

Figure 10.2 Magnetic field generated by a coil for the current entering its base.

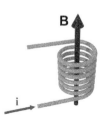

Introductory Electrical Engineering with Math Explained in Accessible Language,
First Edition. Magno Urbano.
© 2020 John Wiley & Sons, Inc. Published 2020 by John Wiley & Sons, Inc.

The magnetic field generated by a coil is basically located inside the coil. Magnetic fields are measured in Tesla and represented by the uppercase letter T in the SI.

Figure 10.3 shows a transversal section of an inductor like the one shown in Figure 10.2 and its respective magnetic field lines. Figure 10.3 also shows how each of the individual magnetic field lines combines to generate a stronger longitudinal magnetic field (B).

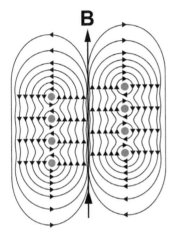

Figure 10.3 Magnetic field in a coil transversal section.

Figure 10.4 shows the symbol used to represent inductors on schematics and in literature in general.

Figure 10.4 Inductor symbol.

10.3 Coils and Magnets

When a magnetic field is created by a coil, it makes the coil a kind of magnet.

In the examples we have given so far, coils were created by winding wires alone in air. But if the coil nucleus is a ferromagnetic rod instead of air, these rods will become magnets when current flows.

Ferromagnetic materials will have their atoms aligned by the magnetic field generated by the coil and behave like real magnets. Therefore, we can create magnets that can be turned on and off.

An invention was born: the electromagnet.

10.4 Inductance

Like we have examined before, capacitors store potential energy in the form of electric field. The higher the capacitance, the higher their storage space.

Inductors have the same ability to store potential energy, but the energy is stored in the form of magnetic field.

> The bigger the number of turns in a coil, the bigger the inductance and its capacity to store energy.

Capacitors hate sudden variations in voltage and generate forces to prevent instant voltage changes. These forces create a delay or inertia between voltage and current variations.

The same happens to inductors. Inductors hate sudden changes in current and will do everything to prevent changes to happen instantly by creating opposite forces, called electromotive forces (EMFs).

The property of creating EMFs opposite to the current flow is called inductance. Inductance is measured in Henry[1] and is represented by the uppercase letter H in the SI.

Inductors are generally represented by the uppercase letter L in schematics, in honor of Heinrich Lenz.[2]

10.5 Variable Inductor

A variable inductor is an inductor that includes a movable element that may be adjusted to different positions to vary its physical dimensions and to change the element's inductance from one value to another. Figure 10.5 shows its symbol.

Figure 10.5 Variable inductor symbol.

---

1 In honor of Joseph Henry (1797–1878), American scientist who discovered electromagnetic induction, independently, at the same time as Michael Faraday (1791–1867) in England.
2 In honor of his pioneering work in electromagnetism.

10.6   Series Inductance

Figure 10.6 Inductors in series.

If we connect two inductors in series, like shown in Figure 10.6, the total inductance measured between A and B will be the sum of its individual inductances, or

$$H = H_1 + H_2$$

For an infinite number of inductors in series, the formula can be written as follows.

SERIES INDUCTANCE

$$H = H_1 + H_2 + \cdots + H_\infty$$

H is the total inductance, in Henries.
$H_1, H_2, \ldots$ is the inductance of each inductor, in Henries.

10.7   Parallel Inductance

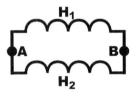

Figure 10.7 Inductors in parallel.

If we connect two inductors in parallel, as seen in Figure 10.7, the total inductance measured between points A and B can be found by using one of the following formulas:

$$\frac{1}{H} = \frac{1}{H_1} + \frac{1}{H_2}$$

or

$$H = \frac{H_1 \times H_2}{H_1 + H_2}$$

For an infinite number of inductors in parallel, we can rewrite the formula as follows.

SERIES INDUCTANCE

$$\frac{1}{H} = \frac{1}{H_1} + \frac{1}{H_2} + \cdots + \frac{1}{H_\infty}$$

H is the total inductance, in Henries.
$H_1, H_2, \ldots$ is the inductance of each inductor, in Henries.

We should notice the similarity between these formulas and the resistor's ones for associations in series or in parallel.

10.8 DC Analysis

We will examine now what happens to an ideal inductor[3] when it is connected to a direct current (DC) voltage/current source.

Figure 10.8 shows a circuit where the inductor is connected in series with a battery and a switch that is initially open.

Figure 10.8 A battery, a switch, and an inductor.

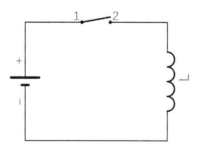

---

3 An ideal inductor has zero electric resistance.

10.8.1   Energizing

As the switch is turned on, current reaches the inductor. Current was 0 before and the inductor was happy. Now, a new nonzero current reaches the inductor, making it very angry. Inductors hate sudden changes in current and will do their best to prevent this change to happen instantly.

To do that, the inductor creates an opposite EMF that manifests itself by generating a voltage across the inductor, which we will call V_L, which is equal to the power supply's but with a negative polarity.

This force will oppose current flow initially, creating a kind of inertia to its passage.

With time, this force will decrease and more current will pass. As the EMF decreases, the inverse voltage across the inductor decreases.

When current reaches its maximum level, the EMF will be 0, so does the voltage across the inductor.[4]

At this point the inductor will be completely energized at its maximum capacity of magnetic field storage. The inductor will then be seen, by the circuit, as a regular wire with no resistance[5] or, in technical terms, a short circuit.

From that point forward, the energy stored inside the inductor will stay there in the form of magnetic field, and magnetic fields hate changes.

Figures 10.9 and 10.10 show current and voltage across an inductor during the energization phase.

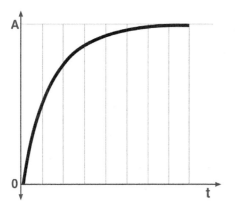

Figure 10.9 Current during the energizing phase.

---

4 Hypothetical situation for an ideal inductor with zero resistance. In real life, voltage will stabilize in a very small value.
5 For the ideal resistor, obviously. In real life, the inductor will be seen as a very small resistor.

Figure 10.10 Voltage during the energizing phase.

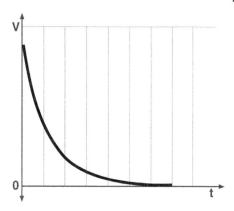

10.8.2 De-energizing

Because inductors hate variations in current, if an inductor is fully energized and receiving current from a power supply and we suddenly disconnect it from the power supply and connect its terminals together, the inductor will see that the current ended and will be very angry.

This time, the inductor will create a favorable EMF to push the current forward and prevent it from stopping. To do that, the inductor will use its entire energy stored, if needed, to keep the current running.

If we are dealing with an ideal inductor with zero resistance, current will flow forever, but in real life, wires have resistance and this resistance will make current decrease slowly until it reaches 0.

Figure 10.11 shows two kinds of inductors using ferromagnetic cores.

Figure 10.11 Cylindrical and toroidal inductors.

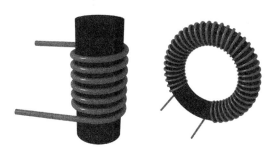

10.9 Electromotive Force

Mathematically, the EMF generated by an inductor in opposition to current, which nowadays is more commonly called "voltage across the inductor," is defined by the following equation.

ELECTROMOTIVE FORCE

$$\varepsilon = -L\frac{di}{dt}$$

ε is the electromotive force, in Volts.
L is the inductance, in Henries.
di/dt is the ratio of current change over time.

The negative sign shows the force is opposite to the current flow.

Voltage across an inductor is proportional to the ratio of current change over time, as seen by the equation.

10.10 Current Across an Inductor

Current across an inductor can be found by the following equation.

CURRENT ACROSS AN INDUCTOR

$$i(t) = \frac{1}{L}\int v(t)dt$$

i(t) is the equation of current across the inductor as a function of time, in Amperes.
v(t) is the equation of voltage across the inductor as a function of time, in Volts.
L is the inductance, in Henries.

10.11 AC Analysis

We will now examine what happens to an inductor when it is subjected to alternating current (AC).

We have seen before that inductors generate opposite forces to prevent sudden increases in current and favorable forces to keep reducing currents flowing.

However, when subjected to AC, inductors will show an internal resistance, called inductive reactance, that will vary with frequency. This inductive reactance is measured in Ohms and normally represented in the technical literature as X_L.

The inductive reactance is ruled by the following formula.

INDUCTIVE REACTANCE

$$X_L = 2\pi fL$$

X_L is the inductive reactance, in Ohms.
f is the current frequency, in Hertz.
L is the inductance, in Henries.

If we substitute $2\pi f$ by ω, we can simplify the formula to the following formula.

INDUCTIVE REACTANCE

$$X_L = \omega L$$

X_L is the inductive reactance, in Ohms.
ω is the angular frequency, in radians per second.
L is the inductance, in Henries.

10.11.1 Alternating Voltage and Current

To understand what happens when an inductor is subjected to AC, we must return to the inductor current equation:

$$i(t) = \frac{1}{L} \int v(t)dt \tag{10.1}$$

Alternating voltages normally follow a sinusoidal pattern like

$$V = V_P \sin(\omega t) \tag{10.2}$$

where V_P is the peak voltage.

If we substitute (10.2) in (10.1), we get

$$i(t) = \frac{V_p}{L} \int \sin(\omega t)dt$$

To solve this integral, we must use the substitution rule:

Let u = ωt.

Thus, du = ωdt

Therefore,

$$\frac{du}{\omega} = dt$$

Substituting this on the main integral, we get

$$i(t) = \frac{V_p}{L} \int \sin(u) \frac{du}{\omega}$$

$$i(t) = \frac{V_p}{\omega L} \int \sin(u) du$$

MATH CONCEPT   The integral of

$$\int \sin(x)dx \longrightarrow -\cos(x)$$

Hence,

$$i(t) = \left(\frac{V_p}{\omega L}\right)(-\cos u)$$

MATH CONCEPT

$$-\cos(\theta) = \sin\left(\theta - \frac{\pi}{2}\right)$$

By applying this concept on the main equation, we get the final equation for the current across the inductor.

CURRENT ACROSS THE INDUCTOR

$$i(t) = \left(\frac{V_p}{\omega L}\right) \sin\left(\omega t - \frac{\pi}{2}\right) \tag{10.3}$$

i(t) is the equation for the current across the inductor as a function of time, in Amperes.

V_p is the peak voltage, in Volts.

ω is the angular frequency, equal to 2πf, in radians per second.

t is the time, in seconds.

10.12 Out of Sync

In this chapter, we have learned that the forces generated by inductors every time current changes for less or for more create a kind of inertia. This inertia creates a delay between current and voltage variations across the inductor. When voltage is maximum, current is minimum and vice versa. Current variations are always delayed compared to voltage in an inductor.

Observing equations (10.2) and (10.3), we see that the first equation describes voltages in terms of $\sin(\omega t)$ and that the second one describes current in terms of $\sin\left(\omega t - \dfrac{\pi}{2}\right)$.

The existence of $-\dfrac{\pi}{2}$, in the current equation, equivalent to $-90°$, tells us that the current is $90°$ delayed compared to voltage.

Figure 10.12 shows a graph of current (second curve) delayed $90°$ in relation to voltage (first curve).

In real life the angle between current and voltage may vary, due to imperfections on the inductors or in the circuits where they are connected.

Figure 10.12 Current behind voltage in an inductor.

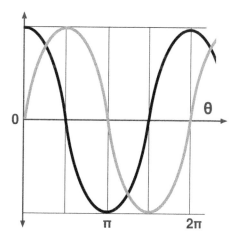

Exercises

1 What is the equivalent inductance of three inductors in series with 100 mH, 1 μH, and 12 mH?

2 What is the equivalent inductance of three inductors in parallel with 100 mH, 1 μH, and 12 mH?

3 A 40 Hz alternating voltage is applied to a 12 mH inductor. Find the inductive reactance for this frequency.

4 A 12 V peak and 50 Hz alternating voltage is applied to a 12 mH inductor. Find the current across the inductor for t = 2s.

Solutions

1 Series inductance can be calculated by the following formula:

$$L_{EQ} = \sum_{n=1}^{\infty} L_n$$

Hence,

$$L_{EQ} = L_1 + L_2 + L_3$$

$$L_{EQ} = 100 \times 10^{-3} + 1 \times 10^{-6} + 12 \times 10^{-3}$$

FINAL RESULT

$$L_{EQ} = 112.001\,mH$$

2 Parallel inductance can be calculated by the following formula:

$$\frac{1}{L_{EQ}} = \sum_{n=1}^{\infty} \frac{1}{L_n}$$

Hence,

$$\frac{1}{L_{EQ}} = \frac{1}{L_1} + \frac{1}{L_2} + \frac{1}{L_3}$$

$$\frac{1}{C_{EQ}} = \frac{1}{100 \times 10^{-3}} + \frac{1}{1 \times 10^{-6}} + \frac{1}{12 \times 10^{-3}}$$

FINAL RESULT

$L_{EQ} = 999.907 \, \text{nH}$

---

3 Inductive reactance can be found by the following formula:

$X_L = 2\pi f L$

Upon substituting the values, we get

$X_L = 2\pi(40)\left(12 \times 10^{-3}\right)$

FINAL RESULT

$X_L = 3.0159 \, \Omega$

---

4 We can find the current using the respective current equation:

$$i(t) = \left(\frac{V_p}{\omega L}\right) \sin\left(\omega t - \frac{\pi}{2}\right)$$

or

$$i(t) = \left(\frac{V_p}{2\pi f L}\right) \sin\left(2\pi f t - \frac{\pi}{2}\right)$$

$$i(2) = \left(\frac{12}{2\pi(50)(12 \times 10^{-3})}\right) \sin\left(2\pi(50)(2) - \frac{\pi}{2}\right)$$

FINAL RESULT

$i(2) = -3.1831 \, \text{A}$

11

Transformers

Not the Movie

11.1 Introduction

In this chapter, we will examine transformers, components created by combining inductors.

11.1.1 Transformers

Inductors have this name because they induce current into other elements.

Suppose we have a coil inside another one, like shown in Figure 11.1.

Both coils are electrically insulated from each other and we apply an alternating current to the first one. This first coil will develop a variable magnetic field, and because the other coil is closer, the second one will absorb part of this magnetic field and produce a variable current. In other words, without any electrical contact, the first coil produced or induced a current on the second one.

A new component was born: the transformer.

Figure 11.1 Two coils, one inside the other.

In a transformer, the first and the second coils are normally called primary and secondary windings.

Introductory Electrical Engineering with Math Explained in Accessible Language,
First Edition. Magno Urbano.
© 2020 John Wiley & Sons, Inc. Published 2020 by John Wiley & Sons, Inc.

11.2 Connected by the Magnetic Field

There is no electrical connection between the primary and secondary coils. However, they are connected by the magnetic field, and they will work by mutual influence.

Like we have explained before, the magnetic field developed on the primary will induce a current in the secondary. Consequently, this secondary current will produce a magnetic field in the secondary, which will induce current on the primary and so on, *ad infinitum.*

The primary and the secondary will induce currents on each other and be locked in an equilibrium state. This connection between the magnetic fields is called flux linkage.

11.3 Faraday's Law

Transformers follow Michael Faraday's law of electromagnetic induction that states: *"The rate of change of flux linkage with respect to time is directly proportional to the induced EMF in a conductor or coil."*

The electrical insulation between the primary and the secondary makes transformers a good component to connect two circuits with different electrical characteristics, keeping one insulated from the other.

11.4 Primary and Secondary

The number of windings of wire on a primary and/or on a secondary of a transformer will dictate how the transformer behaves electrically.

If a transformer is composed of a primary and a secondary with 100 and 50 turns of wire, respectively, any alternating voltage applied to the primary will be reduced in half. If we apply, for example, 500 V across the primary, we will get 250 V across the secondary, because the secondary has half the turns of the primary.

The same is true the other way around: if the secondary has twice the turns of wires of the primary, the secondary voltage will be twice the voltage applied to the primary.

11.5 Real-Life Transformer

In real life, primary and secondary coils are wounded around a rectangular ferromagnetic core with the maximum permittivity possible. This minimizes the losses and maximizes all magnetic fields (see Figure 11.2).

Figure 11.3 shows the symbol used to represent transformers with one primary and one secondary.

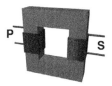

Figure 11.2 A transformer with a primary and a secondary.

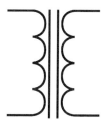

Figure 11.3 Transformer symbol (one primary and one secondary).

11.6 Multiple Secondaries

Nothing prevents a transformer from having multiple secondaries, suitable to obtain multiple output voltages.

Figure 11.4 shows a diagram of a transformer with multiple secondaries, and Figure 11.5 shows its respective symbol.

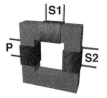

Figure 11.4 A transformer with two secondaries.

Figure 11.5 Symbol used for a transformer with two secondaries.

11.7 Center Tap

There is a kind of transformer with "two secondaries" that is built by connecting, internally, the endings of the secondaries, creating something called center tap.

A center tap is a contact made to a point halfway along a winding of a transformer.

Figure 11.6 shows a transformer with two secondaries or one secondary with a center tap. These secondaries are connected internally, and a center tap secondary is derived (Figure 11.7).

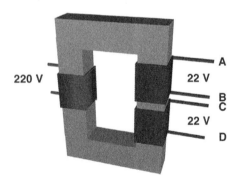

Figure 11.6 Transformer with two secondaries.

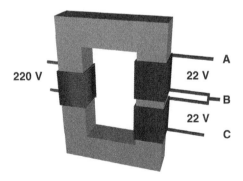

Figure 11.7 Center tap transformer.

Suppose a transformer with two 22 V secondaries like the one in Figure 11.6. Internally we connect two endings of these secondaries (points B and C), making them one point, like shown in Figure 11.7. Now, if we use point B as zero reference, we will measure 22 V between B and A and −22 V between points B and C. This is wonderful to create power supplies with negative voltages. Figure 11.8 shows the symbol for the center tap transformer.

Figure 11.8 Symbol for the center tap transformer.

11.8 Law of Conservation of Energy

The law of conservation of energy states that the total energy of an isolated system remains constant. This means that energy can neither be created nor destroyed, but rather, it can only be transformed or transferred from one form to another.

Power is equal to voltage multiplied to current. Roughly speaking, if we apply a voltage of 100 V and a current of 5 A to the primary of a 2 : 1 transformer, in theory, we will be able to collect 50 V and 10 A at the secondary, because the input power will be 500 W (100 V × 5 A), must be equal to the output power, equal to 50 V multiplied by 10 A, also 500 W.

This is explained by the law of conservation of energy. The power applied to the input of that transformer must be equal to the power collected at the output. In other words, if we apply 500 W of power at the primary, we must be able to collect the same 500 W at the output. So, the product of voltage and current must be equal for the primary and secondary.

This is just a rough calculation. In real life a secondary will always deliver less power than the power applied to the primary, due to losses in the form of heat and leakage flux.

11.9 Leakage Flux

Theoretically, for a 100% efficiency, all magnetic field or flux produced by the primary should be absorbed by the secondary.

In the real world, part of the flux will not be absorbed by the secondary coils and will get lost.

This loss of magnetic field, called "leakage flux," causes a self-resistance of both windings, known as self-reactance or leakage reactance. This leakage reactance causes voltage drops across the primary and secondary.

11.10   Internal Resistance

Windings are created by thousands of wounds of wire and this wire has a specific resistance per meter. That huge number of turns of wire accumulates into a reasonable electrical resistance. Therefore, transformers will also have this internal wire resistance.

11.11   Direct Current

A transformer winding will only induce current on another winding if the current alternates. Therefore, transformers will not work with direct current.

12

Generators

And Motors

12.1 Introduction

In this chapter, we will examine electric generators that are direct applications of electromagnetic induction and their cousins, electric motors.

12.2 Electric Generators

Electric generators are devices that can generate alternating current (AC) by the movement of a wire winding inside a magnetic field, for example.

Figure 12.1 shows the idea behind a simple generator: a winding of wire that rotates inside a magnetic field generated by two static magnets.

Figure 12.1 Electric generator (winding at 0°).

In real life, windings are attached to an axis that rotates by external force like the turbine of a hydroelectric power plant.

As the winding rotates, its turns of wire cut the magnetic field lines, represented by the arrows in Figure 12.1. This process induces electrical current in the winding.

12.2.1 How It Works

Suppose the winding is, initially, at the position 0°, represented by Figure 12.1. The winding is cutting the magnetic field at a position of minimal induced current. The induced current at this point is 0.

Introductory Electrical Engineering with Math Explained in Accessible Language,
First Edition. Magno Urbano.
© 2020 John Wiley & Sons, Inc. Published 2020 by John Wiley & Sons, Inc.

Moved by external forces, the winding rotates anticlockwise and cuts more magnetic field lines. The induced current increases gradually across the winding. When the rotation reaches 90°, as shown in Figure 12.2, current reaches its maximum positive value.

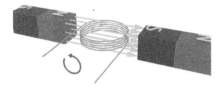

Figure 12.2 Electric generator (winding at 90°).

Rotation continues from 90° to 180°. Current decreases gradually to 0, because the winding moves toward a position where the magnetic field is inverted, compared with what it was at 0°.

Rotation continues and the winding goes from 180° to 270°. Now the winding position is inverted compared with how it was at 90°. Consequently the magnetic field and the current are inverted. The current is at its negative peak.

Rotation continues from 270° to 0° and current will decrease again to 0.

The cycle repeats continuously after that.

In terms of rotation angle, the generator will produce an induced current like the one shown in Figure 12.3.

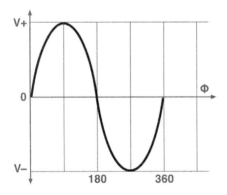

Figure 12.3 Electric generator.

A generator is composed of a static part formed by magnets, called stator, and a rotating part, the winding, called rotor.

12.3   Electric Motor

An electric motor is basically a generator connected backward.

Generators use mechanical movement to rotate a winding inside a magnetic field to produce electricity. Electric motors use electricity to make the winding rotate and produce mechanical movement.

12.3.1   DC Motors

Imagine a winding embedded on a fixed magnetic field between two magnets. The winding is sitting horizontally (0°) as shown in Figure 12.4.

Figure 12.4 Winding of an electric motor.

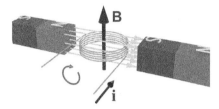

Current is injected on the winding, producing a magnetic field pointing up, as shown in Figure 12.4. The magnetic field produced by the magnets will force this winding magnetic field to be oriented in the same direction. This will force the winding to rotate 90° clockwise. Now, both fields are oriented to the same direction (see Figure 12.5).

Figure 12.5 Electric motor.

The motor is constructed in a way that the current will invert direction when the winding reaches this position.

With the current inverted, the winding magnetic field rotates 180°, forcing the winding to rotate again, to align with the fixed magnetic field.

Current inverts again and the process repeats continuously.

This kind of motor is called induction motor, and because it has only a single winding, it is called a single-phase induction motor.

12.3.2 AC Motors

AC motors use AC and work slightly different.

AC motors can be single or triple phase. Single-phase motors work by having a fixed winding producing an alternating magnetic field. This alternating field induces current on the rotor's winding that creates its own magnetic field that is always trying to align itself with the stationary field.

In triple-phase motors, there are three stationary windings, each one assembled in a 120° to each other. This creates a rotating magnetic field that pushes the rotor.

13

Semiconductors

And Their Junctions

13.1 Introduction

In this chapter, we will examine semiconductors, diodes, transistors, and other components that changed electronics forever.

13.2 It All Started with a Light Bulb

A light bulb is composed of a filament inside a vacuum glass enclosure. This filament is made of a heat-resistant material, like tungsten, that can become incandescent for long periods of time without melting.

The incandescence produces light and emits a cloud of electrons around the filament.

Figure 13.1 shows a light bulb with an incandescent filament and a cloud of electrons, represented by orange circles with the minus sign.

Electrons are negative charges. What would happen if a metallic plate was added to the light bulb and charged with a positive charge? Remember that opposite charges attract themselves.

Figure 13.2 shows a light bulb with a metallic plate at the opposite end. The metallic plate is connected to the positive of a couple of batteries in series.

Figure 13.1 A light bulb.

The single battery is used to make the filament incandescent. The other two batteries are used to make the plate positive.

Introductory Electrical Engineering with Math Explained in Accessible Language,
First Edition. Magno Urbano.
© 2020 John Wiley & Sons, Inc. Published 2020 by John Wiley & Sons, Inc.

As expected, the positive plate will attract the electrons from the filament cloud. The bigger the plate's positive charge, the more intense is the force attracting the electrons from the cloud.

If we make the plate negative, instead of positive, it will repel the electrons from the cloud, because charges with the same polarity repel each other.

A new component was created: the vacuum tube or simply known as tube or valve (in the United Kingdom).

Figure 13.2 A primitive vacuum tube.

This plate is technically known as the anode and the filament as the cathode.

If we have electrons moving from one element to the other, we have current flow in the opposite direction.

However, notice that the electrons move from the filament to the plate and not the other way around. Consequently, current can only move in one direction. If so, we have this device that allows current to flow in one direction (when the anode is positive) but not to the other (when the anode is negative). A new component is born: the diode.

The first tubes were called thermionic cathode tubes or hot cathode tubes, because the cathode was responsible for the emission of electrons. As this idea progressed, a third metallic element was added to the vacuum tube. This third element was placed closer to the filament, to be heated and emit electrons. This element was then named to be the cathode and the filament demoted to be just a heater.

Removing the filament from the equation created an insulated cathode, which could be used by an independent circuit.

Figure 13.3 Diode symbol (vacuum tubes).

Figure 13.3 shows the diode symbol for the vacuum tube, where anode, cathode, and heater can be seen from top down.

The next step on the vacuum tube evolution was placing a metallic grid between the anode and the cathode to control the flux of current between these two elements.

This grid, also called control grid, could be made positive or negative, blocking or letting pass electrons.

A new component was born: the triode.

Figure 13.4 shows the symbol for the triode, where the anode, the grid, the cathode, and the heater can be seen from top down.

Figure 13.4 Triode symbol (vacuum tubes).

13.3 Semiconductors

During the Second World War, vacuum tubes were used as the central element of electronic circuits.

Tubes did the job relatively well, but they were fragile and required high voltages to work – not to mention they require a lot of current and power to operate and their filaments do not last that long.

In 1948, William Shockley, researcher at the legendary Bell Labs, invented something he called "circuit element utilizing semiconductive material", the basis of what we call, today, transistor.

13.3.1 Bipolar Junction

The first transistors were made of germanium (Ge). Today, transistors are made of silicon (Si).

Silicon is a hard and brittle crystalline solid with a blue-gray metallic luster and is classified chemically as a metalloid. Metalloids are chemical elements with properties intermediate between those of typical metals and nonmetals.

Every silicon atom contains four electrons in their valence shell, as illustrated in Figure 13.5.

When a structure of silicon is formed, every atom bonds to its neighbor by sharing one electron from the valence shell. This kind of bond is known as covalent bond, represented in Figure 13.6.

Figure 13.5 A silicon atom.

Figure 13.6 Silicon structure.

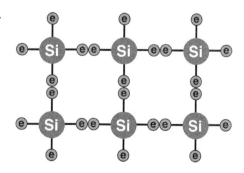

In a structure like this, all electrons are combined to their neighbors, and there are no free electrons to move. This kind of structure can be classified as an insulator with a very high electrical resistance or as an element with a very little electrical conductivity.

13.3.1.1 Making the Structure More Negative

To make a silicon structure a better conductor, we must add more electrons to it.

The solution was to add phosphorus atoms to the silicon structure.

Phosphorus atoms contain five electrons on their valence shell (Figure 13.7).

The technique of adding phosphorus to the silicon structure is called doping. The result is represented in Figure 13.8.

It is now clear that each phosphorus atom will add a free electron to the structure. Millions and millions of these free electrons will now be embedded in the silicon structure, increasing its conductivity.

Figure 13.7 Phosphorus atom.

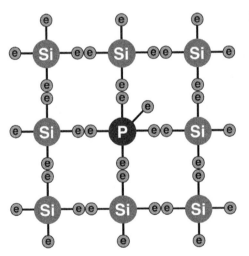

Figure 13.8 N-type silicon structure.

By adding phosphorus, the number of electrons increased. Electrons have negative charge. So, this kind of doping is known as N-type doping (N = negative).

13.3.1.2 Making the Structure More Positive

Boron atoms, on the other hand, differently from phosphorus, have just three electrons on the valence shell. See
Figure 13.9.

Figure 13.9 Boron atom.

If we take silicon and dope it with boron, instead of phosphorus, we will create a structure that have empty spots,
like holes, that represent "parking spots" for electrons. In
other words, a structure that lacks electrons consequently
is less negative or more positive (see Figure 13.10).

Figure 13.10 P-type structure.

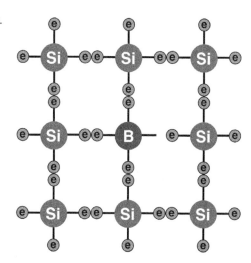

Creating spots or holes in the structure makes it less negative. For this reason,
this type of doping is called P-type doping (P = positive).

13.3.1.3 Pure Magic

Something magical happens when we put together an N-type and a P-type block
of silicon: the excess of electrons from the N-type block migrates to the P-type
block, at the junction between the two materials.
This will make the N-type material less negative
and the P-type material more negative, at the
border.

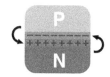

Figure 13.11 shows both materials put together,
P-type and N-type, and the depletion region formed
at their junction.

Figure 13.11 PN junction.

This will create a permanent electric field at the junction, a barrier that will make impossible the migrations of more electrons. This barrier is known as depletion region and represents an insulation barrier between the two materials.

> For silicon junctions, this barrier will have a potential equal to ≈0.7 V. Old germanium junctions have their barrier potentials equal to ≈0.3 V.

In other words, the barrier existence means that current will only flow across the junction where a voltage potential greater than the barrier potential is applied.

13.3.1.4 Reverse Biasing

If we connect the positive and negative poles of a battery to the N-type and P-type blocks of a PN junction, respectively, we will increase the depletion region size.

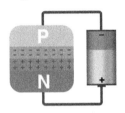

The battery will polarize the blocks in such a way that more holes from the P-type will be attracted to the negative pole and electrons from the N-type will be attracted to the positive pole, enlarging the depletion barrier, raising the barrier's electric potential, and making current circulation impossible. See Figure 13.12. This is called reverse biasing.

Figure 13.12 PN junction – reverse biased.

13.3.1.5 Forward Biasing

If, on the other hand, we invert the battery, we will decrease the barrier potential by attracting electrons from the P-type into the N-type. The battery will force the barrier to contract. In this case, current will flow as soon as the battery can provide a potential bigger than 0.7 V (see Figure 13.13). This is called forward biasing.

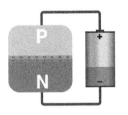

Figure 13.13 PN junction – forward biased.

13.3.1.6 Biasing Curve

The previous description of a junction that can block current flow when reverse biased and allow current flow when forward biased represents the ideal case. In real life, imperfections will prevent junctions to behave like that.

If we forward bias a PN junction using a variable power supply, starting with 0 V and gradually increasing the voltage, the junction barrier will be contracted to a point where it will start to conduct, and this will happen when the power supply is providing around 0.7 V. This is shown in region A in Figure 13.14.

Figure 13.14 PN junction biasing curve.

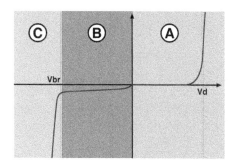

Point Vd is a point beyond which the junction will start to conduct from the P-type to the N-type.

If we invert the power supply and gradually start to increase the voltage, the junction that should block current at all costs will not resist and let current flow, but in the opposite direction, from the N-type to the P-type. Current is negligible, near zero, on the order of microamperes. See region B on Figure 13.14.

If we continue to increase the power supply, keeping the power supply reversed, at some point, the junction will not resist and fail catastrophically, letting current pass at full blast, in the opposite direction. This point is called breakdown voltage or avalanche voltage and is marked as Vbr on region C in Figure 13.14.

13.3.1.7 Thermal Voltage

Atoms are constantly vibrating inside the silicon structure and this vibration produces heat.

If we subject the material to low temperatures, for example, 0 K, all vibrations stop.

As temperature rises, electrons acquire energy, start to vibrate, and become available to move freely across the structure. This proportion of free electrons and temperature is ruled by the Boltzmann constant.[1]

This vibration has a direct relation with temperature and generates an electric potential, called thermal voltage (V_T), expressed by the next formula.

---

1 Equal to $1.380649 \times 10^{-23} J/K$ (Joules per Kelvin), in honor to Ludwig Eduard Boltzmann, an Austrian physicist and philosopher who developed a theory that explains and predicts how the properties of atoms determine the physical properties of matter. The Boltzmann constant was modified recently by the SI. This value was previously equal to $1.38066 \times 10^{-23} J/K$.

THERMAL VOLTAGE

$$V_T = \frac{kT}{q}$$

k is the Boltzmann constant.
T is the junction absolute temperature, in Kelvin.
q is the electron charge, equal to $1.602176634 \times 10^{-19}$ C (Coulombs).[2]

Conversion from Celsius to Kelvin can be calculated by the following formula.

CONVERSION FROM CELSIUS TO KELVIN

$$T_k = T_c + 273.15$$

T_k is the temperature, in Kelvin
T_c is the temperature, in Celsius.

The thermal voltage for a room temperature of, for example, 300 K would be calculated as

$$V_T = \frac{kT}{q}$$

$$V_T = \frac{(1.380649 \times 10^{-23}) \times (300)}{1.602176634 \times 10^{-19}}$$

$$V_T = 0.025198106\,V \approx 26\,mV$$

13.3.1.8 Barrier Voltage

The famous barrier voltage of the PN junction being equal to 0.7V comes from the following formula.

---

2 The electron charge value was modified recently by the SI. This value was previously equal to $1.60217662 \times 10^{-19}$ C.

BARRIER VOLTAGE

$$V_b = V_T \ln\left[\frac{N_A N_D}{n_i}\right]$$

V_B is the barrier voltage, in Volts.
V_T is the thermal voltage, in Volts.
N_A is the acceptor density, or the P-type doping density.
N_D is the donator density, or the N-type doping density.
n_i is the intrinsic carrier density or the number of electrons or holes per volume.

The values of N_A, N_D, and n_i depend on the technology used to produce the N- and P-type compounds.

13.3.1.9 Relation Between Current and Voltage

The relation between current and voltage across a PN junction, or diode, is expressed by the following formula, known as the "diode equation."

RELATION BETWEEN CURRENT AND VOLTAGE ON A DIODE

$$I = I_0 \left(e^{\frac{qV}{kT}} - 1\right)$$

I is the junction current, in Amperes.
I_0 is the junction saturation current, in Amperes.
q is the electron charge.
V is the voltage across the junction, in Volts.
T is the absolute junction temperature, in Kelvin.
k is the Boltzmann constant.

14

Diodes and Transistors

Active Components

14.1 Introduction

In this chapter, we will examine diodes and transistors, the first two practical applications coming from the invention of N-type and P-type materials.

14.2 Diodes

We have explained before that PN junctions allow current to flow in one direction but not on the other. This is the exact definition of a diode.

When a diode is forward biased, current flows and behaves, roughly, like a closed switch or like a resistor of small value.

However, if reverse biased, a diode will block current flow and behave like an open switch or a resistor of large value.

Figure 14.1 shows the symbol used for diodes. The triangle is the anode and the line represents the cathode, equivalent to materials P and N, respectively. The anode is the positive side and the cathode is the negative side.

Figure 14.1 Diode symbol.

14.3 NPN Junction

Suppose we add one more N-type block to the PN structure and create an NPN block like shown in Figure 14.2. What happens now?

Now, we have two electrical barriers between the two N-type and the P-type materials.

Figure 14.2 NPN structure.

Introductory Electrical Engineering with Math Explained in Accessible Language,
First Edition. Magno Urbano.
© 2020 John Wiley & Sons, Inc. Published 2020 by John Wiley & Sons, Inc.

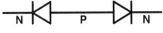

Figure 14.3 NPN junction "diodes."

If the junction PN is equal to a diode, this NPN block is equal to two connected diodes like shown in Figure 14.3.

Instead of NPN, we can also create PNP blocks. PNP blocks behave exactly as NPN blocks, but their current flows in the opposite direction.

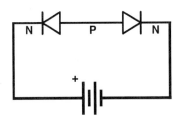

Figure 14.4 NPN structure and a battery.

14.4 Biasing

Suppose that we take our NPN junction and connect it to a battery, like shown in Figure 14.4.

If the battery can provide a voltage superior to 0.7 V, it will make the first N-type element more positive and the second N-type element more negative than normal. This will make the diodes be polarized as illustrated in Figure 14.5.

The cathode of the first diode is now positive and its anode is negative, meaning that it is reverse biased and will not conduct current. However, the second diode has its anode positive and its cathode negative, meaning it is forward biased and can conduct current.

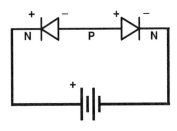

Figure 14.5 NPN structure and a battery.

If we reverse the battery, the first diode will be forward biased and the second one reverse biased.

14.5 The Transistor, Finally!

Figure 14.6 NPN transistor.

The NPN structure we have already shown in Figure 14.3 is, in fact, what we call NPN transistor. Its symbol is shown in Figure 14.6.

In Figure 14.6, we see that transistors are formed by three parts, called collector (C), base (B), and emitter (E). The collector and the emitter are the N-type materials and the base

is the P-type material. The base is the point where both
elements connect.

There is another kind of transistor called PNP transistor.
We will not talk about PNP transistors in this book because
they work similarly to the NPN equivalents. The only difference is that current flows in the opposite directions and their
voltages are reversed. Figure 14.7 shows their symbol.

Figure 14.7 PNP
transistor.

Like shown in Figure 14.7, the symbol for the PNP transistor shows the emitter
arrow pointing to the base, the exact inverse of the NPN symbol.

14.5.1   How Transistors Work?

We will examine in detail how transistors work late in this book, but broadly
speaking, a transistor is a device that lets current flow from collector to emitter
controlled by the current that flows into its base. The larger the current flowing
into the base, the larger the current flowing from collector to emitter.[1]

Current will flow from collector to emitter only if the base voltage is, at least,
0.7 V greater than the emitter voltage.

When a transistor conducts, or enters the "on" state, the collector–emitter
junction is seen, by the circuit, as a resistor of a small value.

When a transistor cuts off, or enters the "off" state, that is, does not conduct,
the collector–emitter junction is seen as an open
circuit.

The amount of current flowing into the base will
control the amount of current flowing from collector to emitter. A small base current will make a
huge current flow from collector to emitter.

How huge a collector current is compared with
the base current is called current gain, known as
h_{FE}[2] or β (beta). Values like 10, 20, 60, 300, 600,
or more for h_{FE} are common, depending on the
transistor.

Figure 14.8 shows a general-purpose low power
transistor in a TO-92 package.

Figure 14.8 Low power
transistor (TO-92 package).

---

1 In the PNP transistor, current flows from emitter to collector.
2 Hybrid parameter forward current gain, common emitter.

15

Voltage and Current Sources

Circuit Analysis

15.1 Introduction

In this chapter, we will examine voltage and current sources and elements used actively to simplify and perform circuit analysis.

15.2 Independent DC Voltage Sources

An ideal voltage source is a two-terminal device that can provide a constant voltage and zero internal resistance and can supply or absorb an infinite amount of power.

In real life, voltage sources will have voltages reduced if the load is too demanding and have a limited capacity to provide or absorb voltage.

Figure 15.1 shows their symbol and a circuit using a DC voltage source.

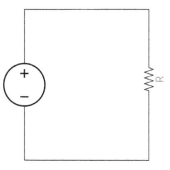

Figure 15.1 DC voltage source and a resistor.

A battery is an example of DC voltage source.

15.3 Independent AC Voltage Sources

An AC voltage source works by the same principles as a DC voltage source but provides alternating voltage that follows a uniform pattern and has a constant frequency.

Introductory Electrical Engineering with Math Explained in Accessible Language,
First Edition. Magno Urbano.
© 2020 John Wiley & Sons, Inc. Published 2020 by John Wiley & Sons, Inc.

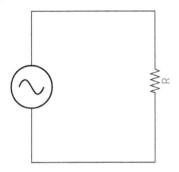

Figure 15.2 AC voltage source and a resistor.

Figure 15.2 shows their symbol and a circuit using an AC voltage source.

15.4 Dependent Voltage Sources

A dependent voltage source is a source that provides voltages which values depend on a voltage or current elsewhere in the circuit.

Dependent voltage sources are useful, for example, in examining the behavior of amplifiers or other kinds of circuits.

Figure 15.3 shows a circuit using a dependent voltage source which values depend on an input voltage.

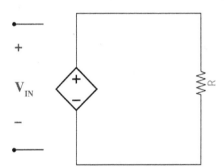

Figure 15.3 Dependent voltage source with voltage control and a resistor.

Figure 15.4 shows a circuit using a dependent voltage source which values depend on an input current.

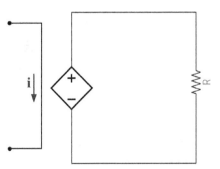

Figure 15.4 Dependent voltage source with current control and a resistor.

15.5   Independent Current Sources

An ideal current source is a two-terminal device that can provide a constant current that will always keep its value and has an infinite internal resistance.

In real life, current sources can see their currents reduced if the load is too demanding and have a limited capacity to provide or absorb current.

Figure 15.5 shows an independent current source and a resistor.

Figure 15.5 Current source and a resistor.

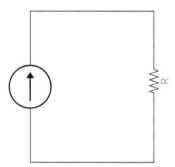

Figure 15.6 shows alternative symbols for current sources.

Figure 15.6 Alternative symbols for current sources.

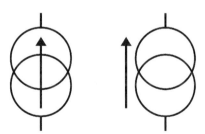

15.6   Dependent Current Sources

Dependent current sources are like independent current sources, but their values depend on voltages or currents elsewhere in the circuit.

Figure 15.7 shows a dependent current source which values depend on the input voltage.

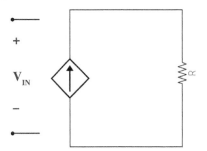

Figure 15.7 Dependent current source with voltage control and a resistor.

Figure 15.8 shows a dependent current source which values depend on an input current.

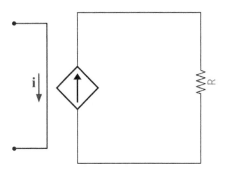

Figure 15.8 Dependent current source with current control and a resistor.

16

Source Transformations

Circuit Analysis

16.1 Introduction

In this chapter, we will use voltage and current sources and perform transformations and simplify circuits, a very powerful technique of circuit analysis.

16.2 The Technique

The technique involves converting voltage sources to current sources and vice versa.

According to the technique:

- A voltage source in series with a resistor can be converted into a current source in parallel with the same resistor and vice versa, as shown in Figure 16.1.

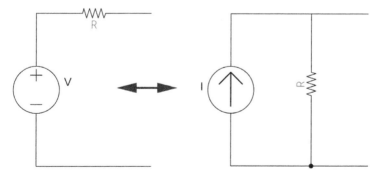

Figure 16.1 Source transformation.

Introductory Electrical Engineering with Math Explained in Accessible Language,
First Edition. Magno Urbano.
© 2020 John Wiley & Sons, Inc. Published 2020 by John Wiley & Sons, Inc.

- The current source arrow should always point to the voltage source's positive pole it is replacing and vice versa.
- The intensities of the voltage and current sources should be found by the following the formula.

SOURCE TRANSFORMATION FORMULA

$$V_{VS} = RI_{CS}$$

V_{VS} is the voltage source value, in Volts.
R is the resistor, in Ohms.
I_{CS} is the current source value, in Amperes.

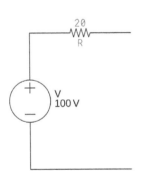

Figure 16.2 Source transformation (circuit converted).

The idea behind a source transformation is to create a substitute circuit that always behaves electrically exactly like the one being replaced.

Consider, for example, a 100 V voltage source in series with a 20 Ω resistor, as shown in Figure 16.2. To convert this voltage source into a current source, we must find the current source value, by using the formula

$$I_{CS} = \frac{V_{VS}}{R}$$

Thus,

$$I_{CS} = \frac{100}{20}$$

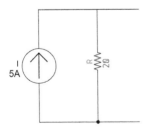

Figure 16.3 Source transformation result.

FINAL RESULT

$$I_{CS} = 5 \text{ A}$$

Hence, the current source will be 5 A, and its arrow should point up, that is, the direction of the positive pole of voltage source it is replacing. The resistor is the same and should be connected in parallel. The result is shown in Figure 16.3.

16.3   Example

Suppose we want to know the value of current I_1 in the circuit shown in Figure 16.4.

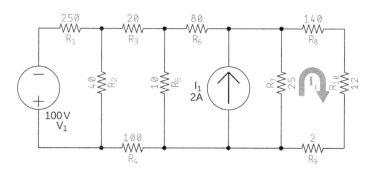

Figure 16.4 Complex circuit for simplification.

16.3.1   Solution

Let us solve this problem by using source transformations.

The circuit starts with a 100 V voltage source in series with a 250 Ω resistor (Figure 16.5).

This can be replaced by a resistor in parallel with a current source with the following value:

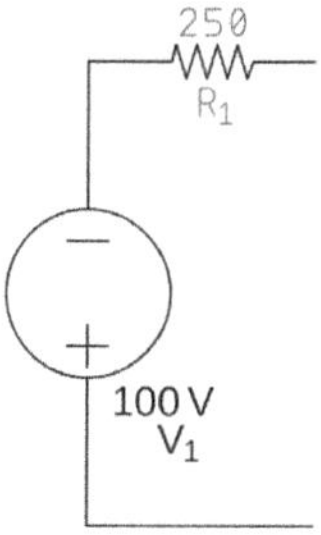

$$I = \frac{V_1}{R_1}$$

$$I = \frac{100}{250}$$

Figure 16.5 Voltage source in series with a resistor.

$$I = 0.4\,A$$

The circuit can then be converted into the one shown in Figure 16.6.

The current source points down because the original voltage source had its positive pole down (see Figure 16.5).

By observing Figure 16.6, we notice two resistors in parallel, R_1 equal to 250 Ω and R_2 equal to 40 Ω.

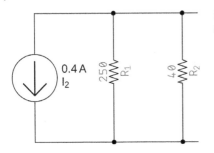

Figure 16.6 First part of the circuit after first conversion.

Resistors in parallel can be replaced by an equivalent resistor with the following value:

$$\frac{1}{R} = \frac{1}{R_1} + \frac{1}{R_2} + \cdots + \frac{1}{R_\infty}$$

$$\frac{1}{R} = \frac{1}{250} + \frac{1}{40}$$

$$R_{EQ1} = 34.48 \ \Omega$$

The circuit is now like it is shown in Figure 16.7.

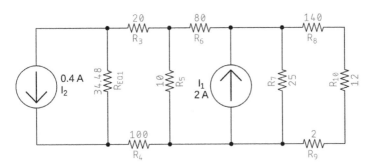

Figure 16.7 Equivalent circuit after second conversion.

Now we see a 0.4 A, current source in parallel with a 34.48 Ω resistor. Using the source transformation rules, we can convert this to a voltage source in series with the same resistor.

We use the formula $V_1 = R_{EQ1}I_2$ to calculate the voltage source value:

$$V_1 = 34.48 \times 0.4$$

$$V_1 = 13.796 \text{ V}$$

The circuit is now transformed into the one seen in Figure 16.8.

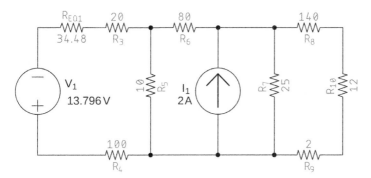

Figure 16.8 Circuit shows a voltage source in series with two resistors.

Notice that the positive pole of the voltage source points down, the same direction as the current source arrow it replaces.

Now we have resistors R_{EQ1} and R_3 in series. In terms of circuits, R_4 can also be considered in series because voltage sources have zero internal resistance.

Thus, the equivalent resistor is the sum of R_{EQ1}, R_3, and R_4:

$$R_{EQ2} = 34.48 + 20 + 100$$

$$R_{EQ2} = 154.48 \ \Omega$$

The circuit is now transformed into the one seen in Figure 16.9.

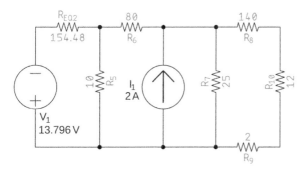

Figure 16.9 Converted circuit.

We see now, again, a voltage source (V_1) in series with a resistor (R_{EQ2}), and we can convert that into a current source in parallel with the same resistor.

The current source value is found by

$$I_2 = \frac{V_1}{R_1}$$

$$I_2 = \frac{13.796}{154.48}$$

$$I_2 = 0.0893 \text{ A}$$

This transformation leads to the following circuit, shown in Figure 16.10.

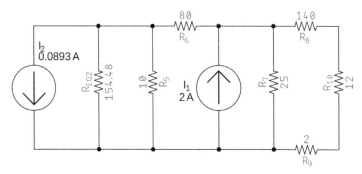

Figure 16.10 Converted circuit shows a current source in parallel with two resistors.

See R_{EQ2} and R_5 in parallel? We can replace them with the equivalent resistance

$$\frac{1}{R_{EQ3}} = \frac{1}{R_{EQ2}} + \frac{1}{R_5}$$

$$\frac{1}{R_{EQ3}} = \frac{1}{154.48} + \frac{1}{10}$$

$$R_{EQ3} = 9.39 \ \Omega$$

The result is shown below (Figure 16.11).

The first current source (I_2) in parallel with R_{EQ3} can be replaced with a voltage source with a series resistor.

The voltage source can be found by

$$V_1 = 9.39 \times 0.0893$$

$$V_1 = 0.8387 \text{ V}$$

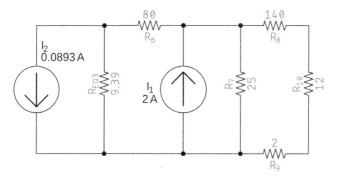

Figure 16.11 Converted circuit.

The result is seen in Figure 16.12.

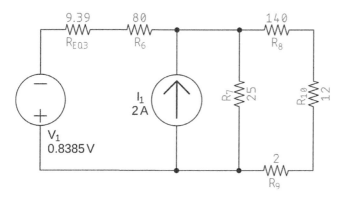

Figure 16.12 Converted circuit.

No need to mention that R_{EQ3} and R_6 are in series and can be replaced by an equivalent resistor equal to their sum, that is, 89.39 Ω.

No need to mention also that this resistor in series with the voltage source can be replaced by a current source in parallel with the same resistor.

The current source will be

$$I_2 = \frac{V_1}{R_{EQ3} + R_6}$$

$$I_2 = \frac{0.8385}{89.39}$$

$$I_2 = 9.3810 \text{ mA}$$

The converted circuit is shown in Figure 16.13.

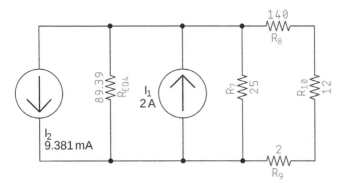

Figure 16.13 Converted circuit.

We now have four elements in parallel: two current sources, I_1 and I_2, and two resistors, R_{EQ4} and R_7.

Parallel resistors can be replaced by their equivalent

$$\frac{1}{R_{EQ5}} = \frac{1}{89.39} + \frac{1}{25}$$

$$R_{EQ5} = 19.53 \ \Omega$$

The current sources can be replaced by their equivalent by adding them. The second current source has a negative value, because it is pointing down.

The equivalent current source will be

$$I = I_1 + I_2$$

$$I = 2 - (0.008697)$$

$$I = 1.9906 \ A$$

The result is shown in Figure 16.14.

Notice that the final current source points up, because the sum of the two current sources was positive.

The last circuit shows another resistor in parallel with a current source, and those can be converted into a voltage source in series with the same resistor.

The voltage source will be

$$V_1 = 19.53 \times 1.9906$$

Figure 16.14 Converted circuit.

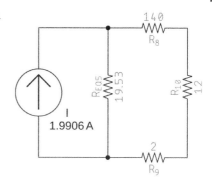

$V_1 = 38.89\ \text{V}$

The final circuit is converted into what is shown in Figure 16.15.

Figure 16.15 Final circuit.

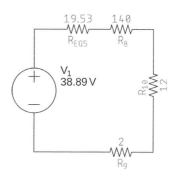

We have simplified the circuit enough. It is now easy to know the current circulating at this point.

By Ohm's law, we get

$$I = \frac{V}{R}$$

$$I = \frac{V}{R_{EQ5} + R_8 + R_9 + R_{10}}$$

$$I = \frac{38.89}{(19.53 + 140 + 12 + 2)}$$

FINAL RESULT

$I = 0.2241\ \text{A}$

Exercises

1 Convert the voltage source in series with the resistor shown in Figure 16.16 into a current source in parallel with a resistor.

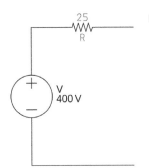

Figure 16.16 Voltage source in series with a resistor.

2 Convert the current source in parallel with the resistor shown in Figure 16.17 into a voltage source in series with a resistor.

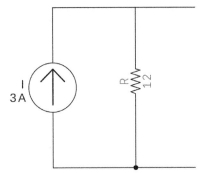

Figure 16.17 Current source in parallel with a resistor.

3 Figure 16.18 shows a circuit formed by several sources and resistors. Convert the circuit using the source transformation techniques to the maximum possible.

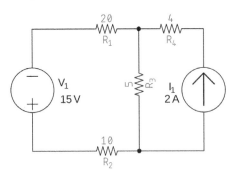

Figure 16.18 Complex circuit for source transformation (exercise).

Solutions

1 If we use the source transformation rules, we can convert this circuit into a current source in parallel with the same resistor.
To find the current source value,

$$I = \frac{V_{FT}}{R}$$

$$I = \frac{400}{25}$$

FINAL RESULT

I = 16 A

Thus, the 400 V voltage source in series with the 25 Ω resistor can be converted into a 16 A current source in parallel with the same resistor.

2 If we use the source transformation rules, we can convert this circuit into a voltage source in series with the same resistor.
To find the voltage source value,

$$V = RI$$

$$V = (12)(3)$$

FINAL RESULT

V = 36 V

Thus, the 3 A current source in parallel with the 12 Ω resistor can be converted into a 36 V voltage source in series with the same resistor.

3 The first thing we notice is that the voltage source is inverted. So, we are dealing with a −15 V voltage source.
Voltage sources have zero internal resistance. If we substitute this source, temporarily, with a wire, we see that R_1 and R_2 are in series and thus, can be substituted by the equivalent resistor, equal to 30 Ω.
The −15 V voltage source in series with a 30 Ω resistor can be replaced by a current source in parallel with the same resistor.
The current source will be

$$I_{EQ1} = \frac{V_1}{R_{EQ1}}$$

$$I_{EQ1} = -\frac{15}{30}$$

$$I_{EQ1} = -0.5 \text{ A}$$

The converted circuit is seen in Figure 16.19

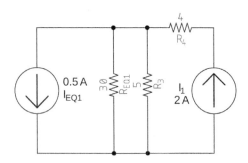

Figure 16.19 Converted circuit.

Resistors R_{EQ1} and R_3 are in parallel and can be substituted by their equivalent

$$\frac{1}{R_{EQ2}} = \frac{1}{R_{EQ1}} + \frac{1}{R_3}$$

$$\frac{1}{R_{EQ2}} = \frac{1}{30} + \frac{1}{5}$$

$$R_{EQ2} = 4.2857 \ \Omega$$

The result is shown in Figure 16.20.

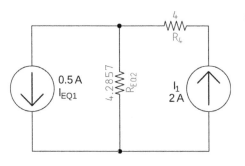

Figure 16.20 Converted circuit.

We now have the current source I_{EQ1} in parallel with resistor R_{EQ2}, and we can convert that into a voltage source in series with the same resistor. The voltage source will be

$$V_{EQ1} = R_{EQ2}I_{EQ1}$$

$$V_{EQ1} = (4.2857)(-0.5)$$

$$V_{EQ1} = -2.1429 \text{ V}$$

The result is shown in Figure 16.21.

Figure 16.21 Converted circuit.

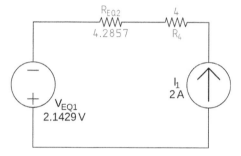

Figure 16.21 shows that V_{EQ1} is in series with R_{EQ2} and R_4. These two resistors can be replaced by their sum, as shown in Figure 16.22.

Figure 16.22 Converted circuit.

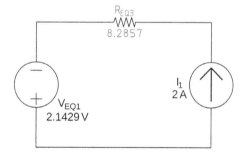

Finally, V_{EQ1} in series with R_{EQ3} can be converted into a current source with the following value:

$$I_2 = \frac{V_{EQ1}}{R_{EQ3}}$$

$$I_2 = \frac{-2.1419}{8.2857}$$

$$I_2 = -0.2586 \text{ A}$$

Figure 16.23 shows the result.

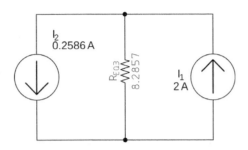

Figure 16.23 Converted circuit.

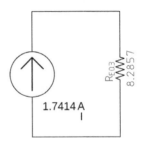

Figure 16.24 Final circuit.

We now have two current sources, with opposite signs, in parallel. We must subtract them to obtain the final circuit, shown in Figure 16.24.

17

Impedance and Phase

Circuit Analysis

17.1 Introduction

In this chapter, we will examine impedance, admittance, and their relations with resistance and reactance, important themes of electrical engineering.

We will use complex numbers extensively. For anyone not familiarized with the concepts behind complex numbers, we advise to read Appendix D, before starting this chapter.

17.2 This Is Just a Phase

In previous chapters, we have examined capacitors and inductors and how these components respond when they are subjected to DC and AC.

Capacitance and inductance are not just properties of capacitors and inductors. In real life, all components, even resistors, have a little bit of both. In most cases these are spurious undesired properties of materials and construction methods.

Simple wires, for example, placed side by side, show capacitance and inductance properties that, in some cases, can prevent more sensitive circuits from working.

We have also examined how alternating current creates reactance in capacitors and inductors and how this reactance creates phase shifts between current and voltage and between input and output signals.

At the end, we have seen two kinds of resistances in a circuit: pure resistance that does not alter the phase of signals and reactance that alters the phase of signals.

Impedance is the sum of both, resistance, and reactance.

Introductory Electrical Engineering with Math Explained in Accessible Language,
First Edition. Magno Urbano.
© 2020 John Wiley & Sons, Inc. Published 2020 by John Wiley & Sons, Inc.

17.3 Impedance

Reactance and resistance cannot be added together directly because they are different entities.

To make the sum possible, we must express impedance as a complex number.

COMPLEX IMPEDANCE

$$Z = R + jX$$

Z is the complex impedance, in Ohms.
R is the resistance, in Ohms.
X is the reactance, in Ohms.
j is the imaginary part.

Complex impedance can represent resistance, reactance, and a phase angle at the same time.

17.3.1 Series Impedance

Like other components, impedances can be combined in series, and the total equivalent impedance (Z) will be the sum of all individual impedances, similar to the one used by resistors in series:

$$Z = Z_1 + Z_2 + Z_3 + \cdots + Z_n$$

or in a generic form shown below.

SERIES IMPEDANCE

$$Z = \sum_{i=0}^{\infty} Z_i$$

Z is the total equivalent impedance, in Ohms.
Z_i is the impedance of each of the series impedances, in Ohms.

17.3.2 Parallel Impedances

Impedances can also be combined in parallel, and the total equivalent impedance (Z) can be found by using the following formula, similar to the one used by resistors in parallel:

$$\frac{1}{Z} = \frac{1}{Z_1} + \frac{1}{Z_2} + \frac{1}{Z_3} + \cdots + \frac{1}{Z_n}$$

Like resistors, if we have just two impedances and parallel, we can use the following formula.

TWO PARALLEL IMPEDANCES

$$Z = \frac{Z_1 \times Z_2}{Z_1 + Z_2}$$

Z is the total equivalent impedance, in Ohms.
Z_1, Z_2 are the impedance of each of the parallel impedances, in Ohms.

or in a generic form shown below.

PARALLEL IMPEDANCE

$$\frac{1}{Z} = \sum_{i=0}^{\infty} \frac{1}{Z_i}$$

Z is the total equivalent impedance, in Ohms.
Z_i is the impedance of each of the parallel impedances, in Ohms.

17.4 Capacitive Impedance

Capacitors have no resistance, just reactance, which is given by

$$X_C = \frac{1}{\omega C}$$

Thus, the impedance of a capacitor is shown below.

CAPACITIVE IMPEDANCE

$$Z_C = \frac{1}{jwC}$$

Z_C is the capacitive impedance, in Ohms.
ω is the angular frequency, equal to $2\pi f$, in radians per second.
C is the capacitance, in Farads.
j is the imaginary part.

Considering that ω = 2πf, we can expand the formula.

CAPACITIVE IMPEDANCE

$$Z_C = \frac{1}{2\pi fCj}$$

Z_C is the capacitive impedance, in Ohms.
C is the capacitance, in Farads.
j is the imaginary part.
f is the frequency, in Hertz.

Mathematically, the j in the denominator is not a good thing, because it will make future calculations harder.

To bring j to the nominator, we must use a mathematical technique – multiply nominator and denominator by j – an operation that will not alter the result.

Hence,

$$Z_C = \frac{1}{2\pi fCj} \times \frac{j}{j}$$

$$Z_C = \frac{j}{2\pi fCj^2}$$

MATH CONCEPT

$j^2 = -1$

By applying this concept, we get

$$Z_C = \frac{j}{2\pi fC(-1)}$$

CAPACITIVE IMPEDANCE

$$Z_C = -\frac{j}{2\pi fC}$$

Z_C is the capacitive impedance, in Ohms.
C is the capacitance, in Farads.
j is the imaginary part.
f is the frequency, in Hertz.

The negative sign shows that a capacitor will have a negative phase angle for the impedance.

17.5 Inductive Impedance

The reactance of an inductor is found by the formula

$$X_L = 2\pi fL$$

Thus, the inductor impedance will be

$$Z_L = jX_L$$

INDUCTIVE IMPEDANCE

$$Z_L = 2\pi fLj$$

Z_L is the inductive impedance, in Ohms.
L is the inductance, in Henries.
j is the imaginary part.
f is the frequency, in Hertz.

The positive sign on the impedance shows that inductors have a positive phase angle for the impedance.

17.6 Examples

17.6.1 Example 1

An alternating voltage source is applied to a resistor–capacitor circuit, as shown in Figure 17.1. How can we calculate the total impedance of this circuit?

17.6.1.1 Solution

The alternating voltage source's peak is 10 V and its frequency is 60 Hz.
The capacitor impedance will be

$$Z_C = -\frac{j}{2\pi fC}$$

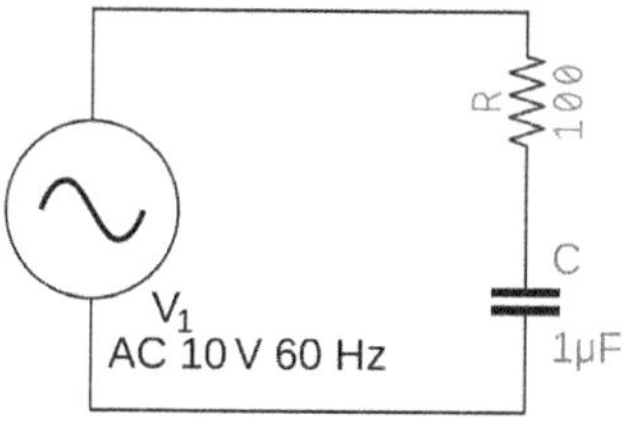

Figure 17.1 Resistor, capacitor, and an AC source.

Thus,

$$Z_C = -\frac{j}{2\pi \times 60 \times (1 \times 10^{-6})}$$

$$Z_C = -j2652.58 \ \Omega$$

Impedance is the real part (resistance) plus the imaginary part (reactance). Hence, this circuit's impedance will be shown below.

FINAL RESULT

$$Z = 100 - j2652.58 \ \Omega$$

If we want to express impedance in the phasor form, we can use the formula in Appendix D:

$$R = |z| = \sqrt{a^2 + b^2}$$

$$R = |z| = \sqrt{100^2 + (-2652.58)^2}$$

$$R = |z| = 2654.46 \ \Omega$$

The real part is positive. Therefore, the impedance's angle can be found by

$$\theta = A \tan\left(\frac{b}{a}\right)$$

$$\theta = A \tan\left(\frac{-2654.46}{100}\right)$$

$$\theta = -87.84°$$

The impedance in phasor notation will be shown below.

FINAL RESULT

$$Z = 2654.46\angle -87.84°\ \Omega$$

17.6.2   Example 2

Find the total impedance of the circuit shown in Figure 17.2.

Figure 17.2 Complex circuit.

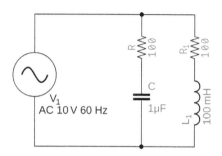

17.6.2.1   Solution

The first branch of this circuit is equal to the one in the last example, to which we have already calculated its impedance as

$$Z_1 = 100 - j2652.58\ \Omega$$

For the second branch, we have the inductor impedance is given by

$$Z_L = 2\pi fLj$$

Thus,

$$Z_L = j\left(2\pi \times 60 \times 100 \times 10^{-3}\right)$$

$$Z_L = j37.699\ \Omega$$

Hence, the impedance of the second branch, as before, is the sum of the real part (resistance) and the imaginary part (reactance), or

$$Z_2 = 100 + j37.699 \ \Omega$$

Both branches are in parallel. The total impedance of these branches can be found by the following formula:

$$Z = \frac{Z_1 \times Z_2}{Z_1 + Z_2}$$

$$Z = \frac{(100 - j2652.58) \times (100 + j37.699)}{(100 - j2652.58) + (100 + j37.699)}$$

$$Z = \frac{10000 + j3769.91 - j265258 - j^2 99999.61}{200 - j2614.88}$$

MATH CONCEPT

$$j^2 = -1$$

$$Z = \frac{10000 + j3769.91 - j265258 - (-1)99999.61}{200 - j2614.88}$$

$$Z = \frac{109999.61 - j261448}{200 - j2614.88}$$

To perform this division, we must multiply nominator and denominator by the denominator's conjugate (see Appendix D):

$$Z = \frac{109999.61 - j261448}{200 - j2614.88} \times \frac{200 + j2614.88}{200 + j2614.88}$$

FINAL RESULT

$$Z = 102.61 + j34.21 \ \Omega$$

---

or in the phasor notation

FINAL RESULT

$Z = 108.15 \angle 18.44\,°\,\Omega$

17.7 The Importance of Impedances in Real Life

Impedance is fundamental for electrical engineering when we must connect different circuits together. Incompatible impedances will prevent circuits from working.

17.7.1 Example

Consider both circuits in Figure 17.3. The first circuit contains a voltage source and two resistors, R_1 and R_2. The second one contains a resistor that represents the load. This load can be anything, a light bulb, a motor, etc.

Figure 17.3 Two independent circuits.

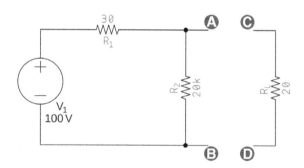

We want to connect both circuits by attaching electrically points A to C and B to D.

After the connection, we need the voltage between points A and B to be closer to the value it has before the connection.

17.7.1.1 Solution

The first thing we must do is to calculate the voltage between A and B in the first circuit alone, without any load.

We calculate the current flowing in the first circuit by using the Ohm's law:

$$V = R \times I$$

$$100 = (20000 + 30) \times I$$

$$I = \frac{100}{20030}$$

$$I = 0.005\,A$$

Hence, voltage between A and B will be

$$V_{AB} = R_2 I$$

$$V_{AB} = (20000)(0.005)$$

FINAL RESULT

$$V_{AB} = 99.85\,V$$

We now connect the second circuit to the first one and see what happens to the voltage between A and B (Figure 17.4).

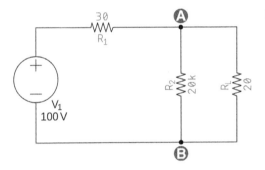

Figure 17.4 Two circuits connected.

After connecting both circuits, we see that R_L is now in parallel with R_2. Therefore, we can substitute both by an equivalent resistor (Figure 17.5):

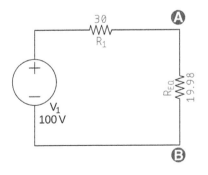

Figure 17.5 Final circuit.

$$R_{EQ} = \frac{R_2 \times R_L}{R_2 + R_L}$$

$$R_{EQ} = \frac{20000 \times 20}{20000 + 20}$$

$$R_{EQ} = 19.98 \ \Omega$$

Now we can calculate the current flowing in the circuit by using Ohm's law:

$$V = RI$$

$$100 = (30 + 19.98) \times I$$

$$I = \frac{100}{49.98}$$

$$I = 2 \ A$$

Voltage between A and B will be

$$V_{AB} = R_{EQ}I$$

$$V_{AB} = (19.98)(2)$$

FINAL RESULT

$$V_{AB} = 39.97 \ V$$

We see that after connecting both circuits, the voltage between A and B dropped from 99.85 to 39.97 V. This is not what we want.

To understand what happened, remember that $V = R \times I$.

Connecting both circuits puts R_2, a 20 kΩ resistor, in parallel with R_L, a 20 Ω resistor, making the resistance between A and B drop almost 1000 times, from 20 kΩ to 19.98 Ω.

This small resistor in combination with a 2 A produces a very small voltage.

17.7.1.2 What About a Bigger Load?

What happens if the load R_L is 100 kΩ instead of 20 Ω?

Connecting both circuits puts R_2, a 20 kΩ resistor, in parallel with R_L, a 100 kΩ resistor, making the equivalent resistance

$$R_{EQ} = \frac{R_2 \times R_L}{R_2 + R_L}$$

$$R_{EQ} = \frac{20000 \times 100000}{20000 + 100000}$$

$$R_{EQ} = 6666.66 \ \Omega$$

The current, according to Ohm's law, will be

$$V = RI$$

$$100 = (30 + 6666.66) \times I$$

$$I = \frac{100}{6696.66}$$

$$I = 0.0149 \ A$$

Hence, the voltage between A and B will be

$$V_{AB} = R_{EQ}I$$

$$V_{AB} = (6696.66)(0.0149)$$

FINAL RESULT

$$V_{AB} = 99.55 \ V$$

The voltage between A and B is practically the same as before, exactly what we want.

17.7.1.3 Conclusion

A circuit with a high output impedance connected to a low impedance load will make the output voltage drop and subject the output to a high current, probably making the circuit stop working.

On the other hand, a circuit with a low output impedance connected to a high impedance load will not create problems for the first circuit. The output voltage will not drop and the load will drain very little current.

Exercises

1 Find the impedance of a circuit composed of a 22 Ω resistor in series with a 10 μF capacitor when an alternating voltage with a frequency of 80 Hz is applied.

2 Find the impedance of a circuit composed of a 47 Ω resistor in series with a 100 mH inductor when an alternating voltage with a frequency of 60 Hz is applied.

3 A circuit with an impedance equal to 4 – j2 Ω is connected in series with another one with an impedance of 2 + j4 Ω. What is the circuit's total impedance? Which circuit shows capacitive properties?

4 What is the total impedance if we connect a circuit with an impedance equal to 4 – j2 Ω in parallel with another one with an impedance of 2 + j4 Ω?

5 What is the phasor representation of an impedance equal to 3 + j5 Ω?

Solutions

1 Capacitor's impedance is given by

$$Z_C = -\frac{j}{2\pi fC}$$

$$Z_C = -\frac{j}{2\pi(80)(10 \times 10^{-6})}$$

$$Z_C = -j198.9437 \ \Omega$$

Circuit's final impedance is the real part (resistance) plus the imaginary part (reactance).

FINAL RESULT

$$Z = 22 - j198.9437 \ \Omega$$

---

2 Inductor's impedance is given by

$$Z_L = 2\pi f L j$$

$$Z_L = j2\pi(60)\left(100 \times 10^{-3}\right)$$

$$Z_L = j37.6991 \ \Omega$$

Circuit's final impedance is the real part (resistance) plus the imaginary part (reactance).

FINAL RESULT

$$Z = 47 + j37.6991 \ \Omega$$

---

3 The final impedance will be the two impedances in series:

$$4 - j2 + 2 + j4$$

FINAL RESULT

$$6 + j2 \ \Omega$$

The result shows inductive properties, because the imaginary part is positive.

The first circuit, with an impedance equal to $4 - j2$, is the one that shows capacitive properties, because the imaginary part is negative.

4 The equivalent impedance for two individual impedances in parallel is found by the following formula:

$$Z_{EQ} = \frac{Z_1 \times Z_2}{Z_1 + Z_2}$$

$$Z_{EQ} = \frac{(4-j2) \times (2+4j)}{(4-j2) + (2+4j)}$$

$$Z_{EQ} = \frac{8 + j16 - j4 - 8j^2}{6 + j2}$$

MATH CONCEPT

$j^2 = -1$

Hence,

$$Z_{EQ} = \frac{8 + j16 - j4 - 8(-1)}{6 + j2}$$

$$Z_{EQ} = \frac{16 + j12}{6 + j2}$$

To perform this division, we must multiply nominator and denominator by the denominator's conjugate (see Appendix D):

$$Z_{EQ} = \frac{16 + j12}{6 + j2} \times \frac{6 - j2}{6 - j2}$$

$$Z_{EQ} = \frac{(16 + j12) \times (6 - j2)}{40}$$

$$Z_{EQ} = \frac{96 - j32 + j72 + 24}{40}$$

$$Z_{EQ} = \frac{120 + j40}{40}$$

FINAL RESULT

$Z_{EQ} = 3 + j \ \Omega$

---

5   To calculate the module of impedance $3 + j5 \ \Omega$, we do

$$R = |z| = \sqrt{a^2 + b^2}$$

$$R = |z| = \sqrt{3^2 + 5^2}$$

FINAL RESULT

$R = |z| = 5.8310 \ \Omega$

The angle can be found by

$$\theta = A\tan\left(\frac{b}{a}\right)$$

$$\theta = A\tan\left(\frac{5}{3}\right)$$

FINAL RESULT

$$\theta = 59.0362°$$

or in the phasor form

$$Z = 5.8310 \angle 59.0362°\ \Omega$$

18

Power
And Work

18.1 Introduction

In this chapter, we will examine several types of electric power.

18.2 Electric Power and Work

Electrical potential is energy and all kinds of energies can create work.

The energy contained in one liter of diesel, for example, can be converted by a machine into mechanical movement or other kinds of work.

The work produced by a car or a tractor, for example, can be measured in Joules. In terms of electrical circuits, work or power is generally measured in Watts.[1]

Electric power is a physical quantity that measures the work produced in a period of time.

An industrial electric oven, for example, has a power greater than a domestic oven because it can produce more work, in that case, heat, in the same amount of time.

Electric power, current, and voltage are related by the following formula.

ELECTRIC POWER

$$P = VI$$

P is the real power, in Watts.
V is the voltage, in Volts.
I is the current, in Amperes.

---

1 In honor of the English scientist James Watt.

Introductory Electrical Engineering with Math Explained in Accessible Language,
First Edition. Magno Urbano.
© 2020 John Wiley & Sons, Inc. Published 2020 by John Wiley & Sons, Inc.

18.3 Powers in Parallel

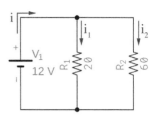

Figure 18.1 Power in parallel.

Consider the circuit shown in Figure 18.1.
Both resistors are in parallel with a 12 V battery and draining a total current i from it. This current splits into i_1 and i_2 that flows across R_1 and R_2, respectively.

Power can be found by using

$$P = VI$$

But Ohm's law states that

$$V = RI$$

If we substitute this relation into the first formula, we get the following.

ELECTRIC POWER

$$P = RI^2$$

P is the real power, in Watts.
I is the current, in Amperes.
R is the resistance, in Ohms.

Back to Figure 18.1, we know that both resistors have the same voltage across, because they are in parallel.
Hence, we can find i_1 and i_2:

$$V_1 = R_1 I_1$$

$$12 = 20 I_1$$

$$I_1 = 0.6 \text{ A}$$

and for R_2,

$$V_2 = R_2 I_2$$

$$12 = 60 I_2$$

$$I_2 = 0.2 \text{ A}$$

We can now calculate the power dissipated by every resistor, by using $P = RI^2$.

For R_1,

$$P_1 = R_1 I_1^2$$

$P_1 = 20 \times (0.6)^2$

$$P_1 = 7.2 \text{ W}$$

and for R_2,

$$P_2 = R_2 I_2^2$$

$P_2 = 60 \times (0.2)^2$

$$P_2 = 2.4 \text{ W}$$

We know that the total current is

$$i = i_1 + i_2$$

Therefore, power will also be the sum of individual powers, or

$$P = P_1 + P_2$$

$P = 7.2 + 2.4$

FINAL RESULT

$$P = 9.6 \text{ W}$$

18.3.1 Conclusion

Total power is the sum of powers in parallel.

18.4 Powers in Series

Consider that we want to find the total power in the circuit shown in Figure 18.2.

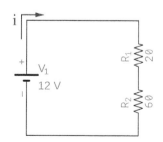

Figure 18.2 Power in series.

In this circuit, we see resistors R_1 and R_2 in series with a battery. Therefore, the sum of the voltage drops across each resistor is equal to the battery's voltage. Total resistance is the sum of R_1 and R_2.
By Ohm's law,

$$V = RI$$

Hence, the current flowing across the circuit is

$$V_1 = (R_1 + R_2)i$$

$$12 = (20 + 60)i$$

$$i = 0.15 \text{ A}$$

Therefore, total power is

$$P = VI$$

$$P = 12 \times 0.15$$

FINAL RESULT

$$P = 1.8 \text{ W}$$

18.4.1 Conclusion

Total power is the sum of all powers in series.

18.5 "Alternating" Power

Measuring the power consumed by a DC circuit is easy.

However, measuring the power of an AC circuit is not easy, because currents and voltages change intensity over time and current changes direction periodically.

18.5.1 Two Ovens

Suppose we have two ovens, one powered by DC and the other powered by AC. How can we compare the work of both ovens? How can we adjust the AC oven to produce the same work as the DC one?

18.5.1.1 The First Oven

Suppose the first oven is powered by a 12 V battery and is draining 5 A of current.

We can find the power consumed by this oven by using

$$P = VI$$

$P = 12 \times 5$

$$P = 12 \times 5 = 60 \text{ W}$$

18.5.1.2 Second Oven

Suppose the second oven is powered by a 12 V alternating voltage and is draining 5 A of current, both peak voltages.

Finding the power consumed by the second oven is a lot more complex. Voltage and current vary with time and current changes direction constantly. The formula $P = VI$ cannot be used anymore, not on a direct way.

18.5.2 The Average Value

To solve the problem, we must find a way to calculate an average value for the alternating current and voltage.

But how exactly is an average value calculated in mathematics?

Suppose we have values 10, 20, and 30. The average value will be the sum of values divided by the number of values.

AVERAGE VALUE

$$A_v = \frac{1}{n} \sum_{i=1}^{n} X_i$$

A_v is the average value.
n is the number of terms.
x_i is every one of the terms.

For the given example, the average value will be

$$A_v = \frac{10 + 20 + 30}{3} = 20$$

18.5.2.1 Average Value of a Sinusoidal Function

Mathematically, an alternating voltage follows a sinusoidal function.

ALTERNATING VOLTAGE

$$V = V_P \sin(\omega t)$$

V is the instantaneous voltage, in Volts.
V_P is the peak voltage, in Volts.
ω is the angular frequency in radians per second.

For the given example, the voltage has a peak value equal to 12 V. The voltage will then follow an equation like

$$V = 12 \sin(\omega t)$$

meaning that its values will oscillate between 12 V and −12 V, like shown in Figure 18.3.

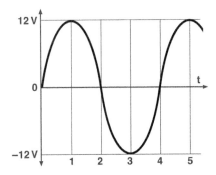

Figure 18.3 Function 12 sin(ωt).

It is clear by Figure 18.3 that the average value is 0.

It is obvious that, in real life, the average voltage is not 0, so this must be a limitation of the method we are using, and we must find another method to calculate the average value of a sinusoidal function.

18.5.3 RMS Value

Mathematically speaking, there is a method called root mean square (RMS) that finds average values of functions.

The RMS value of a function can be found by the following formula.

RMS VALUE

$$RMS = \sqrt{\frac{1}{T}\int_0^T f(t)^2 .dt}$$

T is the wave's period, in seconds.
f(t) is the function.

By applying the RMS formula to

$$V = V_P \sin(\omega t)$$

and to the current

$$I = I_P \sin(\omega t)$$

we obtain the following values for the RMS voltage.

RMS VOLTAGE

$$V_{RMS} = \frac{V_P}{\sqrt{2}} \approx 0.707\ V_P$$

V_{RMS} is the RMS voltage, in Volts.
V_P is the peak voltage, in Volts.

We also obtain the following values for the RMS current.

RMS CURRENT

$$I_{RMS} = \frac{I_P}{\sqrt{2}} \approx 0.707\ I_P$$

I_{RMS} is the RMS current, in Amperes.
I_P is the peak current, in Amperes.

In other words, RMS current and voltage for a sinusoidal function are about 70.71% of their peak values.

In Appendix C, we demonstrate, by using infinitesimal calculus, how we have derived the RMS formula.

18.5.4 Back to the Second Oven

We can now calculate the RMS voltage and current for the given example:

$$V_{RMS} = \frac{12\,V}{\sqrt{2}} \approx 8.48\ V$$

$$I_{RMS} = \frac{5}{\sqrt{2}} \approx 3.53 \text{ A}$$

The average power can be then found by the following formula.

AVERAGE POWER

$$P_{AV} = V_{RMS} \times I_{RMS}$$

P_{AV} is the average power, in Watts.
V_{RMS} is the RMS voltage, in Volts.
I_{RMS} is the RMS current, in Amperes

Substituting the values,

$$P_{AV} = 8.48 \times 3.53 \approx 30 \text{ W}$$

Comparing the average power of 30 W for the AC oven with the first oven's power of 60 W, we see that the AC produces less work (heat) compared with the first one.

18.6   Real, Apparent, and Reactive Power

Formulas like $P = VI$ and $P = RI^2$ relate to power dissipated by purely resistive elements, called real power.

In real life, however, components have inductive and capacitive characteristics that will produce "inductive" and "capacitive" power.

We have seen before that every time we have inductive or capacitive elements, we have reactance and phase shifts.

Therefore, it is obvious that we will have inductive and capacitive power or, in other words, powers that do not belong in the real axis.

Thus, in practice, power will be defined by a real and an imaginary part, according to the following formula.

POWER

$$S = P + jQ$$

S is the apparent power, in VA, Volt-Ampere.
P is the active or real power, in Watts.
Q is the reactive power, in VAR, Volt-Ampere Reactive.
j is the imaginary part.

The given formula defines that the apparent power (S) is the sum of the active power (P), also known as real power, consumed by the resistive part of a system with the reactive power (Q), wasted[2] by the reactive part of the circuit.

> S, P, and Q are the letters normally used to refer to apparent power, real power, and reactive power, respectively.

18.6.1 Reactive Power

The reactive energy of a circuit is required and generated by all electrical components such as capacitors, inductors, transformers, or motors, which use alternating electric or magnetic fields. Reactive energy is needed to keep these components running, but it does not necessarily produce work.

Air conditioners, televisions, computers, refrigerators, etc. are examples of equipment that produce reactive energy.

The only energy that produces work is the energy consumed by the purely resistive parts of the circuits.

Household equipment usually produces inductive reactance. However, inside electronic equipment, there are sources of capacitive and inductive reactive powers.

Since capacitive reactance has a negative sign, capacitive reactive power will also have a negative sign.

CAPACITIVE APPARENT POWER

$$S = P - jQ_C$$

S is the apparent power, in VA, Volt-Ampere.
P is the active or real power, in Watts.
Q is the capacitive reactive power, in VAR, Volt-Ampere Reactive.
j is the imaginary part.

The same is true for inductive power, following the positive sign of inductive reactance.

---

2 Because this power does not produce work.

INDUCTIVE APPARENT POWER

$$S = P + jQ_L$$

S is the apparent power, in VA, Volt-Ampere.
P is the active or real power, in Watts.
Q_L is the inductive reactive power, in VAR, Volt-Ampere Reactive.
j is the imaginary part.

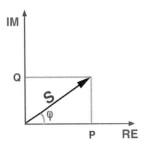

Figure 18.4 Apparent, real, and reactive power.

18.6.1.1 Power Factor

Power is defined in real and imaginary terms by the following complex equation:

$$S = P + jQ$$

If we use the principles of imaginary numbers, we will conclude that the apparent power (S) has a part in the real axis (P) and another on the imaginary axis (Q), as shown in Figure 18.4.

By using trigonometry, we have that

$$S = \sqrt{P^2 + Q^2}$$

and

$$Tan(\varphi) = \left(\frac{Q}{P}\right)$$

Other trigonometric relations can be established, like

$$P = S\cos(\varphi)$$
$$Q = S\sin(\varphi)$$

From Figure 18.4, we see that the greater the angle φ, the greater the reactive power Q, and the smaller the real power P.

Therefore, to measure the efficiency of the circuits, a relation called power factor, defined as the division of the active power by the apparent power, was created.

POWER FACTOR

$$\cos(\varphi) \rightarrow F_p = \frac{P}{|S|}$$

F_p is the power factor.
P is the active power, in Watts.
$|S|$ is the magnitude of the apparent power, in VA, Volt-Ampere.
φ is the angle between the apparent power and the real axis.

Electric companies usually charge electricity consumption based on active power because it is the only one that does work, but the consumer must have a power factor within certain range.

This is needed because if a consumer has an unbalanced system, generating too much reactive power, this consumer is receiving a huge amount of energy and using very little to do work. In addition to being a waste of energy for the utility company and a consequent loss, it causes unnecessary heating on the transmission lines and may cause imbalances in the network.

For this reason, electric companies charge customers if the reactive power value is above certain values.

Exercises

1 Find the power consumed by each one of the resistors shown in Figure 18.5.

Figure 18.5 Power (exercise).

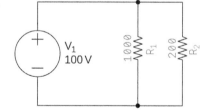

2 A resistor of unknown value has a voltage drop of 25 V and consumes 100 W. What current flows across this resistor?

3 A sinusoidal voltage function equal to V(t) = 30 sin(ωt) is applied to a 100 Ω purely resistive circuit. Find the RMS current.

4   A circuit has the following apparent power:

$S = 200 - j50\,VA.$

- Find the active and the reactive power of this circuit.
- Is this a circuit with inductive or capacitive characteristics?
- What is the power factor?

Solutions

1   Resistors R_1 and R_2 are in parallel; thus, the voltage across them is the same and equal to the power supply.

So, voltage across R_1 is

$V_1 = R_1 I_1$

and voltage across R_2 is

$V_2 = R_2 I_2.$

But $V_1 = V_2$

Therefore,

$R_1 I_1 = R_2 I_2$

$1000\,I_1 = 200\,I_2$

$I_2 = 5I_1$

R_1 and R_2 in parallel have an equivalent resistance equal to

$$\frac{1}{R_{EQ}} = \frac{1}{R_1} + \frac{1}{R_2}$$

$$\frac{1}{R_{EQ}} = \frac{1}{1000} + \frac{1}{200}$$

$R_{EQ} = 166.66\,\Omega$

Total current flowing in the circuit can be found by applying Ohm's law:

$$I = \frac{V}{R}$$

$$I = \frac{100}{166.66}$$

$$I = 0.6 \text{ A}$$

We can now calculate the current across each resistor:

$$I = I_1 + I_2$$

By substituting $I_2 = 5I_1$, we get

$$I = I_1 + 5I_1$$
$$I = 6I_1$$

$$0.6 = 6I_1$$

$$I_1 = 0.1 \text{ A}$$

Calculating I_2,

$$I_2 = 5I_1$$

$$I_2 = 5(0.1)$$

$$I_2 = 0.5 \text{ A}$$

We can now calculate the power consumed by R_1:

$$P_1 = R_1 I_1^2$$

$$P_1 = 1000(0.1)^2$$

FINAL RESULT

$$P_1 = 10 \text{ W}$$

We can also calculate the power consumed by R_2:

$$P_2 = R_2 I_2^2$$

$$P_2 = 200(0.5)^2$$

FINAL RESULT

$$P_2 = 50 \text{ W}$$

Total power will be the sum of both powers, or

$$P = P_1 + P_2$$

$P = 10 + 50$

FINAL RESULT

$P = 60\ \text{W}$

2 Current across the resistor can be found by using the formula:

$$P = VI$$

$100 = 25\,I$

FINAL RESULT

$I = 4\,\text{A}$

3 The RMS voltage can be found by using

$$V_{RMS} = \frac{V_P}{\sqrt{2}} \approx 0.707\ V_P$$

where V_P is the peak voltage.
Thus,

$V_{RMS} = \dfrac{30}{\sqrt{2}}$

$$V_{RMS} = 21.2132\ \text{V}$$

The RMS current is

$21.2132 = 100\,I_{RMS}$

FINAL RESULT

$I_{RMS} = 0.2121\ \text{A}$

4 Apparent power is

$S = 200 - j50$ VA

So, real power is 200 W and reactive power is –50 VAR.
Reactive power has a negative sign; thus, the circuit has capacitive characteristics.
The power factor can be found by

$$\cos(\varphi) \rightarrow F_p = \frac{P}{|S|}$$

where

$$|S| = \sqrt{P^2 + Q^2}$$

$$|S| = \sqrt{200^2 + (-50)^2}$$

$$|S| = 206.1553 \text{ VA}$$

Hence, the power factor will be

$$F_p = \frac{P}{|S|}$$

$$F_p = \frac{200}{206.1553}$$

$$F_p = 0.9701$$

The angle of the apparent power can be found by

$$\tan(\varphi) = \frac{Q}{P}$$

$$\tan(\varphi) = \frac{-50}{200}$$

$$\varphi = A\tan(-0.25)$$

$$\varphi = -14.0362°$$

19

Kirchhoff's Laws

Circuit Analysis

19.1 Introduction

In this chapter, we will examine the Kirchhoff's laws, created by the German physicist Gustav Robert Kirchhoff (1824–1887) and how these laws can be used to find unknowns in electric circuits.

19.2 Kirchhoff's Laws

To use the Kirchhoff's laws, it is necessary to understand two concepts: nodes and meshes.

19.2.1 Nodes or Junctions

It is a point where two or more wires connect.

Figure 19.1 shows a circuit with two nodes, A and B, where all components connect.

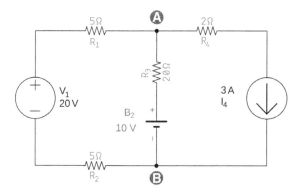

Figure 19.1 Two nodes, A and B.

Introductory Electrical Engineering with Math Explained in Accessible Language,
First Edition. Magno Urbano.
© 2020 John Wiley & Sons, Inc. Published 2020 by John Wiley & Sons, Inc.

In electronic diagrams, nodes are identified by a dot or small circle drawn where two or more lines connect. This indicates that the lines are electrically connected.

19.2.2 Mesh

It is any closed path in which current circulates.

Figure 19.2 shows dashed lines that represent the three meshes possible for the circuit: M1, M2, and M3.

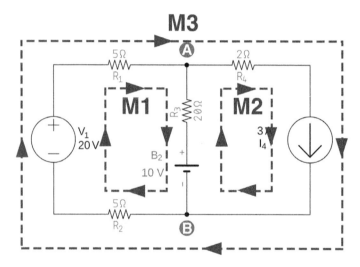

Figure 19.2 Three meshes, M1, M2, and M3.

19.2.3 Kirchhoff's First Law

Kirchhoff's first law, also known as the Kirchhoff's current law (KCL), postulates that at any junction in a circuit, the sum of the currents arriving at the junction is equal to the sum of the currents leaving the same junction.

KIRCHHOFF'S FIRST LAW (KCL)

$$\sum_n I_{in} = \sum_n I_{out}$$

I_{in} are the currents entering the node.
I_{out} are the currents leaving the node.

Figure 19.3 shows a representation of several currents entering and leaving a node.
By applying KCL in the given figure, we get

$$i_3 + i_5 = i_1 + i_2 + i_4$$

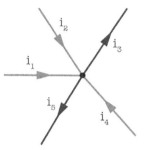

Figure 19.3 Currents leaving and arriving at a node.

19.2.4 Kirchhoff's Second Law

Kirchhoff's second law, also known as Kirchhoff's mesh/loop law, Kirchhoff's Voltage Law, or simply KVL, states that in any closed loop network, the total electromotive force around the loop is equal to the sum of all the electromotive force drops within the same loop. Translating that to popular language, the total voltage around the loop is equal to the sum of all individual voltage drops around the same loop.

KIRCHHOFF'S SECOND LAW (KVL)

$$\sum_k \varepsilon_k = \sum_k R_k I_k \longrightarrow \sum_k V_k = 0$$

ε_k are the electromotive forces or voltages.
$R_k I_k$ is the product of resistance by current of each element in a mesh or, in other words, the voltage drops of each individual element.
V_k are the voltage drops of every element in the mesh.

19.3   Examples

19.3.1   Example 1

Suppose we want to find the several currents flowing across the circuit's branches in Figure 19.4.

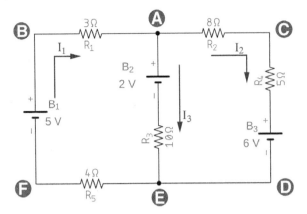

Figure 19.4 Kirchhoff analysis – branch currents.

19.3.1.1 Solution

To begin the analysis, we specify two current meshes in the circuit, I_1 and I_2, both chosen arbitrarily as clockwise, as shown in Figure 19.5.

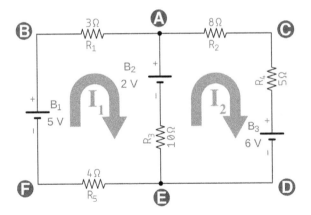

Figure 19.5 Kirchhoff analysis – mesh currents.

19.3.1.2 Meshes

If we start at point A, we can draw two circuit meshes: ACDEA and AEFBA. Any mesh we follow is a closed loop.

19.3.1.2.1 Mesh AEFBA

If we apply Kirchhoff's voltage law (KVL) to the mesh AEFBA, the sum of the voltages of all elements in the mesh must be 0. Therefore, we get the following equation:

$$-V_{B2} - I_1 R_3 + I_2 R_3 - I_1 R_5 + V_{B1} - I_1 R_1 = 0$$

Note that every term in the equation represents a voltage. Voltages across resistors follow Ohm's law, $V = R \times I$.

The components of the equation are explained as follows:

$$-V_{B2}$$

We start at point A. We are following current I_1. The first thing we find along the way is the battery B2. We cross the battery from the positive to the negative pole, going from a pole with greater electrical potential to a pole with less electrical potential. The voltage drops; therefore we have a negative sign for this element.

$$-I_1 R_3$$

The negative sign comes from the fact that the current I_1 crosses R_3, creating a voltage drop.

$$I_2 R_3$$

Current I_2 also creates a voltage drop on R_3 in the direction of its flow. However, current I_2 flows in a direction opposite to I_1, and because we are following I_1, we must account for a voltage raise on R_3 equal to $I_2 R_3$, thus the positive sign.

$$-I_1 R_5$$

The negative sign comes from the fact that the current I_1 crosses R_5, creating a voltage drop.

$$V_{B1}$$

This term gets a positive sign because when passing through B1, we walk from a point of lesser potential (negative pole) to another of greater potential (positive pole), therefore, a potential increase.

$$-I_1 R_1$$

The negative sign of this block occurs because I_1 causes a voltage drop across R_1.

If we substitute the known values and reorganize the equation, we get

$$-V_{B2} - I_1 R_3 + I_2 R_3 - I_1 R_5 + V_{B1} - I_1 R_1 = 0$$

$$-2 - 10 I_1 + 10 I_2 - 4 I_1 + 5 - 3 I_1 = 0$$

$$17 I_1 - 10 I_2 = 3 \qquad (19.1)$$

19.3.1.2.2 Mesh ACDEA

If we start at point A of the mesh ACDEA and apply KVL, we get

$$-I_2 R_2 - I_2 R_4 - V_{B3} - I_2 R_3 + I_1 R_3 + V_{B2} = 0$$

The components of the equation are explained as follows:

$$-I_2 R_2$$

We start at point A. We are following I_2 current. The first component we find is the resistor R_2. Current crosses R_2 and causes a voltage drop. Therefore, the voltage has a negative sign.

$$-I_2 R_4$$

The negative sign of this term is explained by a voltage drop caused by I_2 crossing R_4.

$$-V_{B3}$$

This term gets a negative sign because we are crossing B3 from the positive to the negative pole. Therefore, there is a potential drop, indicated by the negative sign.

$$-I_2 R_3$$

The negative sign of this group happens because I_2 creates a voltage drop across R_3.

$$I_1 R_3$$

While I_2 crosses R_3 from top down, creating a voltage drop, I_1 crosses R_3 in the opposite direction, creating a voltage raise. All voltage raises are positive.

$$V_{B2}$$

This term has a positive sign because the battery is crossed from a point with lesser potential (negative pole) to a point of greater potential (positive pole), thus, a potential increase.

Substituting the known values and rearranging the equation, we get

$$-I_2 R_2 - I_2 R_4 - V_{B3} - I_2 R_3 + I_1 R_3 + V_{B2} = 0$$

$$-8I_2 - 5I_2 - 6 - 10I_2 + 10I_1 + 2 = 0$$

$$10I_1 - 23I_2 = 4 \tag{19.2}$$

We have finished with two equations, (19.1) and (19.2):

$$\begin{cases} 17I_1 - 10I_2 = 3 \\ 10I_1 - 23I_2 = 4 \end{cases}$$

The first equation gives us

$$I_1 = \frac{10I_2 + 3}{17}$$

Substituting I_1 in (19.2),

$$10\left(\frac{10I_2 + 3}{17}\right) - 23I_2 = 4$$

$$100I_2 + 30 - 391I_2 = 68$$

$$-291I_2 = 38$$

$$I_2 = -0.1305\,\text{A}$$

The negative sign of I_2 tells us that this current, in fact, flows in a direction opposite to the one we have chosen arbitrarily.

Substituting I_2 in (19.1),

$$10\,I_2 - 23(-0.13055) = 4$$

$$10\,I_1 + 3.003 = 4$$

$$10\,I_1 = 4 - 3.003$$

$$I_1 = 0.09965\,\text{A}$$

19.3.1.2.3 Branch Currents

Knowing both mesh currents, we can calculate each individual branch current, designated with arbitrary directions as i_1, i_2, and i_3 in Figure 19.4.

By comparing the branch currents with the mesh currents, we get the following.

FINAL RESULT

$$i_1 = I_1 = 0.09965\,\text{mA}$$

$$i_2 = I_2 = -0.1305\,\text{A}$$

Again, the negative sign of i_2 shows that this current really flows in a direction that is opposite to the one we have chosen.

To calculate i_3 we must subtract I_2 from I_1, because these currents flow in opposite directions:

$$i_3 = I_1 - I_2$$

$$i_3 = 0.09965 - (-0.1305)$$

FINAL RESULT

$$i_3 = 0.23015\,\text{A}$$

19.3.2   Example 2

In this example, we will use Kirchhoff's laws to analyze the circuit shown in Figure 19.6. The circuit contains multiple DC voltage sources, drawn on purpose using different symbols, a current source, and a few other components.

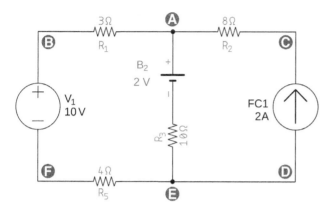

Figure 19.6 Kirchhoff analysis.

We want to find the several currents flowing across the circuit's branches and the voltage drop across R_2 (points A and C).

19.3.2.1   The Analysis
To begin the analysis, we specify the two mesh currents, I_1 and I_2, chosen in a completely arbitrary clockwise direction, as shown in Figure 19.7.

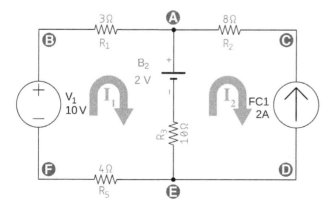

Figure 19.7 Kirchhoff analysis – mesh currents.

Because this circuit has a current source (FC1), we will temporarily ignore mesh ACDEA and focus our attention on mesh AEFBA only.

19.3.2.1.1 AEFBA Mesh

If we start at point A and follow the mesh AEFBA, we get the following equation:

$$-V_{B2} - I_1 R_3 + I_2 R_3 - I_1 R_5 + V_1 - I_1 R_1 = 0$$

The several components of this equation are explained like this:

$$-V_{B2}$$

We cross this battery from the positive to the negative, resulting in a voltage drop and justifying the negative sign for this term.

$$-I_1 R_3$$

We are crossing R_3 in the same direction as I_1, meaning this current will cause a voltage drop across R_3, giving this term a negative sign.

$$I_2 R_3$$

Current I_2 also flows across R_3 in an opposite direction compared with I_1. Because we are traveling the mesh against I_2, this term will not be a voltage drop but a voltage increase. For this reason, it gets a positive sign.

$$-I_1 R_5$$

Current I_1 will cause a voltage drop across R_3, therefore, this term gets a negative sign.

$$V_1$$

We cross the voltage source V_1 from negative to the positive, resulting in a voltage increase, therefore the positive sign.

$$-I_1 R_1$$

Current I_1 will cause a voltage drop across R_1; thus, this term gets a negative sign.

If we substitute the known values and rearrange the equation, we get

$$-V_{B2} - I_1 R_3 + I_2 R_3 - I_1 R_5 + V_1 - I_1 R_1 = 0$$

$$-2 - 10I_1 + 10I_2 - 4I_1 + 10 - 3I_1 = 0$$

$$17 I_1 - 10 I_2 = 8 \qquad (19.3)$$

19.3.2.1.2 Mesh ACDEA

We have assigned I_2 as the current for mesh ACDEA and have chosen a clockwise direction for it. However, we see that this mesh has already a specific 2 A anticlockwise current source.

Thus, we must write I_2 with a negative sign:

$$I_2 = -2\ A$$

Substituting I_2 in (19.3), we get

$$17I_1 - 10(-2) = 8$$

$$I_1 = -\frac{12}{17}\ A$$

$$I_1 = -0.705\ A$$

The negative sign for I_1 shows that, in fact, this current flows in the opposite direction compared with the direction we have chosen arbitrarily.

Now that we have the mesh currents, we can discover the branch currents, i_1, i_2 and i_3, randomly chosen as shown in Figure 19.8.

Comparing i_1 with I_1, we see that they are the same current.

FINAL RESULT

$$i_1 = I_1 = -0.705\ A$$

We can also see that i_2 is equal to I_2.

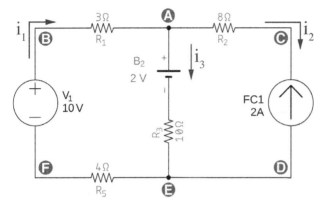

Figure 19.8 Kirchhoff analysis – branch currents.

$$i_2 = I_2 = -2\,A$$

However, two currents flow across R_2: I_1 and I_2 in opposite directions. Hence,

$$i_3 = I_1 - I_2$$

$$i_3 = -\frac{12}{17} - (-2)$$

$$i_3 = \frac{22}{17}A$$

$$i_3 = 1.294\,A$$

Negative signs for i_1 and i_2 show that these currents flow in the opposite direction, as shown in Figure 19.9.

19.3.2.1.3 Voltage A

Current i_3 crosses R_3 and is equal to 1.294 A.
This current will create a voltage drop across R_3 equal to

$$V_{R_3} = R_3 i_3$$

$$V_{R_3} = 10 \times 1.294$$

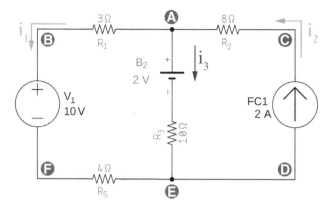

Figure 19.9 Kirchhoff analysis – correct branch currents.

$$V_{R_3} = 12.94\,V$$

Thus, voltage A will be equal to

$$V_A = V_{R_3} + V_{B2}$$

$$V_A = 12.94 + 2$$

FINAL RESULT

$$V_A = 14.94\,V$$

19.3.2.1.4 Voltage C

To find Voltage C we must find the voltage drop across R_2.

Current i_2 flows from the right to the left. This current is equal to 2 A and will create a voltage drop across R_2 equal to

$$V_{R_2} = R_2 i_2$$

$$V_{R_2} = 8 \times 2$$

FINAL RESULT

$$V_{R_2} = 16\,V$$

Notice that i_1 enters V_1 by the positive pole. The same happens for i_3 and B2. Current i_2, on the other hand, is following FC1's direction. This means that V_1 and B2 are absorbing power and that FC1 is the source driving the circuit.

Exercises

1 Find the mesh currents for the circuit shown in Figure 19.10.

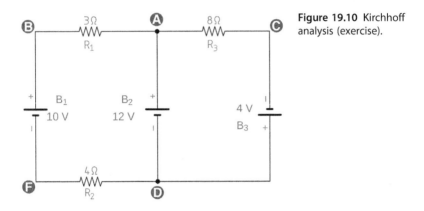

Figure 19.10 Kirchhoff analysis (exercise).

2 Find the mesh currents for the circuit shown in Figure 19.11.

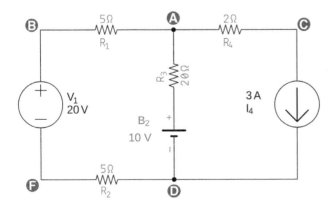

Figure 19.11 Kirchhoff analysis.

Solutions

1 The first thing is to assign mesh currents with arbitrary directions to the circuit, as shown in Figure 19.12.

Figure 19.12 Kirchhoff analysis.

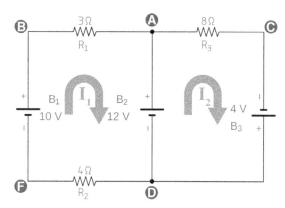

The next step involves analyzing each mesh.

Mesh DFBAD

Starting at point D and following the mesh for I_1 in a clockwise direction, we get the following equation according to the KVL rules:

$$-R_2 I_1 + V_{B1} - R_1 I_1 - V_{B2} = 0$$

Substituting the known values,

$$-4I_1 + 10 - 3I_1 - 12 = 0$$

$$-7I_1 = 2$$

FINAL RESULT

$$I_1 = -\frac{2}{7} A$$

The negative sign of I_1 tells us that this current, in fact, flows in the opposite direction to the one we have chosen.

Mesh DACD

This mesh will give us the following KVL equation:

$$V_{B2} - R_3 I_2 + V_{B3} = 0$$

$$12 - 8I_2 + 4 = 0$$

$$-8I_2 = -16$$

$I_2 = 2 \text{ A}$

2   The first thing we do is to assign mesh currents with arbitrary directions to the circuit as shown in Figure 19.13.

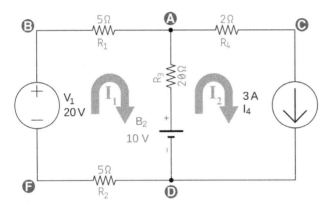

Figure 19.13 Kirchhoff analysis – mesh currents.

The next step involves analyzing each mesh.

Mesh DFBAD

Starting at point D and following mesh DFBAD according to the KVL rules, we get the following equation:

$$-R_2 I_1 + V_1 - R_1 I_1 - R_3 I_1 + R_3 I_2 - V_{B2} = 0$$

In this case we must consider the effects of I_2 across R_3. Current I_1 creates a voltage drop across R_3, but current I_2 creates a voltage raise across the same resistor.

Substituting the known values,

$$-5 I_1 + 20 - 5 I_1 - 20 I_1 + 20 I_2 - 10 = 0$$

$$30 I_1 - 20 I_2 = 10$$

However, I_2 is equal to I_4.

$I_2 = I_4 = 3 \text{ A}$

Substituting the values,

$30I_1 - 20(3) = 10$

$30I_1 = 70$

$I_1 = 2.33$ A

Branch Currents

Having the mesh currents, we can calculate the branch currents, randomly assigned to the circuit and shown in Figure 19.14.

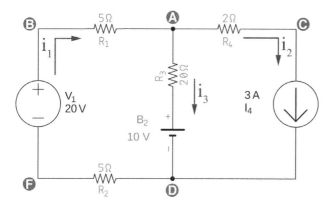

Figure 19.14 Kirchhoff analysis – branch currents.

Comparing Figure 19.13 with Figure 19.14, we discover i_1 and i_2.

$i_1 = I_1 = 2.33$ A

$i_2 = I_2 = I_4 = 3$ A

Current i_2 will be

$i_2 = I_1 - I_2$

$i_2 = 2.33 - 3$

$i_3 = -0.67$ A

20

Nodal Analysis

Circuit Analysis

20.1 Introduction

In this chapter, we will examine a technique of circuit analysis called voltage node analysis or simply nodal analysis.

This technique can be used to find currents and voltages across a circuit by using a zero reference and a peculiar way to use Ohm's law.

20.2 Examples

20.2.1 Example 1

Consider the circuit shown in Figure 20.1, composed of several voltage sources and resistors. We want to find the voltage between points A and B and the several branch currents.

Figure 20.1 Nodal analysis.

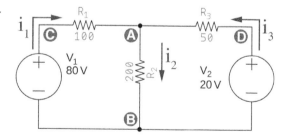

20.2.1.1 Solution

First thing to do, according to the nodal analysis, is to choose a point to be the circuit's zero reference, that is, the ground point. We select point B and redraw the circuit to include a ground point, as shown in Figure 20.2.

Introductory Electrical Engineering with Math Explained in Accessible Language,
First Edition. Magno Urbano.
© 2020 John Wiley & Sons, Inc. Published 2020 by John Wiley & Sons, Inc.

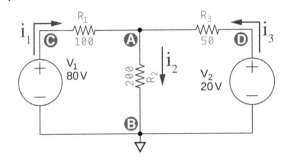

Figure 20.2 Ground reference.

Now that we have a ground point, we can say that the voltages across points C and ground and across D and ground are equal to V_1 and V_2, or 80 and 20 V, respectively.

There is no rule to choose the ground reference point. It is advised to choose the point where the negative poles of more voltage/current sources are connected.

However, the voltage across A, which is between A and ground, is not immediately obvious. To find this voltage we must analyze the circuit.

20.2.1.2 Circuit Analysis

First Ohm's law tells us that $I = V/R$, meaning that current is equal to voltage divided by resistance. This is the principle used by nodal analysis.

20.2.1.3 Kirchhoff's Laws

In Figure 20.2, we have established, randomly, three branch currents, i_1, i_2, and i_3.

According to Kirchhoff's current law (KCL), the sum of the currents that arrive at a node is equal to the sum of the currents that leave the same node.

Thus, we can write the following equation for node A:

$$i_1 + i_3 = i_2$$

20.2.1.3.1 Current i_1

Resistor R_1 is subjected to two voltages: the voltage across point A and the voltage across point C. Therefore, the voltage across R_1 will be the subtraction of these two voltages, or

$$V_{R_1} = V_C - V_A$$

If current is equal to voltage divided by the resistance, or I = V/R, current across R_1 will be

$$i_1 = \frac{V_C - V_A}{R_1} = \frac{80 - V_A}{100}$$

Why have we subtracted V_A from V_C and not the other way around? The answer is simple: i_1 flows from C to A.

20.2.1.3.2 Current i_2

Voltage at point A is the difference between voltages of A and B divided by the value of R_2. However, the voltage at B is 0, because this point is the ground reference.

Therefore, voltage across R_2 is

$$i_2 = \frac{V_A - V_B}{200} = \frac{V_A - 0}{200} = \frac{V_A}{200}$$

20.2.1.3.3 Current i_3

We can calculate i_3 by using the same principle we have used to find i_1:

$$i_3 = \frac{V_D - V_A}{R_3} = \frac{20 - V_A}{50}$$

Again, the subtraction order follows the direction of i_3.

20.2.1.3.4 Bringing It All Together

$$i_1 + i_3 = i_2$$

$$\frac{80 - V_A}{100} + \frac{20 - V_A}{50} = \frac{V_A}{200}$$

Solving this equation will give us the following.

FINAL RESULT

$$V_A = 34.28\,V$$

We can now find the currents

$$i_1 = \frac{80 - V_A}{100}$$

$$i_1 = \frac{80 - 34.28}{100}$$

FINAL RESULT

$$i_1 = 0.4571\,A$$

$$i_2 = \frac{V_A}{200}$$

$$i_2 = \frac{34.28}{200}$$

FINAL RESULT

$$i_2 = 0.1714\,A$$

$$i_3 = \frac{20 - V_A}{50}$$

$$i_3 = \frac{20 - 34.28}{50}$$

FINAL RESULT

$$i_3 = -0.2857\,A$$

The negative sign found for i_3 shows that this current flows in the opposite direction compared with the arbitrarily chosen direction, from point A to D, entering V_2 by the positive pole. Therefore, V_2 is absorbing power from the circuit and V_1 is the source driving the circuit.

20.2.2 Example 2

To prove that the directions chosen for the arbitrary currents are irrelevant, we will now examine an impossible circuit, shown in Figure 20.3.

Figure 20.3 Impossible circuit.

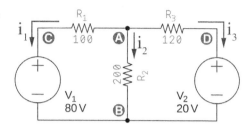

We want to find the voltage across R_2 (points A and B) and the currents, i_1, i_2, and i_3.

However, this circuit has an unusual problem: all currents leave node A, something impossible in real life.

20.2.2.1 Solution

To begin the nodal analysis, we choose point B and establish the zero reference. The circuit is redrawn to include the change, as shown in Figure 20.4.

Figure 20.4 Establishing a ground reference.

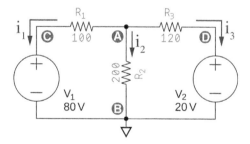

20.2.2.2 Applying Kirchhoff's Laws

According to KCL, the sum of the currents entering a node is equal to the sum of currents leaving the same node, or, in other words, the sum of all the currents entering or leaving a node is equal to 0.

This gives us the following equation:

$$i_1 + i_2 + i_3 = 0$$

20.2.2.2.1 Current i_1

By applying nodal analysis, i_1 is equal to the voltage across R_1 divided by its resistance:

$$i_1 = \frac{V_A - V_C}{R_1} = \frac{V_A - V_1}{R_1} = \frac{V_A - 80}{100}$$

Notice that $V_C = V_1$, equivalent to the power supply V_1.

20.2.2.2.2 Current i_2

By the nodal analysis, i_2 is equal to the voltage across R_2, that is, the voltage at point A subtracted from the voltage at point B, divided by its resistance. However, the voltage of B is 0, since this point is the ground reference, giving us the following equation:

$$i_2 = \frac{V_A - V_B}{R_2} = \frac{V_A - 0}{200} = \frac{V_A}{200}$$

20.2.2.2.3 Current i_3

By using the same principle, current i_3 is equal to the voltage across R_3, that is, voltage at point A subtracted from voltage at point D, divided by its resistance:

$$i_3 = \frac{V_A - V_D}{R_3} = \frac{V_A - 20}{120}$$

20.2.2.2.4 Bringing All Three Currents Together

$$i_1 + i_2 + i_3 = 0$$

$$\frac{V_A - 80}{100} + \frac{V_A}{200} + \frac{V_A - 20}{120} = 0$$

Solving this equation gives us the following.

FINAL RESULT

$$V_A = 41.42\,V$$

We can now find the currents

$$i_1 = \frac{V_A - 80}{100}$$

$$i_1 = \frac{41.42 - 80}{100}$$

FINAL RESULT

$$i_1 = -0.3858\,A$$

$$i_2 = \frac{V_A}{200}$$

$$i_2 = \frac{41.42}{200}$$

FINAL RESULT

$$i_2 = 0.2071\,A$$

$$i_3 = \frac{41.42 - 20}{120}$$

FINAL RESULT

$$i_3 = 0.1785\,A$$

The negative sign for i_1 shows that we got the direction wrong for this current and that it flows the other way around, from C to A.

We see that directions we choose for the currents are irrelevant. The math will always correct our mistakes.

20.2.3 Example 3

Consider the circuit shown in Figure 20.5.

We want to find voltages across A and B and the several currents shown in Figure 20.5.

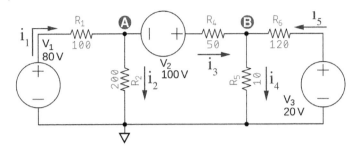

Figure 20.5 Nodal analysis.

20.2.3.1 Node A

Two currents leave and one current arrives at node A.

KCL give us

$$i_1 = i_2 + i_3$$

Each one of the currents can be found by nodal analysis:

$$i_1 = \frac{V_1 - V_A}{R_1} = \frac{80 - V_A}{100}$$

$$i_2 = \frac{V_A}{R_2} = \frac{V_A}{200}$$

$$i_3 = \frac{V_A - V_B + V_2}{R_4} = \frac{V_A - V_B + 100}{50}$$

20.2.3.1.1 *Bringing It All Together*

$$i_1 = i_2 + i_3$$

$$\frac{80 - V_A}{100} = \frac{V_A}{200} + \frac{V_A - V_B + 100}{50}$$

$$\frac{80 - V_A}{2} = \frac{V_A}{4} + \frac{V_A - V_B + 100}{1}$$

$$\frac{80 - V_A}{2} = \frac{V_A + 4V_A - 4V_B + 400}{4}$$

$$320 - 4V_A = 10V_A - 8V_B + 800$$

$$7V_A - 4V_B = -240 \tag{20.1}$$

20.2.3.2   Node B

Node B gives us

$$i_3 + i_5 = i_4$$

Current i_3 is already known.
The other currents can be found by

$$i_5 = \frac{V_3 - V_B}{R_6} = \frac{20 - V_B}{120}$$

$$i_4 = \frac{V_B}{R_5} = \frac{V_B}{10}$$

20.2.3.2.1   *Bringing It All Together*

$$i_3 + i_5 = i_4$$

$$\frac{V_A - V_B + 100}{50} + \frac{20 - V_B}{120} = \frac{V_B}{10}$$

$$\frac{12V_A - 12V_B + 1200 + 100 - 5V_B}{60} = V_B$$

$$12V_A - 12V_B + 1200 + 100 - 5V_B = 60V_B$$

$$12V_A - 77V_B = -1300 \qquad\qquad (20.2)$$

We end with two equations, (20.1) and (20.2), as shown below:

$$\begin{cases} 7V_A - 4V_B = -240 \\ 12V_A - 77V_B = -1300 \end{cases}$$

These equations can be solved by using matrices:

$$\begin{bmatrix} 7 & -4 \\ 12 & -77 \end{bmatrix} \begin{bmatrix} V_A \\ V_B \end{bmatrix} = \begin{bmatrix} -240 \\ -1300 \end{bmatrix}$$

If we label each matrix with a letter

$$\overbrace{\begin{bmatrix} 7 & -4 \\ 12 & -77 \end{bmatrix}}^{A} \overbrace{\begin{bmatrix} V_A \\ V_B \end{bmatrix}}^{B} = \overbrace{\begin{bmatrix} -240 \\ -1300 \end{bmatrix}}^{C}$$

we can say that

$$A \times B = C$$

If we want to find B, we should, in theory, divide C by A, but that is not possible with matrices. The correct way to find B in this case is to multiply C by the inverse of A:

$$B = C \times A^{-1} \tag{20.3}$$

MATH CONCEPT   Given a matrix A

$$A = \begin{bmatrix} a & b \\ c & d \end{bmatrix}$$

The inverse of A can be found by

$$A^{-1} = \begin{bmatrix} a & b \\ c & d \end{bmatrix}^{-1} = \frac{1}{|A|}\begin{bmatrix} d & -b \\ -c & a \end{bmatrix}$$

where $|A|$ is the determinant of A and a,b,c,d are the matrix's components.

Thus, the first thing we need to do is to calculate the determinant of A.

MATH CONCEPT   The determinant of a 2×2 matrix can be found by the following formula:

$$\det(A) = a \times d - b \times c$$

where $\det(A)$ is the determinant and a, b, c, d are the matrix's components.

Therefore,

$$A = \begin{bmatrix} 7 & -4 \\ 12 & -77 \end{bmatrix}$$

$$\det(A) = (7) \times (-77) - (-4) \times (12)$$

$$\det(A) = -491$$

Hence, the inverse of matrix A will be

$$A^{-1} = -\frac{1}{491} \begin{bmatrix} -77 & 4 \\ -12 & 7 \end{bmatrix}$$

If we apply this result in (20.3), we get

$$\begin{bmatrix} V_A \\ V_B \end{bmatrix} = -\frac{1}{491} \begin{bmatrix} -77 & 4 \\ -12 & 7 \end{bmatrix} \begin{bmatrix} -240 \\ -1300 \end{bmatrix}$$

MATH CONCEPT   The product of two 2×2 matrices like

$$\begin{bmatrix} a & b \\ c & d \end{bmatrix} \times \begin{bmatrix} x \\ y \end{bmatrix}$$

is

$$\begin{bmatrix} a*x + b*y \\ c*x + d*y \end{bmatrix}$$

$$\begin{bmatrix} V_A \\ V_B \end{bmatrix} = -\frac{1}{491} \begin{bmatrix} (-77)(-240) + (4)(-1300) \\ (-12)(-240) + (7)(-1300) \end{bmatrix}$$

$$\begin{bmatrix} V_A \\ V_B \end{bmatrix} = -\frac{1}{491} \begin{bmatrix} 13280 \\ -622 \end{bmatrix}$$

$$\begin{bmatrix} V_A \\ V_B \end{bmatrix} = \begin{bmatrix} -27.04 \\ 12.66 \end{bmatrix}$$

FINAL RESULT

$V_A = -27.04 \, V$

$V_B = 12.66 \, V$

We see that V_A is negative, meaning this point has an electric potential less than ground.

We can now calculate the several currents.

FINAL RESULT

$$i_1 = \frac{80 - V_A}{100} = \frac{80 - (-27.04)}{100} = 1.07 \, A$$

$$i_2 = \frac{V_A}{200} = \frac{-27.04}{200} = -0.13 \, A$$

$$i_3 = \frac{V_A - V_B + 100}{50} = \frac{(-27.04) - 12.66 + 100}{50} = 1.20 \, A$$

$$i_4 = \frac{V_B}{10} = \frac{12.66}{10} = 1.26 \, A$$

$$i_5 = \frac{20 - V_B}{120} = \frac{20 - 12.66}{120} = 0.06 \, A$$

Once more, the negative sign of i_2 shows that this current flows inversely, compared with the direction we have chosen for it.

Exercises

1 Find the voltages at points A and C and the several currents assigned with arbitrary directions in the circuit shown in Figure 20.6.

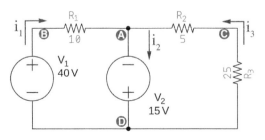

Figure 20.6 Nodal analysis (exercise).

2 Find the voltage across R_3 and the several currents assigned with arbitrary directions in the circuit shown in Figure 20.7.

Figure 20.7 Nodal analysis (exercise).

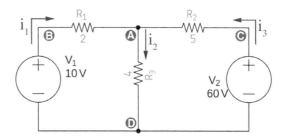

3 Find the voltage at point A and the several currents assigned with arbitrary directions in the circuit shown in Figure 20.8.

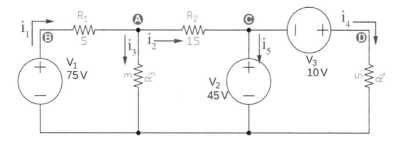

Figure 20.8 Nodal analysis (exercise).

Solutions

1 We choose point D as the ground reference and redraw the circuit to include the change, as shown in Figure 20.9.

Figure 20.9 Nodal analysis (exercise).

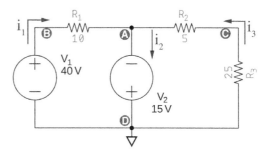

Observing Figure 20.9, we see that voltages B and A, taken in relation to the ground, are equal to V_1 and V_2 or 40 V and -15 V, respectively. We also see that voltage at point D is 0 V.

By applying nodal analysis, we get i_1 and i_3:

$$i_1 = \frac{V_B - V_A}{R_1}$$

$$i_1 = \frac{40 - (-15)}{10}$$

FINAL RESULT

$$i_1 = 5.5\,\text{A}$$

Current i_3 will be equal to

$$i_3 = \frac{V_C - V_A}{R_2}$$

but will be also be equal to

$$i_3 = \frac{V_D - V_C}{R_3}$$

So, we can equal both equations

$$\frac{V_C - V_A}{R_2} = \frac{V_D - V_C}{R_3}$$

By substituting the values we have, we get

$$\frac{V_C - (-15)}{5} = \frac{0 - V_C}{25}$$

$$\frac{V_C + 15}{5} = \frac{-V_C}{25}$$

$$25(V_C + 15) = -5V_C$$

$$25V_C + 375 = -5V_C$$

$$25V_C + 5V_C = -375$$

$$30V_C = -375$$

$$V_C = -12.50\text{A}$$

Therefore, we can calculate i_3

$$i_3 = \frac{V_C - V_A}{R_2}$$

$$i_3 = \frac{-12.50 - (-15)}{5}$$

FINAL RESULT

$i_3 = 0.5\,A$

Current i_2 is found by applying KCL at node A

$$i_2 = i_1 + i_3$$

$$i_2 = 5.5 + 0.5$$

FINAL RESULT

$i_2 = 6\,A$

Current flowing from C to A will cause a voltage drop across R_2 equal to

$$V_{R_2} = V_{AC} = R_2 i_3$$

$$V_{AC} = 5(0.5)$$

FINAL RESULT

$V_{AC} = 2.5\,V$

2 By applying KCL to node A, we get

$$i_1 + i_3 = i_2$$

By applying node analysis for node A, we obtain

$$i_1 = \frac{V_B - V_A}{R_1} = \frac{V_1 - V_A}{R_1} = \frac{10 - V_A}{2}$$

$$i_2 = \frac{V_A}{R_3} = \frac{V_A}{4}$$

$$i_3 = \frac{V_C - V_A}{R_2} = \frac{V_2 - V_A}{R_2} = \frac{60 - V_A}{5}$$

Putting it all together to get the voltage at point A:

$$i_1 + i_3 = i_2$$

$$\frac{10 - V_A}{2} + \frac{60 - V_A}{5} = \frac{V_A}{4}$$

$$\frac{50 - 5V_A + 120 - 2V_A}{10} = \frac{V_A}{4}$$

$$200 - 20V_A + 480 - 8V_A = 10V_A$$

$$680 - 28V_A = 10V_A$$

$$680 = 38V_A$$

FINAL RESULT

$$V_A = 17.8947\,V$$

We can now find the currents

$$i_1 = \frac{10 - V_A}{2}$$

$$i_1 = \frac{10 - 17.8947}{2}$$

FINAL RESULT

$$i_1 = -3.9474\,A$$

$$i_2 = \frac{V_A}{4}$$

$$i_2 = \frac{17.8947}{4}$$

FINAL RESULT

$$i_2 = 4.4737\,A$$

$$i_3 = \frac{60 - V_A}{5}$$

$$i_3 = \frac{60 - 17.8947}{5}$$

$$i_3 = 8.4211\,A$$

3 By applying nodal analysis for node A, we get

$$i_1 = \frac{V_1 - V_A}{R_1} = \frac{75 - V_A}{5}$$

$$i_2 = \frac{V_A - V_C}{R_2} = \frac{V_A - 45}{15}$$

$$i_3 = \frac{V_A}{R_2} = \frac{V_A}{3}$$

By applying KCL to node A,

$$i_1 = i_2 + i_3$$

Thus,

$$\frac{75 - V_A}{5} = \frac{V_A - 45}{15} + \frac{V_A}{3}$$

$$\frac{75 - V_A}{5} = \frac{3V_A - 135 + 15V_A}{45}$$

$$75 - V_A = \frac{3V_A - 135 + 15V_A}{9}$$

$$675 - 9V_A = 3V_A - 135 + 15V_A$$

$$27V_A = 810$$

$$V_A = 30\,V$$

We can now find the currents

$$i_1 = \frac{75 - V_A}{5}$$

$$i_1 = \frac{75 - 30}{5}$$

FINAL RESULT

$$i_1 = 9\,A$$

$$i_2 = \frac{V_A - 45}{15}$$

$$i_2 = \frac{30 - 45}{15}$$

FINAL RESULT

$$i_2 = -1\,A$$

$$i_3 = \frac{V_A}{R_2}$$

$$i_3 = \frac{30}{3}$$

FINAL RESULT

$$i_3 = 10\,A$$

Voltage across point C and ground is equal to V_2, that is, 45 V. The block V_3/R_4 is also connected between C and ground, parallel to V_2. For that reason, V_3/R_4 has the same 45 V across.

If V_3 is equal to –10 V,[1] the voltage across the resistor R_4 will have to be 55 V to make voltage across C and ground equal to 45 V (55 – 10 = 45 V). Thus, current across R_4 must produce a voltage drop of 55 V:

$$i_4 = \frac{V_D}{R_4} = \frac{V_D}{5}$$

$$i_4 = \frac{55}{5}$$

---

1 V_3 is a negative source because its negative pole is connected to C and its positive pole is connected to ground through R_4.

FINAL RESULT

$i_4 = 11\,A$

By applying KCL at node C, we get current i_5:

$i_5 + i_4 = i_2$

$i_5 = i_2 - i_4$

$i_5 = -1 - 11$

FINAL RESULT

$i_5 = -12\,A$

21

Thévenin's Theorem

Circuit Analysis

21.1 Introduction

In this chapter, we will examine the Thévenin's theorem, also known as equivalent circuit theorem, created by Léon Charles Thévenin, a French telegraph engineer who extended Ohm's law to the analysis of complex electrical circuits.

This theorem makes it possible to convert complex circuits into simpler ones that are electrically equivalent and behave in the same way.

21.2 The Theorem

Unlike the source transformation theorem, which aims to replace parts of circuits to convert everything into a voltage or current source with resistors, the Thévenin Theorem's objective is to replace the whole circuit with a voltage source in series with a resistor.

Figure 21.1 shows the goal of the Thévenin's theorem to replace a complete circuit with a voltage source, called the equivalent Thévenin voltage source, or V_{Th} in series with a resistor, called the equivalent Thévenin resistor, or R_{Th}.

Figure 21.1 Thévenin equivalent circuit.

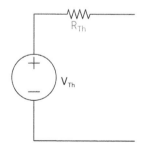

Introductory Electrical Engineering with Math Explained in Accessible Language,
First Edition. Magno Urbano.
© 2020 John Wiley & Sons, Inc. Published 2020 by John Wiley & Sons, Inc.

21.2.1 The Equivalent Thévenin Circuit

The Thévenin circuit always represents the equivalent seen through two terminals of the original circuit that can be the input, the output, or any other points.

Figure 21.2 shows a black box representing a circuit and points A and B representing the input.

Figure 21.2 Two points of a circuit.

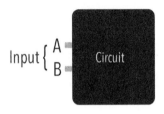

21.2.2 Methodology

To obtain the Thévenin equivalent circuit, we must perform the following steps:

- Remove any loads that the circuit may have.
- Remove all current sources from the circuit.
- Short-circuit all voltage sources in the circuit.
- Calculate the equivalent resistance.
- Calculate the equivalent voltage by using nodal analysis.

21.2.3 Example

Consider the circuit shown in Figure 21.3. This circuit has a few resistors, a voltage source, a current source, and an output represented by points T and U. A load, represented by R_L,[1] is connected to the output.

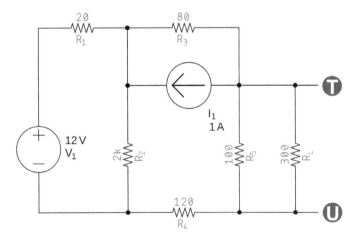

Figure 21.3 Thévenin equivalent circuit.

---

1 A load can be anything, a motor, a light bulb, headphones, etc.

21.2.3.1 Thévenin Equivalent Resistance

To find the Thévenin equivalent circuit, we must follow the theorem's rules. The first rule is to remove all loads from the circuit. Figure 21.4 shows the circuit with the load removed.

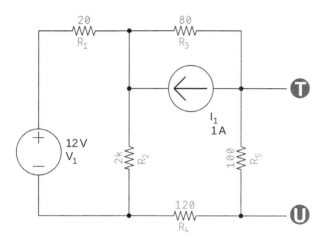

Figure 21.4 Circuit with the load removed.

Next, according to the rules of Thévenin, we must remove all current sources from the circuit and replace all voltage sources with short circuits, that is, a wire. The result is seen in Figure 21.5.

Figure 21.5 All sources removed.

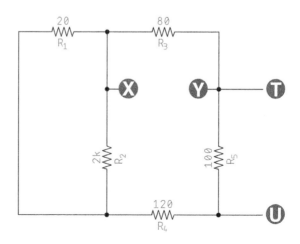

Points X and Y represent where the current source was. There is no more current flowing across these points now.

After removing the sources, we can redraw the circuit in Figure 21.4 as shown in Figure 21.6.

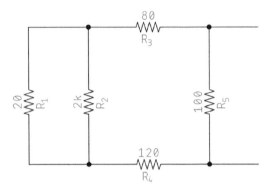

Figure 21.6 Improved version.

In the new circuit, R_1 and R_2 are in parallel and must be replaced by an equivalent resistance equal to

$$\frac{1}{R_E} = \frac{1}{20} + \frac{1}{2000}$$

$$R_E = 19.801\,\Omega$$

The result is seen in Figure 21.7.

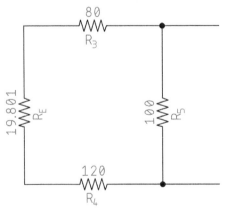

Figure 21.7 Thévenin equivalent circuit without the load.

In this new circuit, we see that R_E, R_4, and R_3 are in series and can be replaced by an equivalent resistance equal to

$R_{E2} = 120 + 80 + 19.801$

$$R_{E2} = 219.801\,\Omega$$

The result is seen in Figure 21.8.

Figure 21.8 Equivalent circuit without the sources.

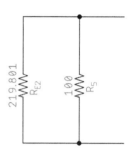

Finally, we have two resistors in parallel, R_{E2} and R_5, and we can replace them with an equivalent resistance:

$$\frac{1}{R} = \frac{1}{219.801} + \frac{1}{100}$$

$$R = 68.73\,\Omega$$

This final resistance is the Thévenin resistance.

FINAL RESULT

$R_{Th} = 68.73\,\Omega$

21.2.3.2 Thévenin Equivalent Voltage

To find the Thévenin equivalent voltage, we bring back the original circuit and choose a point to use as zero reference. The result is seen in Figure 21.9.

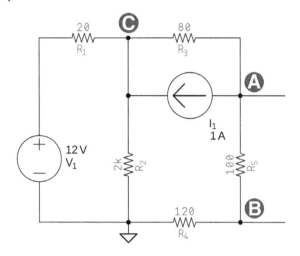

Figure 21.9 Original circuit plus ground.

We must use nodal analysis to find the currents and the voltages across the points we have assigned in Figure 21.10.

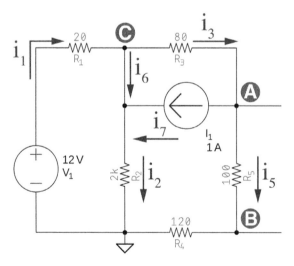

Figure 21.10 Original Circuit plus ground with assigned currents.

21.2.3.3 Node A

By applying Kirchhoff's current law (KCL) in the circuit shown in Figure 21.10, we get the following equation:

$$i_3 = i_5 + i_7 \tag{21.1}$$

Every one of these currents can be described, in terms of nodal analysis, as follows:

$$i_3 = \frac{V_C - V_A}{R_3}$$

Without the load, the current across R_4 and R_5 is the same, that is, i_5. Therefore,

$$i_5 = \frac{V_A}{R_4 + R_5}$$

Current i_7 is the same as I_1, equal to 1 A:

$$i_7 = i_{I_1} = 1\,\mathrm{A}$$

By substituting the values we have already discovered in (21.1), we get

$$\frac{V_C - V_A}{R_3} = \frac{V_A}{R_4 + R_5} + I_{I_1}$$

$$\frac{V_C - V_A}{80} = \frac{V_A}{120 + 100} + 1$$

$$\frac{V_C - V_A}{80} = \frac{V_A + 220}{220}$$

$$22\,V_C - 22\,V_A = 8\,V_A + 1760$$

$$15\,V_A - 11\,V_C = -880 \tag{21.2}$$

21.2.3.4 Node B

The only current that flows across node B is i_5. This current flows across R_5 and then across R_4, because there is no load connected in parallel with R_5.

Therefore, we will ignore this current by now, because it is irrelevant to solve the circuit.

21.2.3.5 Node C

The current diagram has a visualization problem that makes difficult to analyze the currents of node C. The problem is the current source I_1.

To best understand the circuit, we invert the position of R_3 and I_4 and redraw the circuit like shown in Figure 21.11.

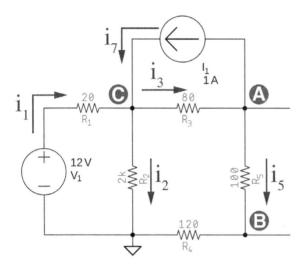

Figure 21.11 Original circuit redrawn.

After redrawing the circuit, we see that current i_6 was not really a current, it was only an illusion created by a drawing that was making it difficult to visualize the real currents in the circuit.

If we now apply KCL to node C, we get

$$i_1 + i_7 = i_2 + i_3 \tag{21.3}$$

By applying nodal analysis to the node, we get each of these currents:

$$i_1 = \frac{V_1 - V_C}{R_1}$$

$$i_7 = i_{I_1}$$

$$i_2 = \frac{V_C}{R_2}$$

$$i_3 = \frac{V_C - V_A}{R_3}$$

21.2.3.5.1 Putting It All Back Together

$$\frac{V_1 - V_C}{R_1} + i_{I_1} = \frac{V_C}{R_2} + \frac{V_C - V_A}{R_3}$$

$$\frac{12 - V_C}{20} + 1 = \frac{V_C}{2000} + \frac{V_C - V_A}{80}$$

$$\frac{12 - V_C + 20}{20} = \frac{80\,V_C + 2000\,V_C - 2000\,V_A}{160000}$$

$$\frac{32 - V_C}{2} = \frac{2080\,V_C - 2000\,V_A}{16000}$$

$$512000 - 16000\,V_C = 2080\,V_C - 4000\,V_A$$

$$25\,V_A - 126\,V_C = -3200 \tag{21.4}$$

We end with two equations, (21.2) and (21.4), which can be solved by using matrices

$$\begin{bmatrix} 15 & -11 \\ 25 & -126 \end{bmatrix} \begin{bmatrix} V_A \\ V_C \end{bmatrix} = \begin{bmatrix} -880 \\ -3200 \end{bmatrix}$$

like we have described before.
The solution gives us the following result.

FINAL RESULT

$V_A = -46.86\,V$

$V_C = 16.09\,V$

21.2.3.6 The Thévenin Voltage
The Thévenin voltage will be the subtraction between the voltage across A and B, or

$$V_{Th} = V_A - V_B$$

This will be the voltage across R_5.
To find this voltage we need to know the current flowing across this resistor. In the first part of this solution, we discovered that

$$i_5 = \frac{V_A}{R_4 + R_5}$$

$$i_5 = -\frac{46.86}{120 + 100}$$

$$i_5 = -0.213\,\text{A}$$

Therefore, the voltage across R_5 is

$$V_{R_5} = R_5 i_5$$

$$V_{R_5} = 100(-0.213)$$

$$V_{R_5} = -21.3\,\text{V}$$

V_{R_5} is exactly the Thévenin voltage.

FINAL RESULT

$$V_{Th} = -21.3\,\text{V}$$

Now that we have the Thévenin voltage and resistance, we can redraw the final Thévenin equivalent circuit, as shown in Figure 21.12, with the load reconnected.

The Thévenin voltage source is drawn upside down because it is negative.

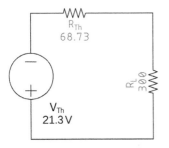

Figure 21.12 Thévenin equivalent circuit.

21.2.3.7   Same Behavior
In this section we will check if the Thévenin circuit behaves like the original.
   Figure 21.13 shows the original circuit with the load connected and its respective currents.

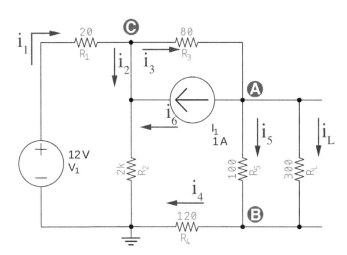

Figure 21.13 Original circuit.

The load R_L and resistor R_5 are in parallel, and we can substitute them by an equivalent resistor with the following resistance:

$$\frac{1}{R} = \frac{1}{100} + \frac{1}{300}$$

$$R = 75\,\Omega$$

We also swap, like we did before, the positions of the current source and R_3. The result is seen in Figure 21.14.

Notice that i_4 does not exist anymore after the conversion, being now equal to i_E.

To compare this circuit with the equivalent Thévenin, we must find the current and the voltage across R_E.

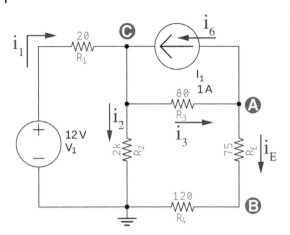

Figure 21.14 Improved original circuit.

21.2.3.7.1 Node A

By applying KCL at node A, we get the following equation:

$$i_3 = i_6 + i_e$$

By applying nodal analysis to node A, we get the following equation for every term of the previous equation:

$$i_3 = \frac{V_C - V_A}{R_3}$$

$i_6 = I_1$, the current source:

$$i_e = \frac{V_A}{R_{RE} + R_4}$$

Putting it all back together,

$$\frac{V_C - V_A}{R_3} = i_{I_1} + \frac{V_A}{R_{R_E} + R_4}$$

$$\frac{V_C - V_A}{80} = 1 + \frac{V_A}{75 + 120}$$

$$195 V_C - 195 V_A = 15600 + 80 V_A$$

$$-275 V_A + 195 V_C = 15600$$

$$-55 V_A + 39 V_C = 3120$$

$$55\,V_A - 39\,V_C = -3120 \tag{21.5}$$

21.2.3.8 Node C

Node C gives the following equation by applying KCL:

$$i_1 + i_6 = i_2 + i_3$$

We already know two terms of this equation. The other two are

$$i_1 = \frac{V_1 - V_C}{R_1}$$

$$i_2 = \frac{V_C}{R_2}$$

Putting it all back together,

$$\frac{V_1 - V_C}{R_1} + i_{I_1} = \frac{V_C}{R_2} + \frac{V_C - V_A}{R_3}$$

$$\frac{12 - V_C}{20} + 1 = \frac{V_C}{2000} + \frac{V_C - V_A}{80}$$

$$\frac{12 - V_C + 20}{20} = \frac{V_C}{2000} + \frac{V_C - V_A}{80}$$

$$12 - V_C + 20 = \frac{V_C}{100} + \frac{V_C - V_A}{4}$$

$$32 - V_C = \frac{4\,V_C + 100\,V_C - 100\,V_A}{400}$$

$$12800 - 400\,V_C = 100\,V_C - 100\,V_A + 4\,V_C$$

$$12800 = 100\,V_C - 100\,V_A + 4\,V_C + 400\,V_C$$

$$12800 = 504\,V_C - 100\,V_A$$

$$25\,V_A - 126\,V_C = -3200 \tag{21.6}$$

We got two equations, (21.5) and (21.6), which can be solved by using matrices

$$\begin{bmatrix} 55 & -39 \\ 25 & -126 \end{bmatrix} \begin{bmatrix} V_A \\ V_B \end{bmatrix} = \begin{bmatrix} -3120 \\ -3200 \end{bmatrix}$$

The solution, by the method we have explained before, gives us

$$V_A = -45.057\,V$$
$$V_C = 16.456\,V$$

We can now calculate i_e:

$$i_e = \frac{V_A}{R_{R_E} + R_4}$$

$$i_e = \frac{-45.057}{75 + 120}$$

$$i_e = -0.2311\,A$$

To calculate the voltage across R_E, we must use the first Ohm's law

$$V_{R_E} = R_E \times I_E$$

$$V_{RE} = 75 \times (-0.2311)$$

$$V_{R_E} = -17.3296\,V \qquad\qquad (21.7)$$

We know that i_e is the current flowing across R_E. However, this resistor does not exist in real life and is just a substitute for R_L in parallel with R_5 (see Figure 21.13).

To compare this circuit with the Thévenin equivalent, we must calculate the current flowing across the load, alone.

R_L and R_5 are in parallel and therefore have the same voltage across. Current across each resistor is their resistance multiplied by their current. We can express both conditions mathematically and obtain a relation between their currents, as follows:

$$V_{R_1} = V_{R_L}$$
$$R_5 I_5 = R_{R_L} I_{R_L}$$

$$100\,I_5 = 300\,I_{R_L}$$

$$I_5 = \frac{300}{100}\,I_{R_L}$$

$$I_5 = 3I_{R_L}$$

Because both resistors are in parallel, we know that

$$i_e = I_5 + I_{R_L}$$

Therefore, the load current is

$$-0.2311 = 3I_{R_L} + I_{R_L}$$

$$I_{R_L} = -0.0578\,\text{A} \qquad (21.8)$$

21.2.3.8.1 Confirmation Using the Thévenin Equivalent

In this section we will perform the same calculation using the Thévenin equivalent circuit, shown again on Figure 21.15, to see if the values match.

Figure 21.15 Thévenin equivalent circuit used previously.

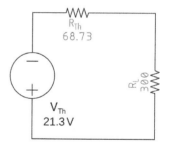

The first thing is to apply the first Ohm's law to find the current flowing across the Thévenin circuit:

$$I = \frac{V}{R}$$

$$I = \frac{-21.3\,\text{V}}{(300 + 68.73)\text{A}}$$

FINAL RESULT

$$I = -0.0578 \, A$$

Notice that this value is the same as obtained in (21.8).
The next step is to find the voltage across R_L, by applying the first Ohm's law, again:

$$V_{R_L} = R_L \times I$$

$$V = 300 \times (-0.0578)$$

FINAL RESULT

$$V_{R_L} = -17.329 \, V$$

Once more, we see that this value is the same as obtained in (21.7).
Therefore, we conclude that the Thévenin equivalent circuit behaves exactly like the original circuit.

Exercises

1 Find the Thévenin equivalent of the circuit connected to a load R_L, shown in Figure 21.16.

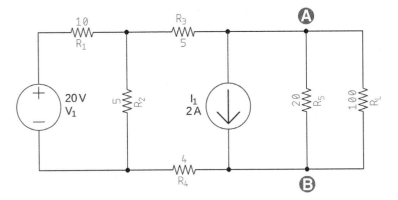

Figure 21.16 Thévenin equivalent circuit (exercise).

Solutions

1 To get the Thévenin resistance, we first remove the load (Figure 21.17).

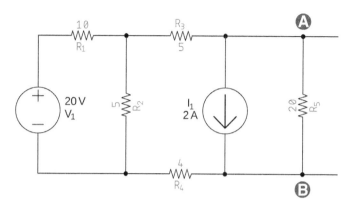

Figure 21.17 Thévenin equivalent circuit without the load.

Then, according to the Thévenin rules, we remove all current sources and replace all voltage sources with short circuits (Figure 21.18).

Figure 21.18 Equivalent circuit (exercise) without the sources.

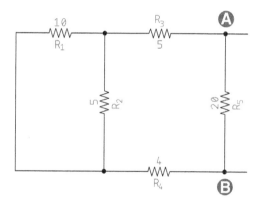

R_1 and R_2 are in parallel and can be replaced by an equivalent resistance equal to (Figure 21.19)

$$\frac{1}{R_{EQ1}} = \frac{1}{R_1} + \frac{1}{R_2}$$

$$\frac{1}{R_{EQ1}} = \frac{1}{10} + \frac{1}{5}$$

$$R_{EQ1} = 3.3333 \, \Omega$$

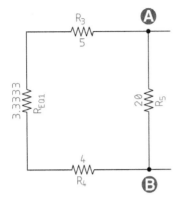

Figure 21.19 Thévenin equivalent circuit (exercise).

R_{EQ1}, R_3, and R_4 are in series and can be replaced by an equivalent resistance equal to the sum of their individual resistances (Figure 21.20):

$$R_{EQ2} = R_{EQ1} + R_3 + R_4$$

$$R_{EQ2} = 3.3333 + 5 + 4$$

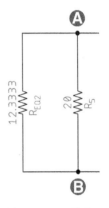

$$R_{EQ2} = 12.3333 \, \Omega$$

The Thévenin resistance will be the equivalent resistance of R_{EQ2} in parallel with R_5:

$$\frac{1}{R_{Th}} = \frac{1}{R_{EQ2}} + \frac{1}{R_5}$$

$$\frac{1}{R_{Th}} = \frac{1}{12.3333} + \frac{1}{20}$$

Figure 21.20 Thévenin equivalent circuit (exercise).

FINAL RESULT

$$R_{Th} = 7.6289 \, \Omega$$

Thévenin Voltage

To find the Thévenin voltage, we go back to the original circuit without the load, add a zero-reference point to it, and assign currents to all branches with directions at random as seen in Figure 21.21.

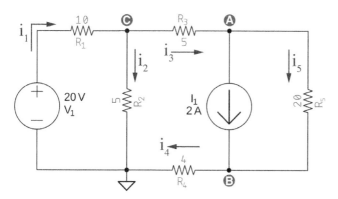

Figure 21.21 Thévenin equivalent circuit – branch currents.

To find the voltage across points A and B, we must apply KCL to all nodes.

Node C

By applying KCL to node C, we get the following equation:

$$i_1 = i_2 + i_3$$

Nodal analysis of node C gives us the following current equations:

$$i_1 = \frac{V_1 - V_C}{R_1}$$

$$i_2 = \frac{V_C}{R_2}$$

$$i_3 = \frac{V_C - V_A}{R_3}$$

Putting it all back together and substituting the known values, we get

$$i_1 = i_2 + i_3$$

$$\frac{V_1 - V_C}{R_1} = \frac{V_C}{R_2} + \frac{V_C - V_A}{R_3}$$

$$\frac{20 - V_C}{10} = \frac{V_C}{5} + \frac{V_C - V_A}{5}$$

$$\frac{20 - V_C}{10} = \frac{V_C + V_C - V_A}{5}$$

$$100 - 5V_C = 20V_C - 10V_A$$

$$100 = 25V_C - 10V_A$$

$$2V_A - 5V_C = -20 \qquad\qquad (21.9)$$

Node B

KCL gives us the following equation:

$$i_4 = I_1 + i_5$$

By nodal analysis, we get that

$$i_5 = \frac{V_A - V_B}{R_5}$$

Therefore,

$$i_4 = I_1 + \frac{V_A - V_B}{R_5}$$

$$i_4 = 2 + \frac{V_A - V_B}{20}$$

$$i_4 = \frac{40 + V_A - V_B}{20}$$

However, i_4 is also

$$i_4 = \frac{V_B}{R_4}$$

$$i_4 = \frac{V_B}{4}$$

Therefore, we can equate both equations for i_4 and obtain a second equation:

$$\frac{40 + V_A - V_B}{20} = \frac{V_B}{4}$$

$$160 + 4V_A - 4V_B = 20V_B$$

$$V_A - 6V_B = -40 \tag{21.10}$$

By applying KCL to the ground node we also get

$$i_1 = i_2 + i_4$$

Therefore, we can get a third equation by substituting the values we have so far:

$$\frac{V_1 - V_C}{R_1} = \frac{V_C}{R_2} + \frac{V_B}{R_4}$$

$$\frac{20 - V_C}{10} = \frac{V_C}{5} + \frac{V_B}{4}$$

$$5V_B + 6V_C = 40 \tag{21.11}$$

We end with three equations, (21.9), (21.10), and (21.11):

$$\begin{cases} 2V_A - 5V_C = -20 \\ V_A - 6V_B = -40 \\ 5V_B + 6V_C = 40 \end{cases}$$

To make it simple, we take the first equation and solve for V_A

$$2V_A - 5V_C = -20$$

$$V_A = \frac{5V_C - 20}{2}$$

$$V_A = 2.5V_C - 10 \tag{21.12}$$

and substitute this value on the second equation:

$$V_A - 6V_B = -40$$
$$2.5V_C - 10 - 6V_B = -40$$
$$2.5V_C - 6V_B = 10 - 40$$

$$2.5V_C - 6V_B = -30$$

We now have two equations instead of three:

$$\begin{cases} -6V_B + 2.5V_C = -30 \\ 5V_B + 6V_C = 40 \end{cases}$$

These equations can be written into the matrix form, which we already know how to solve:

$$\begin{bmatrix} -6 & 2.5 \\ 5 & 6 \end{bmatrix} \begin{bmatrix} V_B \\ V_C \end{bmatrix} = \begin{bmatrix} -20 \\ 40 \end{bmatrix}$$

The solution gives the following.

FINAL RESULT

$$V_B = 5.7732\,V$$
$$V_C = 1.8557\,V$$

If we substitute V_C in (21.12), we obtain the following.

FINAL RESULT

$$V_A = -5.3608\,V$$

The Thévenin voltage will be the voltage across points A and B that is equal to the difference between V_A and V_B, or

$$V_{AB} = V_{Th} = V_A - V_B$$
$$V_{Th} = -5.3608 - 5.7732$$

FINAL RESULT

$$V_{Th} = -11.1340\,V$$

22

Norton's Theorem

Circuit Analysis

22.1 Introduction

In this chapter, we will examine Norton's theorem, also known as Mayer's theorem, developed at the same time, in 1926, by Hans Ferdinand Mayer, researcher of Siemens & Halske in Germany, and by Edward Lawry Norton, engineer and researcher of the legendary Bell Labs[1] in the United States.

22.2 Norton's Theorem

Following the idea of Thévenin's theorem, Norton's theorem proposes the replacement of complex circuits with an equivalent simplified version, known as Norton equivalent circuit.

Thévenin theorem replaces the original circuit with a voltage source in series with a resistor. Norton's theorem, however, replaces the original circuit with a current source in parallel with a resistor, like shown in Figure 22.1.

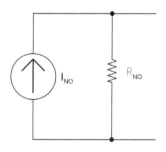

Figure 22.1 Norton equivalent circuit.

---

1 Famous research and development laboratory created in 1870 by Bell Telephone System, company originally founded by Alexander Graham Bell. Researchers working at Bell Labs, awarded with nine Nobel Prizes, are credited with the development of the transistor, LASER, CCDs used in digital photography, radio astronomy, first satellite, photo cell, mobile telephony, and UNIX operating system, including programming languages like C and C++, contributing immensely to the modern society we live today.

Introductory Electrical Engineering with Math Explained in Accessible Language,
First Edition. Magno Urbano.
© 2020 John Wiley & Sons, Inc. Published 2020 by John Wiley & Sons, Inc.

22.2.1 Finding Norton Equivalent Circuit

Like Thévenin, Norton equivalent circuit is always equivalent to two chosen points in a circuit, like the input, the output, etc. Different points in a circuit will generate different Norton circuits.

22.2.2 Methodology

These rules must be followed to obtain the Norton equivalent circuit:

- Create a new circuit version where the load is short-circuited, and calculate the current source using nodal analysis and Kirchhoff's current law (KCL).
- Create a new circuit version without the load, without all current sources, and with all voltage sources replaced with short circuits, and find the equivalent resistance.

22.2.3 Example

Find Norton equivalent circuit for the one shown in Figure 22.2.

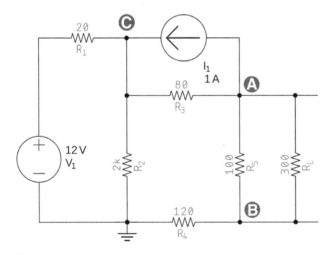

Figure 22.2 Original circuit.

This circuit is the same as used in the last chapter, composed of several resistors, two sources, and a load represented by R_L.

22.2.3.1 Finding the Norton Current Source

To find the Norton current source, that is, the current flowing between points A and B, without the load, we follow the rules defined by Norton. The first one is to replace the load with a short circuit.

The result is seen in Figure 22.3.

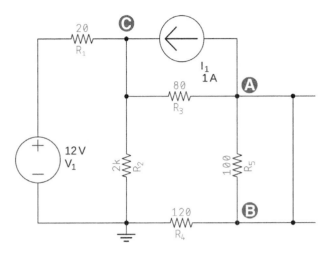

Figure 22.3 Load is short-circuited.

Replacing the load with a short circuit also short-circuited R5 that can now be removed from the circuit.

The result is seen in Figure 22.4.

Figure 22.4 Original circuit.

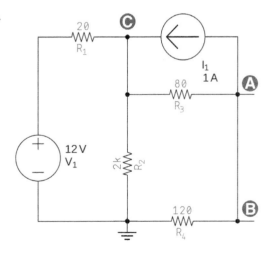

Points A and B makes no sense anymore because they are now the same point.

The Norton current for this circuit is the one flowing between points A and B, without the load. In the given example, the Norton current will be equal to i_4, in the circuit seen in Figure 22.5.

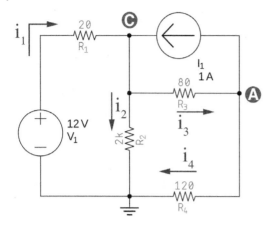

Figure 22.5 Norton current.

To find the Norton current, we must analyze all circuit nodes.

22.2.3.1.1 Node A
By applying KCL on node A, we get

$$i_3 = i_4 + I_1$$

The other unknowns of this equation are

$$i_3 = \frac{V_C - V_A}{R_3}$$

$$i_4 = \frac{V_A}{R_4}$$

Putting it all back together,

$$\frac{V_C - V_A}{R_3} = I_1 + \frac{V_A}{R_4}$$

$$\frac{V_C - V_A}{80} = 1 + \frac{V_A}{120}$$

$$12V_C - 12V_A = 8V_A + 960$$

$$5V_A - 3V_C = -240 \tag{22.1}$$

22.2.3.1.2 Node C

KCL on node C gives us the following equation:

$$i_1 + I_1 = i_2 + i_3$$

We have that

$$i_1 = \frac{V_1 - V_C}{R_1}$$

$$i_2 = \frac{V_C}{R_2}$$

$$i_3 = \frac{V_C - V_A}{R_3}$$

Putting it all back together,

$$\frac{V_1 - V_C}{R_1} + i_{I1} = \frac{V_C}{R_2} + \frac{V_C - V_A}{R_3}$$

$$\frac{12 - V_C}{20} + 1 = \frac{V_C}{2000} + \frac{V_C - V_A}{80}$$

$$-1000\,V_A + 10080\,V_C = 256000$$

$$25\,V_A - 126\,V_C = -3200 \tag{22.2}$$

We end with two equations, (22.1) and (22.2):

$$\begin{cases} 5\,V_A - 3\,V_C = -240 \\ 25\,V_A - 126\,V_C = -3200 \end{cases}$$

We can solve this system of simultaneous equations by using matrices and by using the method we already explained:

$$\begin{bmatrix} 5 & -3 \\ 25 & -126 \end{bmatrix} \begin{bmatrix} V_A \\ V_B \end{bmatrix} = \begin{bmatrix} -240 \\ -3200 \end{bmatrix}$$

The solution gives V_A and V_C:

$$V_A = -37.1891\,V$$

$$V_C = 18.018\,V$$

We can now find the Norton current

$$i_4 = \frac{V_A}{R_4}$$

$$i_4 = \frac{-37.1891}{120}$$

FINAL RESULT

$$I_{No} = i_4 = -0.3099 \, A$$

The negative sign for i_4 shows that this current, in fact, flows in a direction opposite to the one we have arbitrarily chosen.

22.2.3.1.3   Norton Equivalent Resistance

In this next phase of the Norton theorem, we must find the Norton equivalent resistance.

The method used to find this value is the same used by the Thévenin method. Because this circuit is the same as used in the Thévenin chapter of this book, we already know the resistance value, that is, equal to the following value.

FINAL RESULT

$$R_{No} = R_{Th} = 68.73 \, \Omega$$

Therefore, the Norton equivalent circuit will have the final form shown in Figure 22.6.

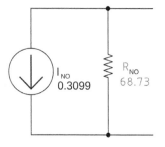

Figure 22.6 Final Norton equivalent circuit.

The current source is drawn upside down because its value is negative.

22.2.3.1.4 Confirmation

In the last chapter we have confirmed that the Thévenin Equivalent Circuit was electrically equivalent to the original circuit.

In this chapter we are using the same circuit as before. If we confirm that this Norton equivalent circuit is the same as the Thévenin equivalent circuit calculated in the last chapter, we know that we have the right values.

To do that, we must use the knowledge we have already acquired about source transformations and convert this Norton equivalent circuit, that is, nothing more than a current source in parallel with a resistor, into a voltage source in series with the same resistor that is equivalent to the Thévenin circuit.

To convert the current source in parallel with a resistor into a voltage source in series with the same resistor, we know that the voltage source can be found by using the following formula:

$$V = R \times I$$

$$V = 68.73 \times -0.3099$$

FINAL RESULT

$$V = -21.299\,A$$

This is the same value found in the last chapter for the voltage source. We conclude that this circuit was calculated correctly.

Exercises

1 Find the Norton equivalent circuit for the circuit shown in Figure 22.7.

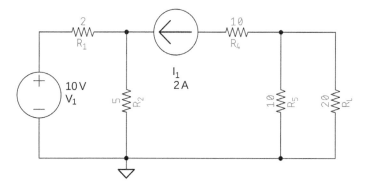

Figure 22.7 Finding the Norton equivalent circuit (exercise).

Solutions

1   Following Norton rules, we replace the load with a short circuit (Figure 22.8).

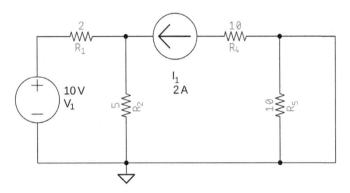

Figure 22.8 Load replaced with a short circuit.

Resistor R_5 was bypassed by the short circuit and can be removed from the circuit (Figure 22.9).

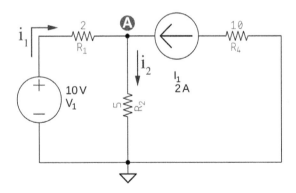

Figure 22.9 Load replaced with a short circuit.

We must apply nodal analysis to node A to find currents i_1 and i_2.

Node A

By applying nodal analysis to node A, we get the following equations:

$$i_1 = \frac{V_1 - V_A}{R_1}$$

$$i_2 = \frac{V_A}{R_2}$$

KCL on node A gives us the next equation:

$$i_1 + I_1 = i_2$$

Therefore,

$$\frac{V_1 - V_A}{R_1} + I_1 = \frac{V_A}{R_2}$$

$$\frac{10 - V_A}{2} + 2 = \frac{V_A}{5}$$

$$\frac{10 - V_A + 4}{2} = \frac{V_A}{5}$$

$$50 - 5V_A + 20 = 2V_A$$

$$7V_A = 70$$

$$V_A = 10\,V$$

Using this value, we can calculate the currents

$$i_1 = \frac{V_1 - V_A}{R_1}$$

$$i_1 = \frac{10 - 10}{2}$$

$$i_1 = 0\,A$$

Interestingly, we see that i_1 is 0! Therefore, there is no current flowing in the branch formed by V_1 and R_1. Current is just flowing in the other half of the circuit (Figure 22.10).

We conclude that current i_4 is then equal to I_1 and equivalent to the Norton current.

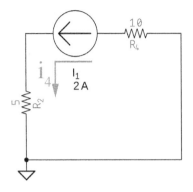

Figure 22.10 Final Norton equivalent circuit.

FINAL RESULT

$i_4 = I_1 = i_{No} = -2\,A$

The Norton current has a negative sign because it flows in the opposite direction compared with the direction we gave to i_1.

Norton Equivalent Resistance

To find the Norton equivalent resistance, we must use the same method used in the Thévenin method.

Therefore, we bring back the original circuit with the load removed (Figure 22.11).

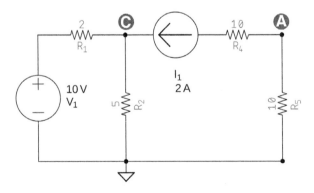

Figure 22.11 Norton equivalent circuit.

To find the Norton equivalent resistance, that is, the resistance between point A and ground, we remove all current sources and replace all voltage sources with short circuits (Figure 22.12).

Figure 22.12 Norton equivalent circuit.

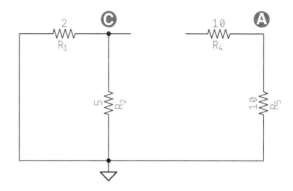

We see that the left part of the circuit is not connected to the right part anymore and can be removed from the circuit. We also see that the left side of R_4 is not connected to anything and can also be removed.

The only thing remaining between A and ground is R_2 and this is the Norton resistance.

FINAL RESULT

$R_{No} = R_5 = 10\,\Omega$

Therefore, the Norton equivalent circuit is equal to the one shown in Figure 22.13.

Figure 22.13 Final Norton equivalent circuit.

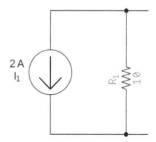

23

Superposition Theorem

Circuit Analysis

23.1 Introduction

In this chapter, we will examine the superposition theorem, another technique for circuit analysis.

23.2 The Theorem

The superposition theorem states that a circuit with multiple voltage and current sources is equal to the sum of simplified circuits using just one of the sources.

A circuit composed of two voltage sources, for example, will be equal to the sum of two circuits, each one using one of the sources and having the other removed.

23.3 Methodology

To simplify a circuit using the superposition theorem, the following steps must be followed:

- Identify all current and voltage sources in the circuit.
- Create multiple versions of the circuit, every version containing just one of the sources. The other sources must be removed using the following rule: voltage sources must be replaced with a short circuit and current sources just removed from the circuit.
- Find the currents and voltages required.
- Sum the results obtained in all circuits.

Introductory Electrical Engineering with Math Explained in Accessible Language,
First Edition. Magno Urbano.
© 2020 John Wiley & Sons, Inc. Published 2020 by John Wiley & Sons, Inc.

23.4   Example

Consider the circuit shown in Figure 23.1, which we have used in the previous chapters. We want to find the current flowing across R_3 and the voltage across points A and B.

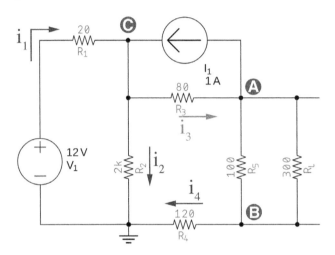

Figure 23.1 Circuit for superposition analysis.

Following the rules of superposition, this circuit will be equal to a first one containing just the voltage source V_1 plus a second one containing just the current source I_1.

23.4.1   First Circuit

To create the first circuit, we remove the load (R_L) and the current source from the original circuit and keep just the voltage source. The result is shown in Figure 23.2.

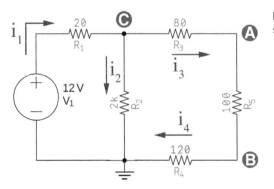

Figure 23.2 First circuit for superposition analysis.

Notice that for this first circuit, after the removal of the current source, i_4 is the same current as i_3.

Next, we will perform a nodal analysis.

23.4.1.1 Node C

By applying Kirchhoff's current law (KCL) to node C, we get the following equation:

$$i_1 = i_2 + i_3$$

Every term of this equation can be discovered by performing a nodal analysis to node C:

$$i_1 = \frac{V_1 - V_C}{R_1}$$

$$i_2 = \frac{V_C}{R_2}$$

$$i_3 = \frac{V_C - V_A}{R_3}$$

Putting it all back together,

$$\frac{V_1 - V_C}{R_1} = \frac{V_C}{R_2} + \frac{V_C - V_A}{R_3}$$

$$\frac{12 - V_C}{20} = \frac{V_C}{2000} + \frac{V_C - V_A}{80}$$

$$25\,V_A - 126\,V_C = -1200 \tag{23.1}$$

23.4.1.2 Relation Between Voltage A and C

By observing Figure 23.2, it is clear that current i_3 that was defined previously as

$$i_3 = \frac{V_C - V_A}{R_3}$$

can also be defined as

$$i_3 = \frac{V_A}{R_4 + R_5}$$

Therefore, we can equate both equations and get V_C:

$$\frac{V_C - V_A}{R_3} = \frac{V_A}{R_4 + R_5}$$

$$\frac{V_C - V_A}{80} = \frac{V_A}{120 + 100}$$

$$220\,V_C - 220\,V_A = 80\,V_A$$

$$V_C = \frac{15}{11}V_A$$

By substituting this value in (23.1), we get

$$25\,V_A - 126\left(\frac{15}{11}\right)V_A = -1200$$

$$25\,V_A - \frac{1890}{11}\,V_A = -1200$$

$$\frac{275\,V_A - 1890\,V_A}{11} = -1200$$

$$-1615\,V_A = -13200$$

$$V_A = 8.1733\,V$$

and we now can find V_C:

$$V_C = \frac{15}{11}V_A$$

$$V_C = \frac{15}{11}(8.1733)$$

$$V_C = 11.1455\,V$$

Therefore, the current i_3 for the first superposition circuit will be

$$i_3 = \frac{V_C - V_A}{R_3}$$

$$i_3 = \frac{11.1455 - 8.1733}{80}$$

FINAL RESULT

$i_3 = 0.03715\,A$

23.4.1.3 Voltage B

Voltage across B is easy to find by simply applying the first Ohm's law

$$V_B = R_4 i_3$$

$$V_B = 120(0.03715)$$

$$V_B = 4.4583\,V$$

We already have V_A and we now need to find V_{AB} that we know is equal to V_A subtracted from V_B.

Therefore,

$$V_{AB} = V_A - V_B$$

$$V_{AB} = 8.1733 - 4.4583$$

and we get V_{AB} for the first superposition circuit.

FINAL RESULT

$V_{AB} = 3.715\,V$

23.4.2 Second Circuit

In this section we analyze and build the second superposition circuit.

For this circuit, we must replace the voltage source, in the given example V_1, with a short circuit and keep the current source.

The result is shown in Figure 23.3.

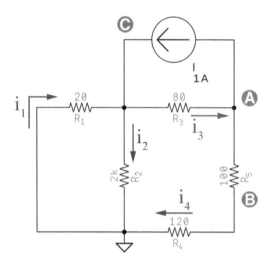

Figure 23.3 Second circuit for superposition analysis.

Replacing V_1 with a wire, put R_1 and R_2 in parallel, and we can substitute them with an equivalent resistor:

$$\frac{1}{R} = \frac{1}{20} + \frac{1}{2000}$$

$$R = 19.801\,\Omega$$

The result is shown in Figure 23.4.

It is now time to perform a nodal analysis.

23.4.2.1 Node A

By applying KCL to node A, we get the following equation:

$$i_3 = I_1 + i_4$$

Figure 23.4 Second circuit for superposition analysis simplified.

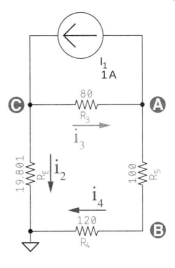

By nodal analysis, we get the unknown terms of this equation:

$$i_3 = \frac{V_C - V_A}{R_3}$$

$$i_4 = \frac{V_A}{R_4 + R_5}$$

Putting it all back together,

$$\frac{V_C - V_A}{R_3} = 1 + \frac{V_A}{R_4 + R_5}$$

$$\frac{V_C - V_A}{80} = 1 + \frac{V_A}{220}$$

$$80\,V_A = 220\,V_C - 220\,V_A + 17600$$

$$15\,V_A - 11\,V_C = -880 \tag{23.2}$$

23.4.2.2   Node C

By applying KCL to node C, we get the following equation:

$$I_1 = i_2 + i_3$$

By nodal analysis, we get the unknown terms of this equation:

$$i_2 = \frac{V_C}{R_{RE}}$$

$$i_3 = \frac{V_C - V_A}{R_3}$$

Putting it all back together,

$$I_{I_1} = \frac{V_C}{R_{RE}} + \frac{V_C - V_A}{R_3}$$

$$1 = \frac{V_C}{19.801} + \frac{V_C - V_A}{80}$$

$$1 = \frac{80\,V_C + 19.801\,V_C - 19.801\,V_A}{(19.801)(80)}$$

$$1584.08 = 80\,V_C + 19.801\,V_C - 19.801\,V_A$$

$$19.801\,V_A - 99.801\,V_C = -1584.08 \tag{23.3}$$

We get two equations, (23.2) and (23.3), which we can solve by using the matrix method we already know:

$$\begin{bmatrix} 15 & -11 \\ 19.801 & -99.801 \end{bmatrix} \begin{bmatrix} V_A \\ V_C \end{bmatrix} = \begin{bmatrix} -880 \\ -1584.08 \end{bmatrix}$$

The solution of this simultaneous equation system will be

$$V_A = -55.0342\,V$$

$$V_C = 4.9533\,V$$

Finding the currents,

$$i_4 = \frac{V_A}{R_4 + R_5}$$

$$i_4 = \frac{-55.0342}{120 + 100}$$

$$i_4 = -0.2501\,A$$

The current we want, i_3, for the second superposition circuit is equal to

$$i_3 = I_1 + i_4$$

$$i_3 = 1 + (-0.2501)$$

$$i_3 = 0.7499\,A$$

We can know calculate V_B:

$$V_B = R_4 i_4$$

$$V_B = 120(-0.2501)$$

$$V_B = -30.0186\,V$$

We know that the voltage across A and B for the second superposition circuit is equal to

$$V_{AB} = V_A - V_B$$

$$V_{AB} = -55.0342 - (-30.0186)$$

$$V_{AB} = -25.0156\,V$$

23.4.2.3 Final Result

In the first superposition circuit, we have found values for V_{AB} and i_3 equal to 3.715 V and 0.03715A , respectively.

In the second superposition circuit, the values for V_{AB} and i_3 found were −25.0156 V and 0.7499 A, respectively.

By following the superposition theorem rules, we must now sum both results to get the final result.

Therefore,

$$V_{AB} = 3.715 + (-25.0156)$$

FINAL RESULT

$$V_{AB} = -21.3006\,\text{V}$$

and current i_3

$$i_3 = 0.03715 + 0.7499$$

FINAL RESULT

$$i_3 = 0.7871\,\text{A}$$

23.4.3 Checking

To confirm the results we have gotten so far, we will perform a nodal analysis of the original circuit without the load, which is shown in Figure 23.5.

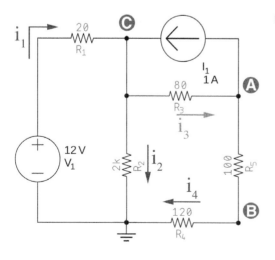

Figure 23.5 Checking.

23.4.3.1 Node A

By applying KCL to node A, we get

$$i_3 = I_1 + i_4$$

The unknown terms of this equation are found by nodal analysis:

$$i_3 = \frac{V_C - V_A}{R_3}$$

$$i_4 = \frac{V_A}{R_4 + R_5}$$

Putting it all back together,

$$\frac{V_C - V_A}{R_3} = \frac{V_A}{R_4 + R_5} + I_{I_1}$$

$$\frac{V_C - V_A}{80} = \frac{V_A}{220} + 1$$

$$\frac{V_C - V_A}{80} = \frac{V_A + 220}{220}$$

$$220\,V_C - 220\,V_A = 80\,V_A + 17600$$

$$220\,V_C - 300\,V_A = 17600$$

$$22\,V_C - 30\,V_A = 1760$$

$$15\,V_A - 11\,V_C = -880 \tag{23.4}$$

23.4.3.2 Node C

KCL gives us the following equation for node C:

$$i_1 + I_1 = i_2 + i_3$$

The unknown terms of these equations are found by nodal analysis:

$$i_1 = \frac{V_1 - V_C}{R_1}$$

$$i_2 = \frac{V_C}{R_2}$$

Putting it all back together,

$$\frac{V_1 - V_C}{R_1} + I_{l_1} = \frac{V_C}{R_2} + \frac{V_C - V_A}{R_3}$$

$$\frac{12 - V_C}{20} + 1 = \frac{V_C}{2000} + \frac{V_C - V_A}{80}$$

$$256000 - 8000\,V_C = 2080\,V_C - 2000\,V_A$$

$$256000 = 10080\,V_C - 2000\,V_A$$

$$25\,V_A - 126\,V_C = -3200 \qquad (23.5)$$

We get Eqs. (23.4) and (23.5) that can be solved using matrices

$$\begin{bmatrix} 15 & -11 \\ 25 & -126 \end{bmatrix} \begin{bmatrix} V_A \\ V_C \end{bmatrix} = \begin{bmatrix} -880 \\ -3200 \end{bmatrix}$$

The result is

$$V_A = -46.8606\,V$$

$$V_C = 16.099\,V$$

23.4.3.3 Voltage Across A and B

Now that we have V_A and V_C, we can calculate i_3:

$$i_3 = \frac{V_C - V_A}{R_3}$$

$$i_3 = \frac{16.099 - (-46.8606)}{80}$$

$$i_3 = \frac{16.099 + 46.8606}{80}$$

FINAL RESULT

$i_3 = 0.7869\,\text{A}$

To find the voltage across A and B, we first need to calculate i_4:

$i_3 = i_4 + I_1$

Therefore,

$i_4 = i_3 - I_1$

$i_4 = 0.7869 - 1$

$i_4 = -0.2130\,\text{A}$

It is now easy to find the voltage across A and B:

$V_{AB} = R_5 i_4$

$V_{AB} = 100(-0.2130)$

$V_{AB} = -21.3\,\text{V}$

The results we have found are the same as we got in previous chapters, confirming that the superposition theory describes precisely the circuit.

Exercises

1 Consider the circuit shown in Figure 23.6. Find the current flowing across R_2 and the voltage across A and B.

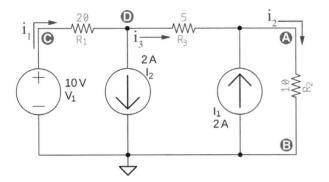

Figure 23.6 Superposition theorem (exercise).

Solutions

1   First Circuit

The first thing to do is to remove all current sources, from the circuit, and keep just the voltage source. The result is shown in Figure 23.7.

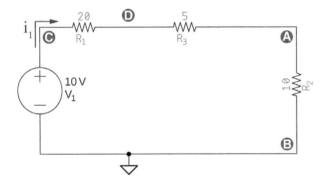

Figure 23.7 Superposition – first circuit (exercise).

All currents have disappeared now, except for i_1 that is equal to

$$i_1 = \frac{V_1}{R_1 + R_2 + R_3}$$

$$i_1 = \frac{10}{20 + 10 + 5}$$

$$i_1 = 0.2857\,\text{A}$$

Voltage across A and B can be found by the first Ohm's law

$$V_{R_2} = V_{AB} = R_2 i_1$$

$$V_{AB} = 10(0.2857)$$

$$V_{AB} = 2.8571\,V$$

Second Circuit

To create the second circuit, we keep current source I_2, remove current source I_1, and short-circuit voltage source V_1. The resulting circuit is shown in Figure 23.8.

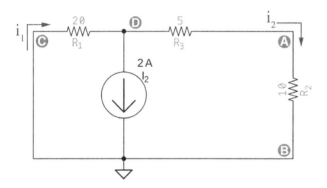

Figure 23.8 Superposition – second circuit (exercise).

To simplify the equations we can invert the direction of i_2. See Figure 23.9.

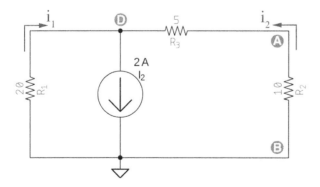

Figure 23.9 Superposition – second circuit simplified (exercise).

The sum of currents, i_1 and i_2, is equal to current source I_2.

$$i_1 + i_2 = I_1$$

The branch on the left, equivalent to resistor R_1, is 20 Ω and the branch in the right, equivalent to the sum of R_2 and R_3, is 15 Ω. If we divide the resistance of the right branch by the left branch, we obtain a relation between them equal to 0.75 or, in other words, 75%. If the resistance of the right branch is 75% of the resistance of the left branch, the current of the left branch, i_1, will be 75% of the current of the right branch, i_2.

$$i_1 = 0.75\, i_2$$

If

$$i_1 + i_2 = I_1$$

then,

$$0.75\, i_2 + i_2 = -2$$
$$1.75\, i_2 = -2$$

$$i_2 = -1.1428\text{A}$$

Consequently,

$$V_A = R_2 i_2$$

$$V_A = 10\,(-1.1428)$$

$$V_A = -11.428\text{V}$$

Third Circuit

To create the third circuit, we keep current source I_2 and remove current source I_1 and short-circuit voltage source V_1.

As in the last circuit, the directions of i_1 and i_3 can be inferred by the direction of I_1. The resulting circuit is shown in Figure 23.10.

The sum of resistors R_1 and R_3 is parallel with R_2. The equivalent resistance of this block is equal

$$\frac{1}{R_{EQ}} = \frac{1}{20 + 5} + \frac{1}{10}$$

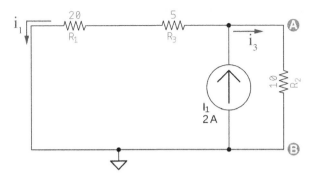

Figure 23.10 – Superposition – third circuit (exercise).

$R_{EQ} = 7.1428 \ \Omega$

Therefore, V_A will be

$$V_A = R_{EQ} I_2$$

$V_A = 7.1428 \,(2)$

$V_A = 14.2856 \ V$

Thus current i_3 is

$$V_A = 10 \, i_3$$

$14.2856 = 10 \, i_3$

$i_2 = 1.4285 \ A$

Consequently, we can now calculate the voltage across A and B.

$$V_{R_2} = V_{AB} = R_2 i_2$$

$V_{AB} = 10 \,(1.4285)$

$V_{AB} = 14.285 \ V$

Final Result

Now that we have found all three superposition circuits, it is time to sum the results of each superposition circuit. Therefore,

$$i_{2_{TOTAL}} = i_{2_{S1}} + i_{2_{S2}} + i_{2_{S3}}$$

$$i_{2_{TOTAL}} = 0.2857 - 1.1428 + 1.4285$$

FINAL RESULT

$$i_{2_{TOTAL}} = 0.5714\,A$$

$$V_{AB_{TOTAL}} = V_{AB_{S1}} + V_{AB_{S2}} + V_{AB_{S3}}$$

$$V_{AB_{TOTAL}} = 2.8571 - 11.428 + 14.285$$

FINAL RESULT

$$V_{AB_{TOTAL}} = 5.7141\,V$$

24

Millman's Theorem

Circuit Analysis

24.1 Introduction

In this chapter, we examine Millman's theorem, another very useful set of techniques for circuit analysis, specifically for circuits containing multiple voltage sources.

24.2 Millman's Theorem

Millman's theorem states that any circuit containing multiple voltage sources, each one in series with its own resistance, can be replaced by one voltage source (V_{EQ}) in series with a resistance (R_{EQ}) (see Figure 24.1).

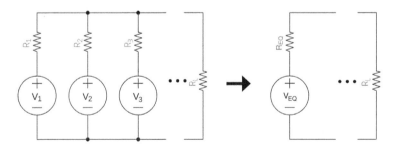

Figure 24.1 Millman's theorem.

24.2.1 The Theory

According to Millman's theorem, each of the voltage sources and their respective resistors produce a current, and the sum of all currents is equal to the total current produced by the circuit, as shown in Figure 24.2.

Introductory Electrical Engineering with Math Explained in Accessible Language,
First Edition. Magno Urbano.
© 2020 John Wiley & Sons, Inc. Published 2020 by John Wiley & Sons, Inc.

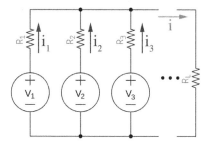

Figure 24.2 Sum of individual currents.

Mathematically, this can be expressed as

$$i = i_1 + i_2 + i_3 + \cdots + i_n$$

or generically as the following postulate.

MILLMAN'S POSTULATE

$$i = \sum_{x=1}^{\infty} i_x$$

i represents the circuit's total current, in Amperes.
i_x represents each of the parallel currents, in Amperes.

The individual element being considered by the theorem is a voltage source in series with a resistor, which we will call a voltage source block.

According to the Millman's theorem, these voltage source blocks are in parallel, each one contributing with its own current to the circuit's total current, as shown in Figure 24.2.

If the individual element is a voltage source block, which is a voltage source in series with a resistor, we can convert each one into a current source in parallel with the same resistor by using the source transformation techniques.

Therefore, we can convert the circuit in Figure 24.2 into the one shown in Figure 24.3.

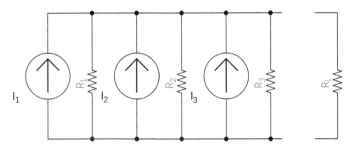

Figure 24.3 Infinite number of current sources and their resistors.

The intensity of each current source can be found by applying the first Ohm's law:

$$I = \frac{V}{R}$$

Therefore, the total current for an infinite number of voltage source blocks in parallel can be expressed as

$$I = \frac{V_1}{R_1} + \frac{V_2}{R_2} + \frac{V_3}{R_3} + \cdots + \frac{V_n}{R_n}$$

After converting an infinite number of voltage source blocks in parallel into current sources in parallel with resistors, like shown in Figure 24.3, we have ended with a lot of resistors in parallel.

We know that resistors in parallel can be replaced with an equivalent resistor, where resistance is

$$\frac{1}{R} = \frac{1}{R_1} + \frac{1}{R_2} + \frac{1}{R_3} + \cdots + \frac{1}{R_n}$$

where R is

$$R = \frac{1}{\dfrac{1}{R_1} + \dfrac{1}{R_2} + \dfrac{1}{R_3} + \cdots + \dfrac{1}{R_n}}$$

Therefore, the total output voltage of a circuit like this, as stated by Millman's theorem, can be expressed by the first Ohm's law:

$$V = R \times I, or$$

$$V_{out} = \left(\overbrace{\frac{1}{\frac{1}{R_1} + \frac{1}{R_2} + \cdots + \frac{1}{R_n}}}^{R} \right) \times \left(\overbrace{\frac{V_1}{R_1} + \frac{V_2}{R_2} + \cdots + \frac{V_n}{R_n}}^{I} \right)$$

Therefore,

$$V_{out} = \left(\frac{\dfrac{V_1}{R_1} + \dfrac{V_2}{R_2} + \cdots + \dfrac{V_n}{R_n}}{\dfrac{1}{R_1} + \dfrac{1}{R_2} + \cdots + \dfrac{1}{R_n}} \right) \tag{24.1}$$

24.2.1.1 Admittance

To simplify the formulas, we will use the concept of admittance.

In electrical engineering, admittance is the reciprocal of impedance, a measure of how easily a circuit or device will allow a current to flow.

The SI unit of admittance is the Siemens, represented by the uppercase letter S.

ADMITTANCE

$$Y = \frac{1}{Z} = Z^{-1}$$

Y is the admittance, in Siemens.
Z is the impedance in Ohms.

Impedance is a sum of resistance with reactance.

Millman circuits are formed exclusively by resistors and there is no reactance involved. Therefore, we can express admittance exclusively by the reciprocal of resistance, or

$$Y = \frac{1}{R} = R^{-1} \tag{24.2}$$

By substituting (24.2) into (24.1), we get the following equation.

MILLMAN VOLTAGE

$$V_M = \frac{V_1 Y_1 + V_2 Y_2 + V_3 Y_3 + \ldots V_n Y_n}{Y_1 + Y_2 + Y_3 + \ldots Y_n}$$

V_M is the Millman voltage.
$V_1, V_2, \ldots$ are every one of the individual voltage sources.
$Y_1, Y_2, \ldots$ are every one of the admittances.

We can define the Millman voltage formula in generic terms as follows.

MILLMAN VOLTAGE

$$V_M = \sum_{i=0}^{\infty} \frac{V_i Y_i}{Y_i}$$

V_M is the Millman voltage.
$V_1, V_2, \ldots$ are every one of the individual voltage sources.
$Y_1, Y_2, \ldots$ are every one of the admittances

24.3   Examples

24.3.1   Example 1

The circuit shown in Figure 24.4 has several voltage sources in series with resistors and every branch is in parallel with a load represented by R_L.

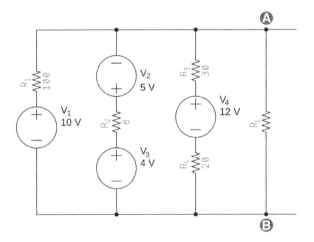

Figure 24.4 Circuit with multiple voltage sources in parallel.

Find the Millman equivalent voltage and resistance.

24.3.1.1   Solution
To find the Millman equivalent circuit, the first thing we must do is to remove the load. The resulting circuit is shown in Figure 24.5.

By looking at Figure 24.5, we can see three branches in parallel. The second one has two voltage sources in series. We can sum these sources and replace them with an equivalent. The same is valid for the two resistors, R_3 and R_4, in the third branch.

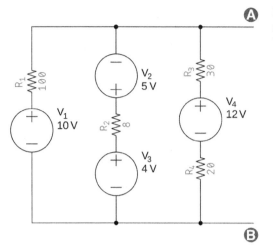

Figure 24.5 Multiple voltage sources in parallel without the load.

The resulting circuit is shown in Figure 24.6.

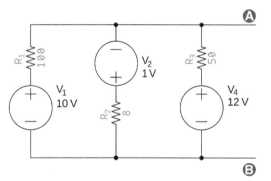

Figure 24.6 Multiple voltage sources in parallel simplified.

We can now apply Millman's formula for the equivalent voltage, but before that, we must find the admittance of each parallel branch, in Siemens:

$$Y_1 = \frac{1}{R_1} = \frac{1}{100}$$

$$Y_2 = \frac{1}{R_2} = \frac{1}{8}$$

$$Y_3 = \frac{1}{R_3} = \frac{1}{50}$$

Therefore,

$$V_M = \frac{V_1Y_1 + V_2Y_2 + V_3Y_3 + \cdots + V_nY_n}{Y_1 + Y_2 + Y_3 + \cdots + Y_n}$$

$$V_M = \dfrac{\dfrac{10}{100} - \dfrac{1}{8} + \dfrac{12}{50}}{\dfrac{1}{100} + \dfrac{1}{8} + \dfrac{1}{50}}$$

$$V_M = \dfrac{0.215}{0.155}$$

FINAL RESULT

$$V_M = 1.38\,\text{V}$$

To find the Millman equivalent resistance, we replace the voltage sources with short circuits. The result is seen in Figure 24.7.

Figure 24.7 Multiple resistors in parallel simplified.

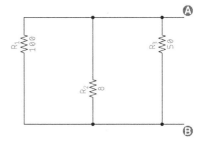

The Millman equivalent resistance will be the equivalent resistance of these three resistors in parallel, or

$$\frac{1}{R} = \frac{1}{R_1} + \frac{1}{R_1} + \frac{1}{R_1} + \cdots + \frac{1}{R_n}$$

$$\frac{1}{R} = \frac{1}{100} + \frac{1}{8} + \frac{1}{50}$$

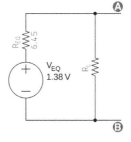

FINAL RESULT

$$R_{EQ} = 6.45\,\Omega$$

Therefore, the Millman equivalent circuit plus the load can be drawn as shown in Figure 24.8.

Figure 24.8 Millman equivalent circuit.

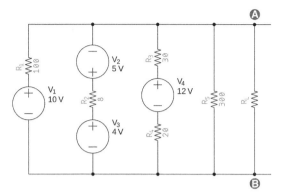

Ⓐ

Figure 24.9 Circuit with multiple voltage sources in parallel and a load.

Ⓑ

24.3.2 Example 2

Consider the circuit shown in Figure 24.9.

This idea behind this circuit is the same of the previous circuit, but there is an additional branch, at the end, with a resistor and no voltage source. In this case, we must consider this branch as having a 0 V voltage source.

We must add the voltage sources and the resistors in the branches with multiple of these elements.

Therefore, the Millman voltage must be calculated like this:

$$V_M = \frac{V_1 Y_1 + V_2 Y_2 + V_3 Y_3 + \cdots + V_n Y_n}{Y_1 + Y_2 + Y_3 + \cdots + Y_n}$$

$$V_M = \frac{\dfrac{10}{100} - \dfrac{1}{8} + \dfrac{12}{50} + \dfrac{0}{300}}{\dfrac{1}{100} + \dfrac{1}{8} + \dfrac{1}{50} + \dfrac{1}{300}}$$

$$V_M = \frac{0.215}{0.1583}$$

FINAL RESULT

$$V_{EQ} = 1.358\,V$$

And the Millman resistance must be calculated as follows:

$$\frac{1}{R} = \frac{1}{R_1} + \frac{1}{R_1} + \frac{1}{R_1} + \cdots + \frac{1}{R_n}$$

$$\frac{1}{R} = \frac{1}{100} + \frac{1}{8} + \frac{1}{50} + \frac{1}{300}$$

FINAL RESULT

$$R_{EQ} = 6.316\,\Omega$$

Exercises

1 Find the Millman equivalent for the circuit shown in Figure 24.10.

Figure 24.10 Multiple voltage sources in parallel.

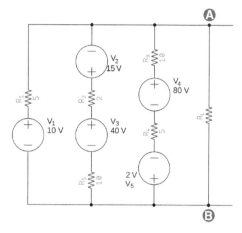

Solutions

1 To find the Millman equivalent circuit, we must remove the load first, as shown in Figure 24.11.

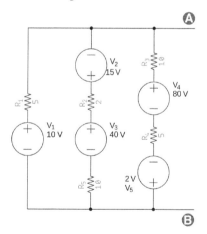

Figure 24.11 Multiple voltage sources in parallel.

The resulting circuit has three branches. The second branch has two voltages sources with reversed polarities that must be subtracted, the third branch has resistors in series that must be added, and the fourth branch has no voltage source or, in other words, is equivalent to a 0 V source in series with R_5.

The resulting simplified circuit is shown in Figure 24.12.

We can now apply these values shown in Figure 24.12 in Millman's formula:

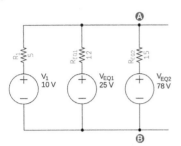

Figure 24.12 Multiple voltage sources in parallel simplified.

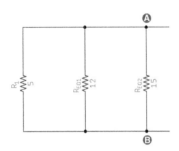

Figure 24.13 Multiple voltage sources in parallel simplified.

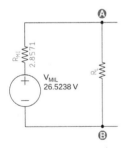

Figure 24.14 Millman equivalent circuit.

$$Y_1 = \frac{1}{R_1} = \frac{1}{5}$$

$$Y_2 = \frac{1}{R_{EQ1}} = \frac{1}{12}$$

$$Y_3 = \frac{1}{R_{EQ2}} = \frac{1}{15}$$

Therefore, the Millman voltage is

$$V_{MIL} = \frac{V_1 Y_1 + V_2 Y_2 + V_3 Y_3 + \cdots + V_n Y_n}{Y_1 + Y_2 + Y_3 + \cdots + Y_n}$$

$$V_{MIL} = \frac{\frac{10}{5} + \frac{25}{12} + \frac{78}{15}}{\frac{1}{5} + \frac{1}{12} + \frac{1}{15}}$$

$$V_{MIL} = \frac{9.2833}{0.35}$$

FINAL RESULT

$$V_{MIL} = 26.5238\,V$$

To find the Millman resistance, we short-circuit all voltage sources, as seen in Figure 24.13.

The Millman resistance can be easily found by

$$\frac{1}{R_{MI}} = \frac{1}{5} + \frac{1}{12} + \frac{1}{15}$$

FINAL RESULT

$$R_{MI} = 2.8571\,\Omega$$

Therefore, the equivalent Millman circuit plus the load can be seen in Figure 24.14.

25

RC Circuits

Voltage and Current Analysis in Circuits Containing Resistors and Capacitors in Series

25.1 Introduction

In this chapter, we will examine the concepts behind resistor–capacitor series circuits, known as RC circuits.

As explained in previous chapters, when a voltage is applied to a completely discharged capacitor, it makes the capacitor charge exponentially to the power supply voltage. As it happens, current will decrease, also exponentially, to 0.

25.2 Charging a Capacitor

Consider the circuit shown on Figure 25.1, composed of a resistor and a capacitor in series with a switch and a power supply. The switch is initially open, so the circuit is not working. The capacitor is completely discharged.

Figure 25.1 RC circuit and an open switch.

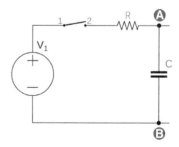

At time t = 0 s, the switch is closed. Current starts flowing from the power supply through the resistor R and reaches the capacitor. The capacitor starts to charge.

Introductory Electrical Engineering with Math Explained in Accessible Language, First Edition. Magno Urbano. © 2020 John Wiley & Sons, Inc. Published 2020 by John Wiley & Sons, Inc.

Current is at its maximum value, just limited by the resistor. As charging starts, the capacitor is seen, by the circuit, as a wire, a short circuit, because it makes no opposition to the current flow.

As time passes and the capacitor charges, current will exponentially decrease to 0, and voltage across the capacitor exponentially increases to the power supply voltage. Variations of current and voltage will then stop and the circuit will stabilize.

> The time interval during which currents and voltages variate is called transient state and the point in time where circuit stabilizes is called steady state.

25.2.1 Charging Voltage

By applying nodal analysis to node A, in the circuit shown in Figure 25.1, we get the following current for the circuit's current:

$$i = \frac{V_1 - V_C(t)}{R} \tag{25.1}$$

However, as defined in this book before, current across a capacitor (i_C) is defined as

$$i_c = C\frac{dV_C(t)}{dt} \tag{25.2}$$

The resistor and the capacitor are in series, so current flowing across one is the same as the current flowing across the other. Therefore, we can equate (25.1) and (25.2):

$$C\frac{dV_C(t)}{dt} = \frac{V_1 - V_C(t)}{R}$$

Rearranging the equation,

$$\frac{dV_C(t)}{V_1 - V_C(t)} = \frac{1}{RC}dt$$

Integrating both sides,

$$\int_{V_0}^{V} \frac{dV_c(t)}{V_1 - V_C(t)} = \int_0^t \frac{1}{RC}dt$$

$$\int_{V_0}^{V} \frac{dV_c(t)}{V_1 - V_C(t)} = \frac{1}{RC} \int_0^t dt$$

First integral goes from the capacitor's voltage at time equal to 0 (V_0) to a generic voltage V at time is equal to t. Second integral goes from 0 to the same time t.

We can compare the first integral to the following math concept.

MATH CONCEPT The integral of

$$\int \frac{1}{ax + b} \, dx \longrightarrow \frac{1}{a} \, \ln |ax + b| + C$$

The constant of integration can be ignored when dealing with definite integrals (see Appendix F).

Compared to this math concept, the first integral

$$\int_{V_0}^{V} \frac{dV_c(t)}{V_1 - V_C(t)}$$

has

$$\begin{cases} b = V_1 \\ a = -1 \\ x = V_C(t) \end{cases}$$

The negative sign of "a" will produce a negative result for the natural log part. For the other integral, we have the following math concept.

MATH CONCEPT   The integral of

$$\int dx \longrightarrow x + C$$

The constant of integration can be ignored when dealing with definite integrals (see Appendix F).

By applying this concept to the second integral and putting it all back together, we get

$$\int_{V_0}^{V} \frac{dV_c(t)}{V_1 - V_C(t)} = \frac{1}{RC} \int_{0}^{t} dt$$

$$-\ln |V_1 - V_C(t)| \Big|_{V_0}^{V} = \frac{1}{RC} t \Big|_{0}^{t}$$

$$-[\ln |V_1 - V_C(t)| - \ln |V_1 - V_0|] = \frac{1}{RC} t$$

$$-\ln |V_1 - V_C(t)| + \ln |V_1 - V_0| = \frac{1}{RC} t$$

Multiplying both sides by –1,

$$\ln |V_1 - V_C(t)| - \ln |V_1 - V_0| = -\frac{1}{RC} t$$

MATH CONCEPT   The subtraction of two natural logarithms

$$\ln a - \ln b = \ln \left(\frac{a}{b} \right)$$

Therefore,

$$\ln \left[\frac{V_1 - V_C(t)}{V_1 - V_0} \right] = -\frac{1}{RC} t$$

We can get rid of the natural logarithm by applying the following rule.

MATH CONCEPT   The exponential of the natural log

$$e^{\ln (a)} = a$$

By applying this rule, we get

$$e^{\ln \left[\frac{V_1 - V_C(t)}{V_1 - V_0} \right]} = e^{-\frac{1}{RC} t}$$

$$\frac{V_1 - V_C(t)}{V_1 - V_0} = e^{-\frac{1}{RC}t}$$

$$V_1 - V_C(t) = (V_1 - V_0)e^{-\frac{1}{RC}t}$$

By passing V_1 to the other side,

$$-V_C(t) = -V_1 + (V_1 - V_0)e^{-\frac{1}{RC}t}$$

and by multiplying both sides by −1, we get the final equation for the voltage across the capacitor during the charge period.

VOLTAGE ACROSS THE CAPACITOR – CHARGE

$$V_C(t) = V_1 + (V_0 - V_1)e^{-\frac{1}{RC}t}$$

$V_C(t)$ is the equation for the voltage across capacitor as a function of time, in Volts.
V_1 is the power supply voltage, in Volts.
V_0 is the capacitor initial voltage, in Volts.
t is an instant in time, in seconds.
R is the resistance, in Ohms.
C is the capacitance, in Farads.

25.2.2 Charge Equation

Charge and voltage across a capacitor have a direct relation through the formula $Q = CV$.

Hence, the charge equation can be obtained by multiplying the voltage equation by C.

CHARGE IN CAPACITOR DURING THE CHARGING PHASE

$$Q_C(t) = C_1V_1 + (Q_0 - C_1V_1)e^{-\frac{1}{RC}t} \tag{25.3}$$

$Q_C(t)$ is the charge equation as a function of time, in Coulombs.
V_1 is power supply's voltage, in Volts.
Q_0 is the capacitor initial charge, in Coulombs.
t is an instant in time, in seconds.
R is the resistance, in Ohms.
C is the capacitance, in Farads.

The same result can be obtained by using Kirchhoff's voltage law (KVL).

KVL states that the sum of the electric potential differences (voltages) around any closed network is 0, or

$$V_1 - V_R - V_C = 0 \tag{25.4}$$

It is known that the voltage drop across the resistor is given by its resistance multiplied by the current, or

$$V_R = Ri$$

and that the voltage across capacitor is equal to the charge divided by capacitance:

$$V_C = \frac{Q}{C}$$

By substituting V_R and V_C in equation (25.4), we get

$$V_1 - Ri - \frac{Q}{C} = 0$$

Current that flows across the capacitor is the same that flows across the resistor and is equal to the ratio between charge over time, or

$$i = \frac{dQ}{dt}$$

If we substitute i in the KVL equation, we obtain

$$V_1 - R\frac{dQ}{dt} - \frac{Q}{C} = 0$$

We organize the equation in a different way:

$$V_1 - \frac{Q}{C} = R\frac{dQ}{dt}$$

$$\frac{1}{R}dt = \frac{dQ}{V_1 - \frac{Q}{C}}$$

and integrating both sides,

$$\frac{1}{R}\int_{0}^{t} dt = \int_{Q_0}^{Q} \frac{dQ}{V_1 - \dfrac{Q}{C}}$$

First integral is defined from time equal to 0 to a generic time t, and the second one goes from an initial value of charge the capacitor may have at time 0 (Q_0) to a generic charge Q when time is equal to t.

To solve the second integral, we use the integral substitution rule:
Let

$$u = V_1 - \frac{Q}{C}$$

Therefore,

$$du = d\left[V_1 - \frac{Q}{C}\right]$$

$$du = d[V_1] - d\left[\frac{Q}{C}\right]$$

MATH CONCEPT The derivative of a constant is equal to 0.

Therefore,

$$du = d[V_1]^0 - d\left[\frac{Q}{C}\right]$$

$$du = -\frac{dQ}{C}$$

$$\frac{du}{dQ} = -\frac{1}{C}$$

$$dQ = -Cdu$$

Upon substituting u and dQ in the main equation, we get

$$\frac{1}{R}\int_0^t dt = \int_{Q_0}^Q \frac{-Cdu}{u}$$

$$\frac{1}{R}\int_0^t dt = -C\int_{Q_0}^Q \frac{du}{u}$$

MATH CONCEPT   The integral of

$$\int \frac{1}{x} dx \longrightarrow \ln|x| + C$$

The constant of integration can be ignored when dealing with definite integrals (see Appendix F).

By applying this concept, we get

$$\frac{1}{R}t\Big|_0^t = -C\ln|u|\Big|_{Q_0}^Q$$

After solving the integrals for the defined ranges, we get

$$\frac{1}{R}t = -C\left[\ln\left|V_1 - \frac{Q}{C}\right| - \ln\left|V_1 - \frac{Q_0}{C}\right|\right]$$

Multiplying both sides by $-1/C$,

$$-\frac{1}{RC}t = \ln\left|V_1 - \frac{Q}{C}\right| - \ln\left|V_1 - \frac{Q_0}{C}\right|$$

MATH CONCEPT   The subtraction of two natural logarithms

$$\ln a - \ln b = \ln\left(\frac{a}{b}\right)$$

Therefore,

$$-\frac{1}{RC}t = \ln\left|\frac{V_1 - \dfrac{Q}{C}}{V_1 - \dfrac{Q_0}{C}}\right|$$

We can get rid of the natural logarithm by applying the following rule.

MATH CONCEPT The exponential of the natural log

$$e^{\ln(a)} = a$$

Therefore,

$$e^{-\frac{1}{RC}t} = e^{\ln\left|\frac{V_1 - \frac{Q}{C}}{V_1 - \frac{Q_0}{C}}\right|}$$

$$e^{-\frac{1}{RC}t} = \frac{V_1 - \dfrac{Q}{C}}{V_1 - \dfrac{Q_0}{C}}$$

By rearranging the equation

$$\left(V_1 - \frac{Q_0}{C}\right)e^{-\frac{1}{RC}t} = V_1 - \frac{Q}{C}$$

$$(CV_1 - Q_0)e^{-\frac{1}{RC}t} = CV_1 - Q$$

$$-CV1 + (CV_1 - Q_0)e^{-\frac{1}{RC}t} = -Q$$

and by multiplying both sides by −1, we get the final equation that is, obviously, identical to the one in (25.3):

$$Q = CV_1 + (Q_0 - CV_1)e^{-\frac{1}{RC}t}$$

25.2.3 Charging Current

We have just obtained the charge equation as

$$Q = CV_1 + (Q_0 - CV_1)e^{-\frac{1}{RC}t}$$

But we know that current is defined as the ratio of charge over time, or in other words the derivative of charge with respect to time, or

$$i = \frac{dQ}{dt}$$

That is the exact definition of derivative, or in this case the derivative of charge with respect to time.

Therefore, current will be the derivative of the charge equation with respect to time:

$$i(t) = \frac{dQ}{dT} = \frac{d}{dt}\left[CV_1 + (Q_0 - CV_1)e^{-\frac{1}{RC}t}\right]$$

We can expand the derivative by using the following concept.

> **MATH CONCEPT** The derivative of a sum is equal to the sum of derivatives.

Therefore,

$$i(t) = \frac{d}{dt}[CV_1] + \frac{d}{dt}\left[(Q_0 - CV_1)e^{-\frac{1}{RC}t}\right]$$

> **MATH CONCEPT** The derivative of a constant is equal to 0.

Hence,

$$i(t) = \frac{d}{dt}[CV_1]^{\,0} + \frac{d}{dt}\left[(Q_0 - CV_1)e^{-\frac{1}{RC}t}\right]$$

$$i(t) = \frac{d}{dt}\left[(Q_0 - CV_1)e^{-\frac{1}{RC}t}\right]$$

We get a result that is equal to a derivative of a product of two functions,

$(Q_0 - CV_1)$ and $e^{-\frac{1}{RC}t}$.

This can be solved by applying the following mathematical concept.

> **MATH CONCEPT** The derivative of a product of two functions is equal to the derivative of the first function multiplied by the second function added to the first function multiplied by the derivative of the second, or
>
> $$(u.v)' = u'v + uv'$$

By applying this rule, we get

$$i(t) = \frac{d}{dt}[(Q_0 - CV_1)]\, e^{-\frac{1}{RC}t} + (Q_0 - CV_1)\frac{d}{dt}\left[e^{-\frac{1}{RC}t}\right]$$

MATH CONCEPT   The derivative of a constant is equal to 0.

Therefore,

$$i(t) = \frac{d}{dt}\overbrace{[(Q_0 - CV_1)]}^{0}\, e^{-\frac{1}{RC}t} + (Q_0 - CV_1)\frac{d}{dt}\left[e^{-\frac{1}{RC}t}\right]$$

$$i(t) = (Q_0 - CV_1)\frac{d}{dt}\left[e^{-\frac{1}{RC}t}\right]$$

MATH CONCEPT   The derivative of an exponential

$$\frac{d}{dt}\, e^{xt} = x\, e^{xt}$$

Thus,

$$i(t) = -\frac{(Q_0 - CV_1)}{RC}\left[e^{-\frac{1}{RC}t}\right]$$

Charge, voltage, and capacitance are related by Q = CV.
Therefore, upon substituting this relationship in the previous formula, we get

$$i(t) = -\frac{(CV_0 - CV_1)}{RC}\left[e^{-\frac{1}{RC}t}\right]$$

By simplifying,

$$i(t) = -\frac{(\cancel{C}V_0 - \cancel{C}V_1)}{R\cancel{C}}\left[e^{-\frac{1}{RC}t}\right]$$

$$i(t) = -\frac{V_0 - V_1}{R}\left[e^{-\frac{1}{RC}t}\right]$$

Hence, after rearranging, we get the final equation for the charging current through the capacitor.

CHARGING CURRENT THROUGH CAPACITOR

$$i(t) = \frac{V_1 - V_0}{R} e^{-\frac{t}{RC}}$$

i(t) is the equation of current as a function of time, in Amperes.
V_1 is the power supply voltage, in Volts.
V_0 is the capacitor initial voltage, in Volts.
t is an instant in time, in seconds.
R is the resistance, in Ohms.
C is the capacitance, in Farads. RC is time constant.

25.3   RC Time Constant

The RC time constant is the time required to charge a capacitor through a resistor, from an initial charge of 0–63.2% of the value of an applied DC voltage or for a discharging capacitor to lose 63.2% of its initial charge.

This value is derived from the mathematical formulas of charge and discharge for the capacitor.

The RC time constant, also called tau (τ) and measured in seconds, is equal to the product of the circuit resistance in Ohms and the circuit capacitance in Farads.

RC TIME CONSTANT

$$\tau = R \times C$$

τ is the RC time constant, in seconds.
R is the resistance, in Ohms.
C is the capacitance, in Farads.

25.3.1.1   Transient and Steady States

The period of time in which an RC circuit exhibits variation of current and voltage, after a voltage pulse is applied to it, is called transient phase.

After the transient phase is over, the circuit stabilizes and will not exhibit any other variations. The stable phase is called steady state.

25.3.1.1.1 How Long Does the Transient Phase Last?

When a capacitor charges, the amount of charge increases following an exponential curve. For this reason, the ratio of charge decreases with time. Hence, an infinite amount of time is required to reach the maximum charge.

In practice, engineers cannot wait indefinitely for a capacitor to charge. In electrical engineering, a capacitor is considered charged after a time equal to five time constants.

RC TIME CONSTANT – STEADY STATE

$$t \geq 5\tau \rightarrow t \geq 5 \times RC$$

τ is the time constant, in seconds.
R is the resistance, in Ohms.
C is the capacitance, in Farads.

This happens because after each time constant, the voltage or charge will raise or decay according to the following table.

RC time constant interval	Percentage of total voltage (%)
1τ	62.3
2τ	86.5
3τ	95.0
4τ	98.2
5τ	99.3

The table shows that after five time constants, the voltage of a charging capacitor will be 99.3% of the charging value or the voltage of a discharging capacitor will have decayed by the same amount, which is reason why five time constants is the number chosen to represent a fully charged/discharged capacitor.

25.3.2 General Formula

Suppose the capacitor on Figure 24.1 has an initial charge and, consequently, an initial voltage (V_0) before the switch is turned on. When the switch is closed, the capacitor will start charging, not from 0 but from its initial voltage to the power supply voltage.[1]

Because it takes an infinite time for the capacitor to charge to the power supply's voltage, this voltage at infinite time is called V_∞ or steady-state voltage.

---

1 The capacitor will discharge if its voltage is higher than power supply voltage.

The formulas we have described so far are represented on the international literature, as shown below.

VOLTAGE ACROSS THE CAPACITOR DURING CHARGE (GENERAL FORMULA)

$$V_C(t) = V_\infty + (V_0 - V_\infty)e^{-\frac{t}{RC}}$$

$V_C(t)$ is the equation of the voltage across the capacitor as a function of time, in Volts.
V_∞ is the power supply voltage, in Volts.
V_0 is the capacitor initial voltage, in Volts.
t is an instant in time, in seconds.
R is the resistance, in Ohms.
C is the capacitance, in Farads.

CURRENT ACROSS THE CAPACITOR DURING CHARGE (GENERAL FORMULA)

$$i(t) = \frac{V_\infty - V_0}{R}e^{-\frac{t}{RC}}$$

i(t) is the equation of the current across the capacitor as a function of time, in Amperes.
V_∞ is power supply's voltage, in Volts.
V_0 is the capacitor initial voltage, in Volts.
t is an instant in time, in seconds.
R is the resistance, in Ohms.
C is the capacitance, in Farads.

25.3.3 Discharging a Capacitor

Figure 25.2 shows a DC power supply, a resistor, a capacitor, and a two-position switch.

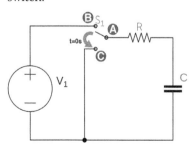

Figure 25.2 RC circuit initially working.

In the following circuit, the switch has been sitting for a long time, connecting points A and B.

The resistor R and the capacitor C have been connected to a constant voltage source, V_1, for a long time. The circuit is, therefore, on the steady-state phase, the capacitor is completely charged, and no current is flowing in the circuit anymore.

At time t = 0 s, the switch is moved from position AB to AC, disconnecting the resistor and the capacitor from the power supply and connecting them in parallel.

The capacitor has now a path to discharge and will do it slowly through the resistor.

25.3.3.1 Charge During Discharge

A capacitor will always discharge following an exponential curve.

During discharge, voltage, current, and charge all decay from their initial values to a lower ones following an exponential curve. This phase where current, charge, and voltage vary is called transient phase.

25.3.3.1.1 Considerations

Circuits with transient phases will always have equations for voltages, currents, and charges that are the sum of what happens during the transient and during the steady-state phases:

$$f(t) = f_{tr} + f_{ss}$$

In the example of a capacitor discharging, current, voltage, and charge will all decay from a given value to a lower value and then stabilize.

25.3.3.1.2 Back to the Equations

Charge, voltage, and capacitance are related by V = Q/C.

The first Ohm's law states that I = V/R.

Upon substituting the first equation into the second, we get

$$I = \frac{\frac{Q}{C}}{R}$$

$$I = \frac{Q}{RC} \tag{25.5}$$

Current across a capacitor is defined as the ratio of charge over time, or

$$I = -\frac{dQ}{dt} \tag{25.6}$$

taken with a negative sign because charge is decreasing.

Equations (25.5) and (25.6) represent the same current. Thus, they can be equated:

$$\frac{Q}{RC} = -\frac{dQ}{dt}$$

$$-\frac{1}{RC}dt = \frac{1}{Q}dQ$$

and integrating both sides,

$$-\int_0^t \frac{1}{RC}dt = \int_{Q_0}^Q \frac{1}{Q}dQ \tag{25.7}$$

First integral goes from time equal 0 to a generic time t. Second integral goes from an initial value of charge (Q_0) to a generic value of charge (Q) when time is equal to t.

To solve the second integral, we must use the following mathematical concept.

MATH CONCEPT   The integral of

$$\int \frac{1}{x}dx \longrightarrow \ln |x| + C$$

The constant of integration can be ignored when dealing with definite integrals (see Appendix F).

By applying this concept on the second integral of equation (25.6), we get

$$-\frac{t}{RC}\Bigg|_0^t = \ln |Q| \Bigg|_{Q_0}^Q$$

Solving for the interval,

$$\left[-\frac{t}{RC} - \left(-\frac{0}{RC}\right)\right] = \ln|Q| - \ln|Q_0|$$

$$-\frac{t}{RC} = \ln|Q| - \ln|Q_0|$$

MATH CONCEPT The subtraction of two natural logarithms

$$\ln a - \ln b = \ln\left(\frac{a}{b}\right)$$

By applying this math concept, we get

$$-\frac{t}{RC} = \ln\left|\frac{Q}{Q_0}\right|$$

We can get rid of the natural logarithm by applying the following mathematical rule.

MATH CONCEPT The exponential of the natural log

$$e^{\ln(a)} = a$$

$$e^{-\frac{t}{RC}} = e^{\ln\left|\frac{Q}{Q_0}\right|}$$

$$e^{-\frac{t}{RC}} = \frac{Q}{Q_0}$$

We get the final equation for the charge during the decaying transient phase.

CHARGE EQUATION (DISCHARGE)

$$Q(t) = Q_0 e^{-\frac{t}{RC}}$$

Q(t) is the charge equation as a function of time, in Coulombs.
Q_0 is the capacitor initial charge, in Coulombs.
t is an instant in time, in seconds.
R is the resistance, in Ohms.
C is the capacitance, in Farads.

Charge is given as Q = CV. Thus, the previous equation can also be expressed by the following equation.

CHARGE EQUATION (DISCHARGE)

$$Q(t) = CV_0 e^{-\frac{t}{RC}} \qquad (25.8)$$

Q(t) is the charge equation as a function of time, in Coulombs.
V_0 is the capacitor initial voltage, in Volts.
t is an instant in time, in seconds.
R is the resistance, in Ohms.
C is the capacitance, in Farads.

25.3.3.2 Voltage During Discharge

The relation between charge, voltage, and capacitance given by Q = CV can be used to obtain the voltage equation as a function of time. All we have to do is to divide (25.8) by C.

VOLTAGE ACROSS THE CAPACITOR DURING DISCHARGE

$$V(t) = V_0 e^{-\frac{t}{RC}}$$

V(t) is the equation for the voltage across the capacitor as a function of time, in Volts.
V_0 is the capacitor initial voltage, in Volts.
t is an instant in time, in seconds.
R is the resistance, in Ohms.
C is the capacitance, in Farads.

25.3.4 Current During Discharge

Charge is defined as the derivative of charge with respect to time:

$$i(t) = -\frac{dQ}{dt}$$

Hence, the current equation will be the derivative of the charge equation (25.7) with respect to time:

$$i(t) = -\frac{dQ(t)}{dt} = \frac{d}{dt}\left[CV_0 e^{-\frac{t}{RC}}\right]$$

$$i(t) = -CV_0 \frac{d}{dt}\left[e^{-\frac{t}{RC}}\right]$$

MATH CONCEPT   The derivative of an exponential

$$\frac{d}{dt} e^{xt} = xe^{xt}$$

By applying this math concept and simplifying,

$$i(t) = -CV_0\left[-\frac{1}{RC}e^{-\frac{t}{RC}}\right]$$

$$i(t) = -\mathcal{C}V_0\left[-\frac{1}{R\mathcal{C}}e^{-\frac{t}{RC}}\right]$$

and we get the current equation as a function of time during discharge.

CURRENT ACROSS THE CAPACITOR DURING DISCHARGE

$$i(t) = \frac{V_0}{R}e^{-\frac{t}{RC}}$$

i(t) is the equation of the current across the capacitor as a function of time, in Amperes.
V_0 is the capacitor initial voltage, in Volts.
t is an instant in time, in seconds.
R is the resistance, in Ohms.
C is the capacitance, in Farads.

25.4   Examples

25.4.1   Example 1

Suppose the circuit shown on Figure 25.3, containing a direct voltage power supply (V_1), a resistor, a capacitor, and an open switch.

In the circuit in Figure 25.3, the switch has been opened for a long time. There is no current flowing and C is completely discharged.

At time t = 0 s, the switch is closed, and current flows and reaches the capacitor that starts charging.

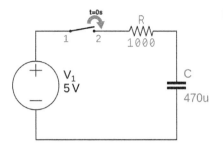

Figure 25.3 RC circuit initially not working.

We want to find the current flowing in the circuit the moment the switch closes and the voltage and current across the capacitor 1 s after the switch is closed.

25.4.1.1 Current Flowing as the Switch Is Turned On

By Kirchhoff's law,

$$V_1 = V_R + V_C$$

We know that when the switch is turned on, the voltage across the capacitor is 0 and the capacitor will offer no resistance to the current flow, behaving like a regular wire or short circuit, letting maximum current pass. Therefore, the capacitor can be removed from the circuit, and the current flowing in the circuit when the switch is turned on will be, according to the first Ohm's law, equal to the power supply voltage divided by the resistor, or

$$i = \frac{V_1}{R}$$

$$i(0^+) = \frac{5}{1000}$$

FINAL RESULT

$$i(0^+) = 5\,mA$$

0^+ represents the instant in time just after the switch is closed. By the same principle, 0^- represents the instant in time before the switch is closed.

25.4.1.2 Current and Voltage After 1 s

Voltage at any given time will be given by the formula

$$V_C(t) = V_1 + (V_0 - V_1)e^{-\frac{1}{RC}t}$$

If we plug in the known values, we get

$$V_C(1\,s) = 5 + (0-5)e^{-\frac{1}{1000 \times 470 \times 10^{-6}}}$$

$$V_C(1\,s) = 4.404\,V$$

Current at any given time can be calculated by the following formula:

$$i(t) = \frac{V_1 - V_0}{R}e^{-\frac{t}{RC}}$$

By plugging in the known values, we get

$$i(1\,s) = \frac{5-0}{1000}e^{-\frac{1}{1000 \times 470 \times 10^{-6}}}$$

$$i(1\,s) = 0.000595579$$

$$i(1\,s) = 595.579\,\mu A$$

25.4.2 Example 2

Figure 25.4 shows a DC power supply, a resistor, a capacitor, and a two-position switch.

Figure 25.4 RC circuit initially working.

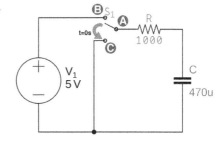

The switch has been closed for a long time connecting points A and B. Thus, R and C have been connected across a constant voltage source, V_1, for a long time. The circuit is on the steady-state phase, C is completely charged, and there is no more current flowing.

At time t = 0s, the switch is moved from position AB to AC, disconnecting R and C from the power supply and connecting them parallel.

- What is the current flowing in the circuit before the switch is moved?
- What is the current flowing in the circuit as the switch is moved?
- What is the voltage across the capacitor 2 s after the switch is moved?
- What is the current flowing in the circuit 2 s after the switch is moved?

25.4.2.1 Current Before the Switch Is Moved

The circuit was working for a long time before the switch is moved. The capacitor was fully charged and no more current was flowing.

FINAL RESULT

$$i(0^-) = 0 \, \text{mA}$$

25.4.2.2 Current as the Switch Is Moved

The moment the switch moves from AB to AC, there is a potential difference across the resistor. This drives a current to flow. The capacitor has now a path to discharge and will do it slowly through the resistor.

The flow of current will start out big and gradually decay exponentially to 0. The initial big current can be calculated by simple Ohm's law:

$$i = \frac{V_c}{R}$$

$$i = \frac{5}{1000}$$

FINAL RESULT

$$i(0^+) = 5 \, \text{mA}$$

25.4.2.3 Voltage After 2 s

As soon as the switch is moved, the capacitor starts to discharge. The following formula can be used to calculate its voltage 2 s into discharge:

$$V(t) = V_0 e^{-\frac{t}{RC}}$$

By plugging in the known values, we get

$$V(2\,s) = 5e^{-\frac{2}{1000 \times 470 \times 10^{-6}}}$$

$$V_C(2\,s) = 70.9428\,mV$$

We can obtain the same value by using the general formula:

$$V_C(t) = V_\infty + (V_0 - V_\infty)e^{-\frac{t}{RC}}$$

where V_0 is equal to capacitor initial voltage, which is equal to the power supply voltage (5 V), and V_∞ is 0, because at an infinite time the capacitor will be completely discharged.

Upon substituting these values on the general formula, for t = 2 s, we obtain the same value as before:

$$V_C(2\,s) = 0 + (5 - 0)e^{-\frac{2}{1000 \times 470 \times 10^{-6}}}$$

$$V_C(2\,s) = 70.9428\,mV$$

25.4.2.4 Current After 2 s

To calculate the current flowing in the circuit two seconds into discharge, we can use the following formula:

$$i(t) = -\frac{V_0}{R}e^{-\frac{t}{RC}}$$

By plugging in the known values, we get

$$i(2\,s) = -\frac{5}{1000}e^{-\frac{2}{1000 \times 470 \times 10^{-6}}}$$

$$i(2\,s) = -\frac{0.07094}{1000}$$

$$i(2\,s) = -70.94\,\mu V$$

The negative sign means that current is flowing in the opposite direction during discharge, compared to the direction it flows during charge.

25.4.3 Example 3

Figure 25.5 shows a circuit with two DC power supplies, two resistors, a capacitor, and a two-position switch.

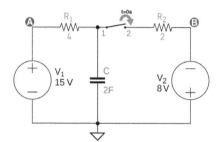

Figure 25.5 RC circuit with two voltage sources.

In the following circuit, V_1, R_1, and C make the main circuit. This part of the circuit is working for a long time, C is fully charged, and there is no current flowing in that part of the circuit anymore. Voltage across the capacitor is the same as power supply voltage, V_1, which is 15 V.

The secondary part of the circuit is composed of R_2, V_2, and a switch that is initially open, making this part of the circuit electrically isolated from other part. There is no current flowing in that part of the circuit initially.

At time t = 0 s, the switch is closed, connecting the secondary circuit to the main circuit.

We want to know the voltage across the capacitor after the switch is closed and circuit stabilizes.

25.4.3.1 After the Switch Is Closed

As the switch is closed, components R_1/V_1 and C are put parallel with R_2/V_2. Notice that V_2's polarity is reversed compared with V_1's.

A new transient phase begins. The circuit tries to stabilize for the new values of resistance and voltages. During this new transient phase, current will flow as shown in Figure 25.6.

Figure 25.6 RC circuit tries to stabilize.

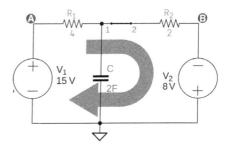

To best understand the circuit, consider the voltages of points A and B in relation to ground as seen in Figure 25.7.

Figure 25.7 RC circuit tries to stabilize.

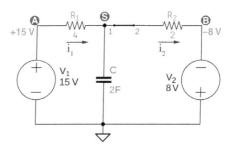

Figure 25.7 shows, clearly, that voltages of points A and B, in relation to ground, are +15 V and −8 V, respectively.

By subtracting these two voltages, we can get the difference in potential between points A and B:

$$V_{AB} = 15 - (-8) = 15 + 8 = 23\,V$$

To calculate the voltage across the capacitor when the circuit stabilizes, all we must do is to calculate the voltage at point S, removing the capacitor from the circuit.

Currents across R_1 and R_2 are

$$i_1 = \frac{V_1 - V_S}{R_1}$$

$$i_2 = \frac{V_S - V_2}{R_2}$$

Without the capacitor, i_1 and i_2 are the same current; thus we can equate them and obtain the voltage across the capacitor when the circuit stabilizes:

$$\frac{V_1 - V_S}{R_1} = \frac{V_S - V_2}{R_2}$$

$$\frac{15 - V_S}{4} = \frac{V_S - (-8)}{2}$$

$$30 - 2V_S = 4V_S + 32$$

$$6V_S = -2$$

FINAL RESULT

$$V_S = -333\,\text{mV} \rightarrow V_\infty$$

V_S is the same as the voltage across the capacitor or V_∞ (voltage of the steady state).

25.4.4 Example 4

Figure 25.8 shows a circuit with an alternating current power supply, a resistor, a capacitor, and a switch permanently closed. The circuit is working like this for a long time.

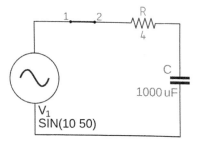

Figure 25.8 AC powered RC circuit.

V_1 is a sinusoidal alternating voltage source. Its amplitude is 10 V and its frequency is 50 Hz.

This is what we want to know about the circuit:

- The circuit's impedance and its phase angle.
- The root mean square (RMS) current flowing in the circuit.
- The circuit's power factor.
- The apparent, the real, and the reactive power.

25.4.4.1 Circuit's Impedance

The impedance of a resistor (Z_R) is its own value of resistance (R), because resistors have no reactance.

Thus,

$$Z_R = R$$

For a series R_C circuit, the total impedance will be the resistor's impedance added to the capacitor's impedance, or

$$Z = Z_R + Z_C$$

Capacitor's impedance is related to its reactance by

$$Z_C = -jX_C$$

where X_C is the capacitor's reactance and $-j$ is the impedance's imaginary part.

Capacitor's reactance can be found by plugging in the known values in the following formula:

$$X_C = \frac{1}{2\pi fC}$$

$$X_C = \frac{1}{2\pi(50)(1000 \times 10^{-6})}$$

$$X_C = 3.1831 \, \Omega$$

Hence, capacitor's impedance is

$$Z_C = -jX_C$$

$$Z_C = -j3.1831 \, \Omega$$

and the circuit's total impedance is

$$Z = R - jX_C$$

CIRCUIT'S IMPEDANCE

$$= 4 - j3.1831\,\Omega$$

By converting the complex impedance to phasor notation, we get the impedance's magnitude:

$$Z = \sqrt{R^2 + Z_C^2}$$

$$Z = \sqrt{4^2 + (-3.1831)^2}$$

$$Z = 5.112\,\Omega$$

and for the phase angle,

$$\tan(\phi) = \frac{Z_C}{R}$$

$$\tan(\phi) = \frac{-3.1831}{4} = -0.7958$$

$$\phi = A\tan(-0.7958)$$

$$\phi = -38.5119°$$

CIRCUIT'S IMPEDANCE – PHASOR NOTATION

$$Z \approx 5.112\,\Omega \angle -38.5119°$$

See Appendix D for more information about complex number transformations.

Figure 25.9 shows the impedance triangle. Horizontal axis represents impedance's real part (Re), in this case equal to resistance R, and the capacitor reactance X_C is the imaginary part (Im). The impedance magnitude Z is the triangle's hypotenuse. Its angle to the real axis is the phase angle ϕ.

Capacitive circuits will always have negative imaginary terms.

Figure 25.9 Impedance triangle.

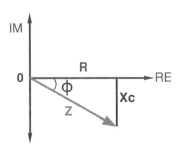

25.4.4.2 RMS Current

V_{RMS} can be calculated by using the following formula:

$$V_{RMS} = \frac{V_{MAX}}{\sqrt{2}}$$

By plugging in the power supply maximum voltage V_{MAX} of 10 V, we get

$$V_{RMS} = \frac{10}{\sqrt{2}}$$

$$V_{RMS} = 7.071 \, V$$

The magnitude of impedance, already calculated as 5.112 Ω, can be used to find the RMS current:

$$I_{RMS} = \frac{V_{RMS}}{Z}$$

$$I_{RMS} = \frac{7.071}{5.112}$$

RMS CURRENT

$$I_{RMS} = 1.3832 \, A$$

25.4.4.3 Power Factor and Phase Angle

The power factor of a series RC circuit powered by alternating current is equal to the resistance divided by the impedance magnitude.

POWER FACTOR – SERIES RC CIRCUIT POWERED BY AC

$$\cos(\varphi) \rightarrow F_p = \frac{R}{Z}$$

$\cos(\varphi), F_p$ is the power factor.
R is the resistance, in Ohms.
Z is the impedance module, in Ohms.

By plugging in

$$\begin{cases} Z = 5.112\,\Omega \\ R = 4\,\Omega \end{cases}$$

we get

$$\cos(\varphi) \rightarrow F_p = \frac{4}{5.112}$$

POWER FACTOR

$$\cos(\varphi) \rightarrow F_p = 0.7825$$

25.4.4.4   The Apparent, the Real, and the Reactive Power

The apparent, the real and the reactive powers can be calculated by the following formulas.

REAL POWER – CIRCUIT POWERED BY AC

$$P = V_{RMS}\, I_{RMS} \cos(\varphi)$$

P is the real or active power, in Watts.
V_{RMS} is the power supply RMS voltage, in Volts.
I_{RMS} is the power supply RMS current, in Amperes.
$\cos(\varphi)$ is the power factor.

REACTIVE POWER – CIRCUIT POWERED BY AC

$$Q = V_{RMS}\, I_{RMS} \sin(\varphi)$$

Q is the reactive power, in VAR.
V_{RMS} is power supply RMS voltage, in Volts.
I_{RMS} is power supply RMS current, in Amperes.
$\sin(\varphi)$ is the sine of the phase angle.

APPARENT POWER – CIRCUIT POWERED BY AC

$$S = \sqrt{P^2 + Q^2}$$

S is the apparent power, in VA.
P is the real power, in Watts.
Q is the reactive power, in VAR.

The real power is then calculated by plugging in the known values for the given circuit:

$$P = V_{RMS} I_{RMS} \cos(\varphi)$$

$$P = (7.071)(1.3832)(0.7825)$$

REAL POWER

$$P = 7.6534\,W$$

For the reactive power,

$$Q = V_{RMS} I_{RMS} \sin(\varphi)$$

$$Q = (7.071)(1.3832)(\sin(-38.5119))$$

$$Q = -6.0902\,VAR$$

The negative sign represents a capacitive reactive power and is, generally, never used to represent Q. Thus, the result is shown below.

REACTIVE POWER

$$Q = 6.0902\,VAR$$

The apparent power can be calculated as

$$S = \sqrt{P^2 + Q^2}$$

$$S = \sqrt{7.6534^2 + 6.0902^2}$$

APPARENT POWER

$$S = 9.7809\,VA$$

Exercises

1   Consider the circuit shown in Figure 25.10, composed of a resistor and a capacitor in series with a switch and a direct voltage power supply. The switch is initially open, so the circuit is not working. The capacitor is completely discharged.

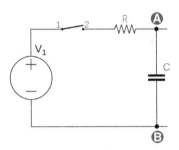

Figure 25.10 RC circuit and an open switch.

The components in the circuit are

$$\begin{cases} R = 220\,\Omega \\ C = 100\,\mu F \\ V_1 = 12\,V \end{cases}$$

At time t = 0 s, the switch is closed.
The following is what we want to find:

- What is the current flowing in the circuit before the switch is closed?
- What is the current flowing in the circuit after the switch is closed?
- What is the current flowing in the circuit 5 ms after the switch is closed?
- What is the voltage across capacitor (V_{AB}) 3 ms after the switch is closed?

2   Figure 25.11 shows a DC power supply, a resistor, a capacitor, and a two-position switch.

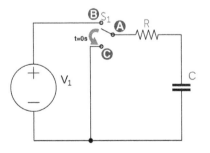

Figure 25.11 RC circuit and a two-pole switch.

The switch has been closed for a long time, connecting points A and B. For that reason, the resistor and the capacitor have been connected to the power supply for a long time, and the capacitor is fully charged.

The components in the circuit are

$$\begin{cases} R = 330\,\Omega \\ C = 47\,\mu F \\ V_1 = 10\,V \end{cases}$$

At time t = 0 s, the switch is moved from AB to AC, putting the capacitor and the resistor in parallel.

The following is what we want to find:

- What is the current flowing in the circuit before the switch is moved?
- What is the current flowing in the circuit after the switch is moved?
- What is the current flowing in the circuit 4 ms after the switch is moved?
- What is the voltage across capacitor 5 ms after the switch is moved?

3   Figure 25.12 shows a circuit with two DC power supplies, two resistors, a capacitor, and a switch.

Figure 25.12 RC circuit with two voltage sources.

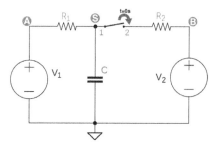

V_1, R_1, and C make the main circuit. This part of the circuit is working for a long time like this. For that reason, the capacitor is fully charged.

The secondary circuit is composed of R_2, V_2, and a switch in series, and the switch is initially open.

The components in the circuit are

$$\begin{cases} V_1 = 40\,V \\ V_2 = 30\,V \\ R_1 = 100\,\Omega \\ R_2 = 22\,\Omega \\ C = 1000\,\mu F \end{cases}$$

At time t = 0 s, the switch is closed, connecting the second circuit to the first.

The following is what we want to find:

- What is the voltage across capacitor after the circuit stabilizes?
- What is the voltage across capacitor 3 ms after the switch is closed?

4 Figure 25.13 shows a circuit with an alternating current power supply, a resistor, a capacitor, and a switch in series. The switch is permanently closed. The circuit is working like this for a long time.

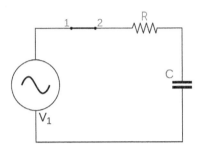

Figure 25.13 AC powered RC circuit.

The components in the circuit are

$$\begin{cases} V_1(\text{amplitude}) = 48\,\text{V} \\ V_1(\text{frequency}) = 60\,\text{Hz} \\ R = 470\,\Omega \\ C = 330\,\mu\text{F} \end{cases}$$

The following is what we want to find:

- What are the circuit's impedance and its phase angle?
- What is the RMS current flowing in the circuit?
- What is the circuit's power factor?
- What are the apparent, the real and the reactive power?

Solutions

1 What is the current flowing in the circuit before the switch is closed?
Before the switch is closed, there is no current flowing in the circuit, because the switch is open and there is no circuit continuity.

FINAL RESULT

$i(0^-) = 0\,\text{mA}$

What is the current flowing in the circuit after the switch is closed?

The moment the switch is turned on, voltage across the capacitor (V_C) is 0, and the capacitor will offer zero resistance to the current flow, behaving like a regular wire or short circuit, letting maximum current pass.

By Kirchhoff's law,

$$V_1 = V_R + V_C$$

We know that

$$V_C = 0$$

$$V_1 = Ri \longrightarrow i = \frac{V_1}{R}$$

Upon substituting these values on the equation, we get

$$V_1 - Ri - \cancel{V_C(t)}^0 = 0$$

$$i = \frac{V_1}{R}$$

By plugging in the known values

$$\begin{cases} R = 220\,\Omega \\ V_1 = 12\,V \end{cases}$$

we get

$$i(0^+) = \frac{12}{220}$$

FINAL RESULT

$$i(0^+) = 54.545\,\text{mA}$$

What is the current flowing in the circuit 5 ms after the switch is closed?

In an RC series circuit, current at any given time can be found by the following formula:

$$i(t) = \frac{V_1 - V_0}{R} e^{-\frac{t}{RC}}$$

If we plug in the known values,

$$\begin{cases} R = 220\,\Omega \\ C = 100\,\mu F \\ V_0 = 0\,V \\ V_1 = 12\,V \\ t = 5\,ms \end{cases}$$

we get

$$i(5\,ms) = \frac{12-0}{220}\,e^{-\frac{5\times 10^{-3}}{220\times 100\times 10^{-6}}}$$

FINAL RESULT

$$i(5\,ms) = 43.456\,mA$$

What is the voltage across capacitor (V_{AB}) 3 ms after the switch is closed?
Voltage at any given time can be found by using the following formula with the known values we have:

$$\begin{cases} R = 220\,\Omega \\ C = 100\,\mu F \\ V_0 = 0\,V \\ V_1 = 12\,V \\ t = 3\,ms \end{cases}$$

$$V_C(t) = V_1 + (V_0 - V_1)e^{-\frac{1}{RC}t}$$

$$V_C(3\,ms) = 12 + (0-12)e^{-\frac{3\times 10^{-3}}{220\times 100\times 10^{-6}}}$$

FINAL RESULT

$$V_C(3\,ms) = 1.5296\,V$$

2   **What is the current flowing in the circuit before the switch is moved?**
The main part of the circuit was working for a long time before the switch is moved. The capacitor was fully charged and no more current was flowing.

FINAL RESULT

$$i(0^-) = 0\,mA$$

What is the current flowing in the circuit after the switch is moved?
The moment the switch moves from AB to AC, the capacitor is fully charged, and this charge will produce a difference in potential across the resistor. Voltage across the capacitor is equal to V_1. This drives a current to flow. The capacitor has now a path to discharge and will do it slowly through the resistor.
The flow of current will start out big and gradually decay exponentially to 0. The initial current can be calculated by using the values we have:

$$\begin{cases} R = 330\,\Omega \\ V_C = 10\,V \end{cases}$$

and by applying the first Ohm's law,

$$i = \frac{V_C}{R}$$
$$i = \frac{10}{330}$$

FINAL RESULT

$$i(0^+) = 30.3030\,mA$$

What is the current flowing in the circuit 4 ms after the switch is moved?
To calculate the current flowing in the circuit 4 ms after the switch is moved, we use the following formula:

$$i(t) = -\frac{V_0}{R}e^{-\frac{t}{RC}}$$

By plugging in the known values

$$\begin{cases} R = 330\,\Omega \\ C = 47\,\mu F \\ V_0 = 10\,V \\ t = 4\,ms \end{cases}$$

we get

$$i(4\,ms) = -\frac{10}{330}\,e^{-\frac{4\times 10^{-3}}{330\times 47\times 10^{-6}}}$$

FINAL RESULT

$$i(4\,\text{ms}) = -23.414\,\text{mA}$$

What is the voltage across capacitor 5 ms after the switch is moved?

As soon as the switch is moved, the capacitor starts to discharge. The following formula can be used to calculate its voltage 5 ms into discharge:

$$V(t) = V_0 e^{-\frac{t}{RC}}$$

By plugging in the known values

$$\begin{cases} R = 330\,\Omega \\ C = 47\,\mu\text{F} \\ V_0 = 10\,\text{V} \\ t = 5\,\text{ms} \end{cases}$$

we get

$$V(5\,\text{ms}) = 10e^{-\frac{5 \times 10^{-3}}{330 \times 47 \times 10^{-6}}}$$

FINAL RESULT

$$V_C(5\,\text{ms}) = 7.2442\,\text{V}$$

3 What is the voltage across capacitor after the circuit stabilizes?

V_1, R_1, and C make the main circuit. This part of the circuit is working for a long time, C is fully charged, and there is no current flowing anymore on that part. Voltage across the capacitor is the same as the power supply voltage V_1, which is 40 V.

As the switch is closed, components R_1/V_1 are put parallel with R_2/V_2.

A new transient phase begins, as the circuit tries to stabilize for the new values of resistance and voltages.

Voltage at point S

Points A and B have 40 V and 30 V, respectively. Resistors R_1 and R_2 are connected between A and B. By subtracting these voltages, we conclude that R_1 and R_2 are subjected to a difference in potential of 10 V.

The way these resistors are connected between A and B makes them a simple voltage divider. We can use Ohm's law to find the voltage at point S, which is exactly the voltage across the capacitor when the circuit stabilizes.

Currents flowing across R_1 and R_2 are

$$I_{R_1} = \frac{V_1 - V_S}{R_1}$$

$$I_{R_2} = \frac{V_S - V_2}{R_2}$$

Without capacitor, currents flowing across R_1 and R_2, which are I_{R_1} and I_{R_2}, respectively, are the same current; thus, we can equate their equations:

$$\frac{V_1 - V_S}{R_1} = \frac{V_S - V_2}{R_2}$$

By plugging in the known values

$$\begin{cases} R_1 = 100\,\Omega \\ R_2 = 22\,\Omega \\ V_1 = 40\,V \\ V_2 = 30\,V \end{cases}$$

we get the voltage at point S or, in other words the voltage across the capacitor when the circuit stabilizes (V_∞):

$$\frac{40 - V_S}{100} = \frac{V_S - 30}{22}$$

$$880 - 22\,V_S = 100\,V_S - 3000$$

$$122\,V_S = 3880$$

FINAL RESULT

$$V_S = 31.8032\,V \longrightarrow V_\infty$$

What is the voltage across capacitor 3 ms after the switch is closed?

Voltage across the capacitor before the switch is closed is 40 V. As soon as the switch is closed, a new transient phase begins and the voltage across the capacitor slowly decays to 31.8032 V.

If we substitute both voltage sources with wires, it shows that the capacitor is in parallel with both resistors, R_1 and R_2.

The equivalent resistance of these resistors is

$$\frac{1}{R} = \frac{1}{R_1} + \frac{1}{R_2}$$

$$\frac{1}{R} = \frac{1}{100} + \frac{1}{22}$$

$$R = 18.0327 \ \Omega$$

Hence, as soon as the switch is closed, the capacitor will be put in parallel with a resistance equal to 18.0327 Ω, meaning that it will discharge from 40 to 31.8032 V through this resistance.

To calculate the voltage across the capacitor after the time required, we must use the following equation:

$$V_C(t) = V_C(\infty) + [V_C(0) - V_C(\infty)]e^{-\frac{t}{RC}}$$

by plugging in the known values

$$\begin{cases} R = 18.0327 \ \Omega \\ C = 1000 \ \mu F \\ V_0 = 40 \ V \\ V_C(\infty) = 31.8032 \ V \\ t = 3 \ ms \end{cases}$$

we get,

$$V(3ms) = 31.8032 + [40 - 31.8032] \ e^{-\frac{3 \times 10^{-3}}{18.0327 \times 1000 \times 10^{-6}}}$$

FINAL RESULT

$$V(3 \ ms) = 38.7437 \ V$$

4   What are the circuit's impedance and its phase angle?

The impedance of a resistor (Z_R) is its own value of resistance (R), because resistors have no reactance.

Thus,

$$Z_R = R$$

For a series RC circuit, the total impedance will be the resistor's impedance added to the capacitor's impedance, or

$$Z = Z_R + Z_C$$

Capacitor's impedance is related to its reactance by

$$Z_C = -jX_C$$

where X_C is the capacitor's reactance and $-j$ is the impedance's imaginary part. Capacitor's reactance can be found by plugging in the known values

$$\begin{cases} f = 60\,Hz \\ C = 330\,\mu F \end{cases}$$

in the following formula:

$$X_C = \frac{1}{2\pi fC}$$

$$X_C = \frac{1}{2\pi(60)(330 \times 10^{-6})}$$

$$X_C = 8.038\,\Omega$$

Hence, capacitor's impedance is

$$Z_C = -jX_C$$

$$Z_C = -j8.038\,\Omega$$

and the circuit's total impedance is

$$Z = R - jX_C$$

FINAL RESULT

$$Z = 470 - j8.038\,\Omega$$

By converting the complex impedance to phasor notation, we get the impedance's magnitude:

$$Z = \sqrt{R^2 + Z_C^2}$$

$$Z = \sqrt{470^2 + (-8.038)^2}$$

$$Z = 470.0687\,\Omega$$

and for the phase angle,

$$\tan(\phi) = \frac{Z_C}{R}$$

$$\tan(\phi) = \frac{-8.038}{470} = -0.0171$$

$$\phi = A\tan(-0.0171)$$

$$\phi = -0.9797°$$

$$Z \approx 470.0687\,\Omega \angle -0.9797$$

The phase angle is very small, almost horizontal, because resistance is very big compared with reactance.

What is the RMS current flowing in the circuit?
If we extend Ohm's law to impedances, current will be defined as

$$V_{RMS} = \frac{V_{MAX}}{\sqrt{2}}$$

By plugging in power supply maximum voltage V_{MAX} of 48 V, we get

$$V_{RMS} = \frac{48}{\sqrt{2}}$$

$$V_{RMS} = 33.9411\,V$$

We can finally calculate the RMS current by plugging in V_{RMS} into equation

$$I_{RMS} = \frac{V_{RMS}}{Z}$$

$$I_{RMS} = \frac{33.9411}{470.06}$$

FINAL RESULT

$$I_{RMS} = 72.2059\,mA$$

What is the circuit's power factor?

The power factor of a series RC circuit powered by alternating current is equal to the resistance divided by the impedance magnitude, or

$$\cos(\varphi) \to F_p = \frac{R}{Z}$$

By plugging in the known values, we get

$$\cos(\varphi) \to F_p = \frac{470}{470.06}$$

FINAL RESULT

$$\cos(\varphi) \to F_p = 0.9998 \longrightarrow F_p \approx 1$$

What are the apparent, the real, and the reactive powers?

The real power can be calculated by

$$P = V_{RMS}I_{RMS}\cos(\varphi)$$

$$P = 33.9411 \times 0.0722059 \times 0.9998$$

FINAL RESULT

$$P = 2.4502\,W$$

The reactive power is

$$Q = V_{RMS}I_{RMS}\sin(\varphi)$$

$$Q = 33.9411 \times 0.0722059 \times \sin(-0.9797)$$

FINAL RESULT

$$Q = 0.0419\,VAR$$

The apparent power is

$$S = \sqrt{P^2 + Q^2}$$

$$S = \sqrt{2.4502^2 + 0.0419^2}$$

FINAL RESULT

$$S = 2.4505\,VA$$

26

RL Circuits

Voltage and Current Analysis in Circuits Containing Resistors and Inductors in Series

26.1 Introduction

In this chapter, we will examine RL series circuits formed by resistors and capacitors in series and what happens when these circuits are powered on or off.

26.2 Energizing

We have seen previously in this book that inductors hate current variations and will do everything to oppose the change by trying to block current increasing or pushing forward current decreasing.

Consider the circuit shown in Figure 26.1, composed of an inductor, a resistor, a voltage source, and a switch in series. The switch is initially open and the inductor is completely de-energized.

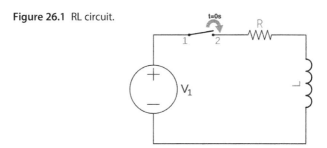

Figure 26.1 RL circuit.

At time t = 0 s, the switch is closed, and current flows from the voltage source and reaches the inductor.

However, like we have learned before, inductors hate sudden variations in current and create an electromotive force to oppose this new current to flow. This electromotive force will make a voltage rise across inductor with a polarity inverted in relation to V_1. At the exact time the switch is closed, no current will

Introductory Electrical Engineering with Math Explained in Accessible Language,
First Edition. Magno Urbano.
© 2020 John Wiley & Sons, Inc. Published 2020 by John Wiley & Sons, Inc.

flow across the inductor, and this component is seen by the circuit as an open circuit or as a resistor with a very large resistance.

As time passes, current gradually wins the battle against the electromotive force until the point where the electromotive force and the respective voltage across the inductor disappear. At this point, current reaches its maximum value possible, just limited by the resistor and the first Ohm's law.

26.2.1 Current During Energizing

Like explained previously in this book, voltage across an inductor is equal to the derivative of current with respect to time, according to the following formula:

$$V_L(t) = L\frac{di(t)}{dt}$$

Kirchhoff's voltage law states that the sum of all voltages in a loop is equal to 0. Therefore, we can write this equation for the RL series circuit:

$$V_1 = Ri(t) + L\frac{di(t)}{dt}$$

This equation can be rewritten as

$$\frac{di(t)}{V_1 - i(t)R} = \frac{1}{L}dt$$

To find the current, we integrate both sides

$$\int_{I_0}^{I} \frac{di(t)}{V_1 - i(t)R} = \int_{0}^{t} \frac{1}{L}dt$$

First integral is defined from an initial current I_0 that may be flowing in the circuit to a generic current I at time equal to t. The second integral is defined from time 0 to a generic time t.

To solve the first integral, we use the substitution rule:

$$\text{let } u = V_1 - i(t)R$$
$$du = d[V_1 - i(t)R]$$

MATH CONCEPT   The derivative of a sum is equal to the sum of derivatives.

$$du = d[V_1] - d[i(t)R]$$

MATH CONCEPT   The derivative of a constant is equal to 0.

Therefore,

$$du = d[V_1]^0 - [Rdi]$$
$$du = -Rdi$$
$$di = -\frac{du}{R}$$

By substituting this on the first integral, we get

$$\int_{I_0}^{I} \frac{di(t)}{V_1 - i(t)R}$$

$$\int_{I_0}^{I} \frac{-\dfrac{du}{R}}{u}$$

$$-\frac{1}{R} \int_{I_0}^{I} \frac{du}{u}$$

MATH CONCEPT   The integral of

$$\int \frac{1}{x} dx \longrightarrow \ln|x| + C$$

The constant of integration can be ignored when dealing with definite integrals (see Appendix F).

By applying this concept on the first integral, we can solve the whole equation as

$$-\frac{1}{R}\left[\ln|u|\right]_{I_0}^{I} = \frac{1}{L}t\Big|_0^t$$

Substituting u,

$$-\frac{1}{R}\left[\ln\left|\frac{V_1}{R}-i(t)\right|\right]_{I_0}^{I} = \frac{1}{L}t\Big|_0^t$$

$$-\frac{1}{R}\left[\ln\left|\frac{V_1}{R}-I\right|-\ln\left|\frac{V_1}{R}-I_0\right|\right]_{I_0}^{I} = \frac{1}{L}t\Big|_0^t$$

We solve for the intervals

$$-\frac{1}{R}\left[\ln\left|\frac{V_1}{R}-I\right|-\ln\left|\frac{V_1}{R}-I_0\right|\right] = \frac{1}{L}[t-0]$$

We multiply both sides by $-R$:

$$\left[\ln\left|\frac{V_1}{R}-I\right|-\ln\left|\frac{V_1}{R}-I_0\right|\right] = -\frac{R}{L}t$$

MATH CONCEPT   The subtraction of two natural logarithms

$$\ln a - \ln b = \ln\left(\frac{a}{b}\right)$$

Therefore, the first part can be converted to

$$\ln\left|\frac{\frac{V_1}{R}-I}{\frac{V_1}{R}-I_0}\right| = -\frac{R}{L}t$$

$$\ln\left|\frac{V_1-IR}{V_1-I_0R}\right| = -\frac{R}{L}t$$

To get rid of the natural log, we can use the following math concept.

MATH CONCEPT   The exponential of the natural log

$e^{\ln{(a)}} = a$

Therefore, we apply the exponential to both sides:

$$e^{\ln{\left|\frac{V_1 - IR}{V_1 - I_0R}\right|}} = e^{-\frac{R}{L}t}$$

and we get

$$\frac{V_1 - IR}{V_1 - I_0R} = e^{-\frac{R}{L}t}$$

If we pass the fraction denominator to the other side,

$$V_1 - IR = (V_1 - I_0R)e^{-\frac{R}{L}t}$$

We can isolate the current on one side by rewriting the equation

$$-IR = -V_1 + (V_1 - I_0R)e^{-\frac{R}{L}t}$$

Multiplying both sides by –1,

$$IR = V_1 - (V_1 - I_0R)e^{-\frac{R}{L}t}$$

$$I = \frac{V_1 + (I_0R - V_1)e^{-\frac{R}{L}t}}{R}$$

and we finally get the current equation for the inductor during the energizing phase.

CURRENT EQUATION – INDUCTOR ENERGIZING

$$i(t) = \frac{V_1 + (I_0R - V_1)e^{-\frac{R}{L}t}}{R} \tag{26.1}$$

i(t) is the current equation as a function of time, in Amperes.
I_0 is the initial current, in Amperes.
V_1 is the voltage applied to the inductor, in Volts.
t is the instant in time, when energizing starts, in seconds.
R is the resistance, in Ohms.
L is the inductance, in Henries.

If the initial current I_0 is 0, the formula simplifies to

$$i(t) = \frac{V_1 + \left(\cancel{I_0}R - V_1\right)e^{-\frac{R}{L}t}}{R}$$

$$i(t) = \frac{V_1 - V_1 e^{-\frac{R}{L}t}}{R}$$

CURRENT EQUATION – INDUCTOR ENERGIZING (INITIAL CURRENT ZERO)

$$i(t) = \frac{V_1}{R}\left(1 - e^{-\frac{R}{L}t}\right)$$

---

i(t) is the current equation as a function of time, in Amperes.
V_1 is the voltage applied to the inductor, in Volts.
t is the instant in time, when energizing starts, in seconds.
R is the resistance, in Ohms.
L is the inductance, in Henries.

26.2.2 Voltage During Energizing

Like mentioned before, voltage across an inductor follows the equation

$$V_L(t) = L\frac{di(t)}{dt}$$

This equation tells us that voltage is equal to the derivative of current with respect to time.

Therefore, if we take the derivative of the equation we have just obtained in (26.1) with respect to time, we get the voltage equation:

$$V_L(t) = L\frac{d}{dt}\left[\frac{V_1 + (I_0 R - V_1)e^{-\frac{R}{L}t}}{R}\right]$$

MATH CONCEPT   The derivative of a sum is equal to the sum of derivatives.

$$V_L(t) = L\left\{\frac{d}{dt}\left[\frac{V_1}{R}\right] + \frac{d}{dt}\left[\frac{(I_0 R - V_1)e^{-\frac{R}{L}t}}{R}\right]\right\}$$

The first term is a derivative of a constant that mathematically is 0:

$$V_L(t) = L\left\{\frac{d}{dt}\left[\frac{V_1}{R}\right]^0 + \frac{d}{dt}\left[\frac{(I_0R - V_1)e^{-\frac{R}{L}t}}{R}\right]\right\}$$

$$V_L(t) = L\left\{\frac{d}{dt}\left[\frac{(I_0R - V_1)e^{-\frac{R}{L}t}}{R}\right]\right\}$$

We are before the derivative of a product of two functions (u) and (v), as shown in the following equation:

$$V_L(t) = L\left\{\frac{d}{dt}\left[\frac{\overbrace{(I_0R - V_1)}^{u}\;\overbrace{e^{-\frac{R}{L}t}}^{v}}{R}\right]\right\}$$

and we can apply the following math concept.

MATH CONCEPT The derivative of a product of two functions is equal to the derivative of the first function multiplied by the second function added to the first function multiplied by the derivative of the second, or

$$(u.v)' = u'v + uv'$$

Therefore, for the given case,

$$u = \frac{(I_0R - V_1)}{R}$$

$$v = e^{-\frac{R}{L}t}$$

We see that u is a constant. Therefore, its derivative is 0 or

$$u' = 0$$

Thus, the mathematical concept simplifies to

$$(u.v)' = u'v + uv'$$

$$(u.v)' = u'^0 v + uv'$$

$$(u.v)' = uv'$$

In other words, we see that all we must do is to find the derivative of v and multiply by u.

We know that

$$v = e^{-\frac{R}{L}t}$$

Therefore, its derivative can be found by using the following math concept.

MATH CONCEPT The derivative of an exponential

$$\frac{d}{dt}e^{xt} = xe^{xt}$$

$$v = e^{-\frac{R}{L}t}$$

$$v' = -\frac{R}{L}e^{-\frac{R}{L}t}$$

Therefore, by putting it all back together back into the equation, we get

$$V_L(t) = L\left\{\frac{d}{dt}\left[\frac{(I_0R - V_1)e^{-\frac{R}{L}t}}{R}\right]\right\}$$

$$V_L(t) = L\left\{\frac{(I_0R - V_1)}{R}\left(-\frac{R}{L}e^{-\frac{R}{L}t}\right)\right\}$$

$$V_L(t) = -\frac{R}{L}L\left(\frac{(I_0R - V_1)}{R}\right)\left(e^{-\frac{R}{L}t}\right)$$

$$V_L(t) = -\frac{\cancel{R}}{\cancel{L}}\cancel{L}\left(\frac{(I_0R - V_1)}{\cancel{R}}\right)e^{-\frac{R}{L}t}$$

$$V_L(t) = -(I_0R - V_1)e^{-\frac{R}{L}t}$$

$$V_L(t) = (V_1 - I_0R)e^{-\frac{R}{L}t}$$

We get the voltage equation for the inductor during the energizing phase, considering the initial current I_0 equal to zero:

VOLTAGE ACROSS THE INDUCTOR – ENERGIZING PHASE

$$V_L(t) = V_1(t)e^{-\frac{R}{L}t}$$

$V_L(t)$ is the equation for the voltage across the inductor as a function of time, in Volts.
V_1 is the voltage applied to the inductor, in Volts.
t is the instant in time, when energizing starts, in seconds.
R is the resistance, in Ohms.
L is the inductance, in Henries.

26.3 De-energizing

Figure 26.2 shows a direct voltage source, a resistor, an inductor, and a switch.

 **Figure 26.2** RL circuit.

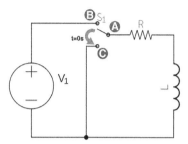

This circuit is working for a long time with the switch connecting points A and B, or in other words, with the power supply in series with the resistor and the inductor. The circuit is stabilized and the inductor is fully energized.

At time t = 0 s, the switch is moved from position AB to AC, disconnecting the resistor and the inductor from the power supply and putting them in parallel.

Current flowing across the circuit was at the maximum value when the switch was moved from AB to AC. The inductor was happy. Now, the switch moves and puts the resistor and the inductor in parallel. The inductor sees the current ceasing to exist and hates it.

The inductor will then try to prevent the current from vanishing and will use its own energy, stored as magnetic field, to generate an electromotive force to push current forward and keep it flowing. This electromotive force manifests itself as a voltage across the inductor that has the inverse polarity, when compared to the power supply.

26.3.1 Current During De-energizing

The switch moving from AB to AC puts the resistor in parallel with the inductor. If we apply the Kirchhoff's laws this circuit, we get

$$V_R + V_L = 0 \qquad (26.2)$$

We know that the voltage across the inductor is

$$V_L(t) = L\frac{di(t)}{dt}$$

and that the voltage across the resistor is

$$V_R(t) = Ri(t)$$

Therefore, we can substitute both terms in (26.2) and get

$$Ri(t) + L\frac{di(t)}{dt} = 0$$

or the following equation after isolating the current on one side:

$$\frac{di}{i(t)} = -\frac{R}{L}dt$$

If we integrate both sides, we can get the current equation as a function of time:

$$\int_{i=V_1/R}^{I} \frac{di}{i(t)} = -\int_{0}^{t} \frac{R}{L}dt$$

The first integral goes from the initial current, defined by the first Ohm's law as V_1/R to a generic current and the second integral goes from a time equal to 0 to a generic time.

We can apply the following math concept to the first integral.

MATH CONCEPT   The integral of

$$\int \frac{1}{x}dx \longrightarrow \ln|x| + C$$

The constant of integration can be ignored when dealing with definite integrals (see Appendix F).

$$\ln|i|\Big|_{V_1/R}^{I} = -\frac{R}{L}[t]\Big|_{0}^{t}$$

If we solve for the intervals,

$$\ln|I| - \ln\left|\frac{V_1}{R}\right| = -\frac{R}{L}[t-0]$$

$$\ln|I| - \ln\left|\frac{V_1}{R}\right| = -\frac{R}{L}t$$

MATH CONCEPT   The subtraction of two natural logarithms

$$\ln a - \ln b = \ln\left(\frac{a}{b}\right)$$

Thus,

$$\ln\left|\frac{I}{\frac{V_1}{R}}\right| = -\frac{R}{L}t$$

$$\ln\left|\frac{RI}{V_1}\right| = -\frac{R}{L}t$$

We can get rid of the natural log by applying the following math concept.

MATH CONCEPT   The exponential of the natural log

$$e^{\ln(a)} = a$$

Therefore, we apply the exponential to both sides:

$$e^{\ln\left|\frac{RI}{V_1}\right|} = e^{-\frac{R}{L}t}$$

$$\frac{RI}{V_1} = e^{-\frac{R}{L}t}$$

We get the final equation for the current across the inductor during the de-energizing phase.

CURRENT ACROSS INDUCTOR – DE-ENERGIZING PHASE

$$i(t) = \frac{V_1}{R}e^{-\frac{R}{L}t} \tag{26.3}$$

i(t) is the current equation as a function of time, in Amperes.
V_1 is the initial voltage feeding the inductor, in Volts.
t is the instant in time, in seconds.
R is the resistance, in Ohms.
L is the inductance, in Henries.

26.3.2 Voltage During De-energizing

Voltage across an inductor is defined as

$$v(t) = L\frac{di}{dt}$$

Like the energizing case, this equation tells us that voltage is the inductance multiplied by the derivative of the current with respect to time.

Therefore, we can take the derivative of Eq. (26.3) with respect to time to obtain the current equation:

$$v(t) = L\frac{d}{dt}\left[\frac{V_1}{R}e^{-\frac{R}{L}t}\right]$$

$$v(t) = L\frac{V_1}{R}\frac{d}{dt}\left[e^{-\frac{R}{L}t}\right]$$

MATH CONCEPT   The derivative of an exponential

$$\frac{d}{dt}e^{xt} = xe^{xt}$$

Therefore,

$$v(t) = L\frac{V_1}{R}\left[-\frac{R}{L}e^{-\frac{R}{L}t}\right]$$

$$v(t) = -L\frac{V_1 R}{R L}\left[e^{-\frac{R}{L}t}\right]$$

Simplifying,

$$v(t) = -\cancel{L}\frac{V_1 \cancel{R}}{R \cancel{L}}\left[e^{-\frac{R}{L}t}\right]$$

We get the current equation for the inductor during the de-energizing phase.

VOLTAGE ACROSS AN INDUCTOR – DE-ENERGIZING PHASE

$$v(t) = -V_1 e^{-\frac{R}{L}t}$$

v(t) is the equation for the voltage across the inductor as a function of time, in Volts.
V_1 is the voltage applied to the inductor, in Volts.
t is the instant in time, when energizing starts, in seconds.
R is the resistance, in Ohms.
L is the inductance, in Henries.

26.3.2.1   RL Time Constant

The RL time constant is the amount of time that it takes to conduct 63.2% of the current that results from a voltage applied across an energizing inductor or to reach 36.8% of its initial current for a de-energizing inductor.

This value is derived from the mathematical formulas of charge and discharge for the inductor.

The RL time constant, also called tau (τ) and measured in seconds, is equal to the inductance in Henries, divided by the resistance in Ohms.

RL TIME CONSTANT

$$\tau = \frac{L}{R}$$

τ is the RL time constant, in seconds.
R is the resistance, in Ohms.
L is the inductance, in Henries.

26.3.2.2   Transient and Steady States

RC circuits exhibit a transient phase when the voltage changes, because capacitors hate sudden changes in voltage and the circuit takes a longer time to stabilize to the new voltage applied.

RL circuits exhibit the same transient phase when current changes, because in this case inductors hate sudden changes in current and the circuit will take a while to stabilize to the new current applied.

In electrical engineering, an RL series circuit is considered stabilized and enters the steady-state phase, after a time equal to five time constants.

RL TIME CONSTANT – STEADY STATE

$$t \geq 5\tau \rightarrow t \geq 5 \times \frac{L}{R}$$

τ is the RL time constant, in seconds.
R is the resistance, in Ohms.
L is the inductance, in Henries.

This happens because after each time constant, the current raises or decays, exponentially, according to the following table.

RL time constant interval (τ)	Percentage of total voltage (%)
1	62.3
2	86.5
3	95.0
4	98.2
5	99.3

The table shows that after five time constants, the current of an energizing inductor will be 99.3% of the final current or that the current of a de-energizing inductor will have decayed by the same amount, which is reason why five time constants is the number chosen to represent a fully energized/de-energized inductor.

26.4　Examples

26.4.1　Example 1

Consider the circuit shown in Figure 26.3.

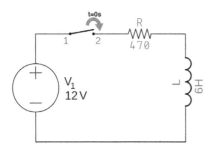

Figure 26.3 RL circuit initially off.

In the circuit in Figure 26.3, the switch has been opened for a long time. There is no current flowing and the inductor is completely de-energized.

At time t = 0 s, the switch is closed, and current flows and reaches the inductor that starts to energize.

We want to find the current flowing in the circuit 10ms after the switch is closed and after the circuit stabilizes.

26.4.1.1 Current After 10 ms

The current at any time across an inductor, during energizing, can be found by using the following formula:

$$i(t) = \frac{V_1}{R}\left(1 - e^{-\frac{R}{L}t}\right)$$

$$i(10\,\text{ms}) = \frac{12}{470}\left(1 - e^{-\frac{470}{6}(0.01)}\right)$$

FINAL RESULT

$i(10\,\text{ms}) = 13.8668\ \text{mA}$

26.4.1.2 Final Current

At the time the switch is closed, no current flows across the inductor.

As time passes, current increases until it reaches the maximum current possible. At that time, inductors behave like a wire with zero or near-zero resistance.

Therefore, the maximum current will be only limited by the resistor and by the first Ohm's law, where

$$i(t) = \frac{V}{R}$$

$$i(\infty) = \frac{12}{470}$$

FINAL RESULT

$i(\infty) = 25.5319\ \text{mA}$

The current at time equal to infinity means the current after the transient phase is over and the circuit stabilizes (steady state).

26.4.2 Example 2

Figure 26.4 shows a DC power supply, a resistor, an inductor, and a two-position switch.

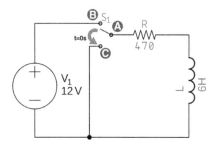

Figure 26.4 RL circuit initially working.

The switch has been closed for a long time connecting points A and B. Thus, R and L have been connected across a constant voltage source, V_1, for a long time. The circuit is on the steady-state phase, and L is completely charged.

At time t = 0 s, the switch is moved from position AB to AC, disconnecting R and L from the power supply and connecting them in parallel.

- What is the current flowing in the circuit before the switch is moved?
- What is the current flowing in the circuit as the switch is moved?
- What is the voltage across the inductor 5ms after the switch is moved?
- What is the current flowing in the circuit 5ms after the switch is moved?

26.4.2.1 Current Before the Switch Is Moved

The circuit was working for a long time before the switch is moved. Current was at its maximum, just limited by the first Ohm's law:

$$i(t) = \frac{V}{R}$$

$$i(0^-) = \frac{12}{470}$$

FINAL RESULT

$$i(0^-) = 25.5319 \text{ mA}$$

26.4.2.2 Current as the Switch Is Moved

The moment the switch moves from AB to AC, the resistor and the inductor are connected in parallel.

At that time, the inductor tries to prevent the current from vanishing by creating an electromotive force to push the current forward and keep it flowing.

Therefore, the current at the time the switch is moved is still the same as before.

FINAL RESULT

$$i(0^+) = 25.5319 \text{ ma}$$

26.4.2.3 Voltage After 5 ms

Voltage across an inductor at any time during the de-energizing phase can be found by using the following formula:

$$v(t) = -V_1 e^{-\frac{R}{L}t}$$

$$v(5 \text{ ms}) = -12 e^{-\frac{470}{6}(0.005)}$$

FINAL RESULT

$$v(5 \text{ ms}) = -8.1111 \text{ V}$$

26.4.2.4 Current After 5 ms

Current across an inductor at any time during the de-energizing phase can be found by using the following formula:

$$i(t) = \frac{V_1}{R} e^{-\frac{R}{L}t}$$

$$i(5 \text{ ms}) = \frac{12}{470} e^{-\frac{470}{6}(0.005)}$$

FINAL RESULT

$$i(5 \text{ ms}) = 17.2577 \text{ mA}$$

26.4.3   Example 3

Figure 26.5 shows a circuit with an alternating current power supply, a resistor, an inductor, and a switch permanently closed. The circuit is working like this for a long time.

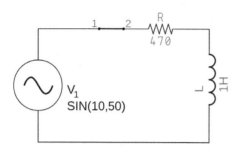

Figure 26.5 AC powered RL circuit.

The power supply has an amplitude of 10 V and a frequency of 50 Hz. This is what we want to know about the circuit:

- The circuit's impedance and the phase angle.
- The root mean square (RMS) current flowing in the circuit.
- The circuit's power factor.
- The apparent, the real, and the reactive power.

26.4.3.1   The Circuit's Impedance

The impedance of a resistor (Z_R) is its own value of resistance (R), because resistors have no reactance.

Thus,

$$Z_R = R$$

For a series RL circuit, the total impedance will be the resistor's impedance added to the inductor's impedance, or

$$Z = Z_R + Z_L$$

The inductor's impedance is related to its reactance by

$$Z_L = jX_L$$

and its reactance is

$$X_L = 2\pi fL$$

$$X_L = 2\pi(50)(1)$$

$$X_c = 100\,\pi\,\Omega$$

Therefore, the total impedance is

$$Z = R + jX_L$$

CIRCUIT'S IMPEDANCE

$$Z = 470 + j(100\,\pi)$$

If we convert this into phasor notation, we have

$$|Z| = \sqrt{470^2 + (100\pi)^2}$$

$$\theta = A\tan\left(\frac{100\pi}{470}\right)$$

CIRCUIT'S IMPEDANCE – PHASOR NOTATION

$$Z = 565.33\ \Omega \angle 33.76°$$

See Appendix D for more information about complex number transformations.

26.4.3.2   RMS Current

V_{RMS} can be calculated by using the following formula:

$$V_{RMS} = \frac{V_{MAX}}{\sqrt{2}}$$

By plugging in the power supply maximum voltage V_{MAX} of 10 V, we get

$$V_{RMS} = \frac{10}{\sqrt{2}}$$

$$V_{RMS} = 7.071 \text{ V}$$

The magnitude of impedance, already calculated as 565.33 Ω, can be used to find the RMS current:

$$I_{RMS} = \frac{V_{RMS}}{Z}$$

$$I_{RMS} = \frac{7.071}{565.33}$$

FINAL RESULT

$$I_{RMS} = 0.0125 = 12.50 \text{ mA}$$

26.4.3.3 Power Factor

The power factor of a series RL circuit powered by alternating current is equal to the resistance divided by the impedance magnitude.

POWER FACTOR – AC POWERED RL SERIIES CIRCUIT

$$\cos(\varphi) \rightarrow F_p = \frac{R}{Z}$$

$\cos(\varphi), F_p$ is the power factor.
R is the resistance, in Ohms.
Z is the impedance module, in Ohms.

By plugging in

$$\begin{cases} Z = 565.33 \ \Omega \\ R = 470 \ \Omega \end{cases}$$

we get

$$\cos(\varphi) \rightarrow F_p = \frac{470}{565.33}$$

POWER FACTOR

$$\cos(\varphi) \rightarrow F_p = 0.8314$$

26.4.3.4 The Apparent, the Real, and the Reactive Power

The apparent, the real, and the reactive powers can be calculated by the following formulas.

REAL POWER – CIRCUIT POWERED BY AC

$$P = V_{RMS} I_{RMS} \cos(\varphi)$$

P is the real or active power, in Watts.
V_{RMS} is the power supply RMS voltage, in Volts.
I_{RMS} is the power supply RMS current, in Amperes.
$\cos(\varphi)$ is the power factor.

REACTIVE POWER – CIRCUIT POWERED BY AC

$$Q = V_{RMS} I_{RMS} \sin(\varphi)$$

Q is the reactive power, in VAR.
V_{RMS} is power supply RMS voltage, in Volts.
I_{RMS} is power supply RMS current, in Amperes.
$\sin(\varphi)$ is the sine of the phase angle.

APPARENT POWER – CIRCUIT POWERED BY AC

$$S = \sqrt{P^2 + Q^2}$$

S is the apparent power, in VA.
P is the real power, in Watts.
Q is the reactive power, in VAR.

Real power is then calculated by plugging in the known values for the given circuit:

$$P = V_{RMS} I_{RMS} \cos(\varphi)$$

$$P = (7.071)(0.0125)(0.8314)$$

REAL POWER

$$P = 73.4861 \, \text{mW}$$

For the reactive power,

$$Q = V_{RMS} I_{RMS} \sin(\varphi)$$

$$Q = (7.071)(0.0125)(\text{Sin}(33.76))$$

FINAL RESULT

$$Q = 0.0491 \, \text{VAR}$$

The apparent power can be calculated as

$$S = \sqrt{P^2 + Q^2}$$

$$S = \sqrt{0.0735^2 + 0.0491^2}$$

FINAL RESULT

$$S = 0.0884 \, \text{VA}$$

Exercises

1 Consider the circuit shown in Figure 26.6, composed of a resistor, an inductor, a voltage source, and a switch, in series.

The switch has been opened for a long time. There is no current flowing and L is completely de-energized.

At time t = 0 s, the switch is closed, and current flows and reaches the inductor that starts to energize.

The components in the circuit are

Figure 26.6 RL circuit initially off.

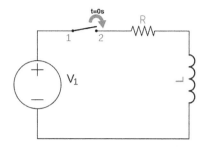

$$\begin{cases} R = 100\ \Omega \\ L = 2\ H \\ V_1 = 12\ V \end{cases}$$

The following is what we want to find:

- The current flowing in the circuit before the switch is closed.
- The current flowing in the circuit as soon as the switch is closed.
- The current flowing in the circuit 5 ms after the switch is closed.
- The voltage across the inductor 3 ms after the switch is closed.

2   Figure 26.7 shows a DC power supply, a resistor, an inductor, and a two-position switch.

Figure 26.7 RL circuit initially off.

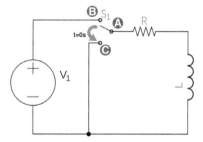

The switch has been closed for a long time, connecting points A and B. For that reason, the resistor and the inductor have been connected to the power supply for a long time, and the inductor is fully energized.

The components in the circuit are

$$\begin{cases} R = 270\ \Omega \\ L = 3\ H \\ V_1 = 20\ V \end{cases}$$

At time t = 0 s, the switch is moved from position AB to AC, putting the resistor and the inductor in parallel.

The following is what we want to find:

- The current flowing in the circuit before the switch is closed.
- The current flowing in the circuit as soon as the switch is closed.
- The current flowing in the circuit 4 ms after the switch is closed.
- The voltage across the inductor 5 ms after the switch is closed.

3 Figure 26.8 shows a circuit with two DC power supplies, two resistors, an inductor, and a switch.

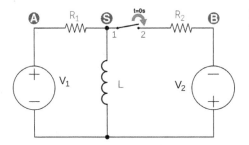

Figure 26.8 RL circuit with double voltage sources.

V_1, R_1, and L make the main circuit. This part of the circuit is working for a long time like this. For that reason, the inductor is fully energized. Notice that V_2 has the polarity inverted.

The components in the circuit are

$$\begin{cases} R_1 = 200 \ \Omega \\ R_2 = 50 \ \Omega \\ L = 2 \ H \\ V_1 = 20 \ V \\ V_2 = 15 \ V \end{cases}$$

At time t = 0 s, the switch is closed, connecting the second circuit to the first.

The following is what we want to find:

- What is the voltage across the inductor after the circuit stabilizes?
- What is the voltage across the inductor 3 ms after the switch is closed?

4 Figure 26.9 shows a circuit containing an alternating current power supply, a resistor, an inductor, and a switch in series. The switch is permanently closed. The circuit is working like this for a long time.

The components in the circuit are

$$\begin{cases} V_1 (\text{amplitude}) = 24 \ V \\ V_1 (\text{frequency}) = 50 \ Hz \\ R = 120 \ \Omega \\ L = 3 \ H \end{cases}$$

Figure 26.9 AC powered RL circuit.

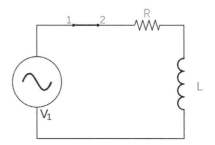

The following is what we want to find:

- What are the circuit's impedance and its phase angle?
- What is the RMS current flowing in the circuit?
- What is the circuit's power factor?
- What are the apparent, the real, and the reactive power?

Solutions

1 **The current flowing in the circuit before the switch is closed.**
Before the switch is closed, there is no current flowing in the circuit, because the switch is open and there is no circuit continuity.

FINAL ANSWER

$i(0^-) = 0$ mA

The current flowing in the circuit as soon as the switch is closed.
The moment the switch is closed, current reaches the inductor. The inductor sees that as a change in current, from 0 to a new value. Inductors hate current variations with passion and will prevent that current from flowing.
Therefore, the moment the switch is closed, current will be 0.

FINAL ANSWER

$i(0^+) = 0$ mA

The current flowing in the circuit 5ms after the switch is closed.
Current across an inductor, at any time, can be found by using the following formula:

$$i(t) = \frac{V_1}{R}\left(1 - e^{-\frac{R}{L}t}\right)$$

By plugging in the known values we have,

$$\begin{cases} R = 100\ \Omega \\ L = 2\,H \\ V_1 = 12\,V \\ t = 5\ ms \end{cases}$$

we get

$$i(5\,ms) = \frac{12}{100}\left(1 - e^{-\frac{100}{2}(5 \times 10^{-3})}\right)$$

FINAL RESULT

$$i(5\,ms) = 26.5439\,mA$$

The voltage across the inductor 3 ms after the switch is closed.
Voltage across an inductor, at any given time, can be found by using the following formula:

$$V_L(t) = V_1(t)e^{-\frac{R}{L}t}$$

If we plug in the values we have

$$\begin{cases} R = 100\ \Omega \\ L = 2\,H \\ V_1 = 12\ V \end{cases}$$

we get

$$V_L(3\,ms) = 12e^{-\frac{100}{2}(3 \times 10^{-3})}$$

FINAL RESULT

$$V_C(3\,ms) = 10.3284\ V$$

2   The current flowing in the circuit before the switch is closed.
The main part of the circuit was working for a long time before the switch is moved. The inductor was fully energized and current was flowing at maximum intensity, just limited by the first Ohm's law, as

$$i(t) = \frac{V}{R}$$

$$i(0^-) = \frac{20}{270}$$

FINAL RESULT

$i(0^-) = 74.074 \text{ mA}$

The current flowing in the circuit as soon as the switch is closed.
The moment the switch moves from AB to AC, it puts the inductor, fully energized, in parallel with the resistor.
At that time, the inductor sees the current ceasing to exist and unhappy with this fact and tries to use its stored magnetic energy to push current forward and keep it flowing.
Therefore, the moment the switch is closed, the current is still the same as before.

FINAL RESULT

$i(0^+) = 74.074 \text{ mA}$

The current flowing in the circuit 4 ms after the switch is closed.
The current in an inductor, at any given time after that, can be found by using the following formula:

$$i(t) = \frac{V_1}{R} e^{-\frac{R}{L}t}$$

and by plugging in the values we have for the given circuit,

$$\begin{cases} R = 270 \ \Omega \\ L = 3 \ H \\ V_1 = 20 \ V \\ t = 4 \ ms \end{cases}$$

we get

$$i(4\,ms) = \frac{20}{270}e^{-\frac{270}{3}(4 \times 10^{-3})}$$

FINAL RESULT

$i(4\,ms) = 51.6793 \text{ mA}$

The voltage across the inductor 5 ms after the switch is closed.

The moment the switch is closed and the current ceases to exist, the inductor uses its energy stored as a magnetic field and creates an electromotive force to push the current to continue flowing. This electromotive force manifests itself by creating a voltage across the inductor with the same value as the power supply but with reversed polarity, or for the given example, –20 V.

This value will decrease with time because the resistor will reduce the current gradually.

The voltage across the inductor at any time can be found by using the following formula:

$$V_L(t) = -V_1 e^{-\frac{R}{L}t}$$

and by plugging in the values we have,

$$\begin{cases} R = 270 \ \Omega \\ L = 3\,H \\ V_1 = 20\,V \\ t = 5\,ms \end{cases}$$

we get

$$V_L(5\,ms) = -20e^{-\frac{270}{3}(5 \times 10^{-3})}$$

FINAL RESULT

$V_L(5\,ms) = -12.7525 \text{ V}$

---

3 What is the voltage across the inductor after the circuit stabilizes?
In the following circuit, V_1, R_1, and C make the main circuit. This part of the circuit was working for a long time, L was fully energized, and the current was flowing at the maximum intensity.

When an ideal inductor is fully charged, it behaves like a wire with zero resistance. Thus, the voltage across the inductor was 0.

As the switch is closed, components R_1/V_1 and L are put in parallel with R_2/V_2.

The new components in parallel will make the circuit pass through another transient phase, but after that phase is over and the circuit enters on steady state, the voltage across the inductor will be 0 again.

FINAL RESULT

$$V_L(\infty) = 0\,V$$

What is the voltage across the inductor 3 ms after the switch is closed?
Before the switch is closed, the inductor is fully energized and the voltage across it is 0.

After the switch is closed, R_2/V_2 is connected in parallel with the circuit and a certain voltage will appear across point S. This is the energizing voltage for the inductor during the transient phase.

We know that before the switch is moved, current was flowing at its maximum, just limited by the first Ohm's law, or

$$i(t) = \frac{V_1}{R_1}$$

$$i(0^-) = \frac{20}{200}$$

$$i(0^-) = 0.1\,A$$

This current is the one flowing across the inductor before the switch is closed; therefore, we can substitute the inductor with a current source of that intensity and redraw the circuit to be like the one shown in Figure 26.10 to illustrate that point in time.

The current source points down because current flows from V_1, crossing R_1 from A to S to ground.

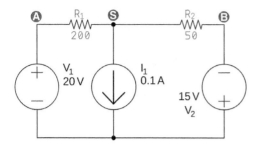

Figure 26.10 Inductor as a current source.

We can use the source transformation techniques to convert V_1/R_1 into a current source in parallel with the same resistor.

In that case, the current source will have an intensity equal to V_1 divided by R_1, that is,

$$I_3 = \frac{V_1}{R_1}$$

$$I_3 = \frac{20}{200}$$

$$I_3 = 0.1\,A$$

The result is the circuit shown in Figure 26.11.

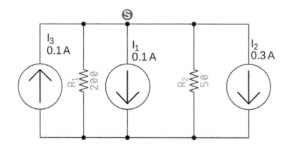

Figure 26.11 Inductor as a current source.

We can now replace all resistors in parallel with their equivalent:

$$\frac{1}{R} = \frac{1}{R_1} + \frac{1}{R_2}$$

$$\frac{1}{R} = \frac{1}{200} + \frac{1}{50}$$

$$R = 40 \, \Omega$$

and all current sources with their sum:

$$I = I_1 + I_2 + I_3$$

$$I = -0.1 - 0.3 + 0.1$$

$$I = -0.3 \, A$$

We end with a current source of -0.3 A in parallel with a resistor of 40 Ω, which can be converted into a voltage source of -12 V (-0.3 A $\times$ 40 Ω) in series with the same resistor.

Therefore, -12 V is the energizing voltage for the inductor, in series with 40 Ω resistance, during the transient phase.

We can now apply the inductor formula and find the voltage across the inductor after 3 ms:

$$V_L(t) = V_1(t)e^{-\frac{R}{L}t}$$

$$V_L(3\,ms) = -12e^{-\frac{40}{7}(3 \times 10^{-3})}$$

FINAL RESULT

$$V_L(3\,ms) = -11.3011 \, V$$

4 **What are the circuit's impedance and its phase angle?**
The impedance of a resistor (Z_R) is its own value of resistance (R), because resistors have no reactance.
Thus,

$$Z_R = R$$

For a series RL circuit, the total impedance will be the resistor's impedance added to the inductor's impedance, or

$$Z = Z_R + Z_L$$

Inductor's impedance is related to its reactance by

$$Z_C = jX_L$$

where X_L is the capacitor's reactance and j is the impedance's imaginary part. Capacitor's reactance can be found by plugging in the known values (f = 50 Hz and L = 3 H) in the following formula:

$$X_L = 2\pi fL$$

$$X_L = 2\pi(50)(3)$$

$$X_L = 942.4777 \ \Omega$$

Therefore its impedance is

$$Z_L = jX_L$$

$$Z_L = j942.4777 \ \Omega$$

and the circuit's total impedance is

$$Z = R + jX_L$$

FINAL RESULT

$$Z = 120 + j942.4777 \ \Omega$$

This impedance can be expressed in terms of phasors by calculating its magnitude as

$$Z = \sqrt{R^2 + Z_C^2}$$

$$Z = \sqrt{120^2 + 942.4777^2}$$

$$Z = 950.0865 \ \Omega$$

and the phase angle as

$$\tan(\varphi) = \frac{Z_L}{R}$$

$$\tan(\varphi) = \frac{942.4777}{120} = 7.8539$$

$$\varphi = A\tan(7.8539)$$

$$\varphi = 82.7439°$$

FINAL RESULT

$$Z = 950.0865 \ \Omega \angle 82.7439°$$

What is the RMS current flowing in the circuit?

V_{RMS} can be calculated by using the following formula:

$$V_{RMS} = \frac{V_{MAX}}{\sqrt{2}}$$

By plugging in the power supply maximum voltage V_{MAX} of 10 V, we get

$$V_{RMS} = \frac{24}{\sqrt{2}}$$

$$V_{RMS} = 16.9705 \ V$$

The magnitude of impedance, already calculated as 5.112 Ω, can be used to find the RMS current:

$$I_{RMS} = \frac{V_{RMS}}{Z}$$

$$I_{RMS} = \frac{16.9705}{950.0865}$$

FINAL RESULT

$$I_{RMS} = 17.8621 \ mA$$

What is the circuit's power factor?

The power factor of a series RL circuit powered by alternating current is equal to the resistance divided by the impedance magnitude.

POWER FACTOR – AC POWERED RL SERIES CIRCUIT

$$\cos(\varphi) \rightarrow F_p = \frac{R}{Z}$$

$\cos(\varphi), F_p$ is the power factor.
R is the resistance, in Ohms.
Z is the impedance module, in Ohms.

By plugging in

$$\begin{cases} Z = 950.0865 \ \Omega \\ R = 120 \ \Omega \end{cases}$$

we get

$$\cos(\varphi) \rightarrow F_p = \frac{120}{950.0865}$$

POWER FACTOR

$$\cos(\varphi) \rightarrow F_p = 0.1263$$

What are the apparent, the real, and the reactive power?

The real power can be calculated by the following formula:

$$P = V_{RMS} \, I_{RMS} \cos(\varphi)$$

$$P = 16.9705 \times 0.0178621 \times 0.1263$$

FINAL RESULT

$$P = 38.2867 \, mW$$

The reactive power is

$$Q = V_{RMS} \, I_{RMS} \sin(\varphi)$$

$$Q = 16.9705 \times 0.0178621 \times \sin(82.7439)$$

FINAL RESULT

$Q = 0.3007\,\text{VAR}$

The apparent power is

$$S = \sqrt{P^2 + Q^2}$$

$S = \sqrt{0.0382867^2 + 0.3007^2}$

FINAL RESULT

$S = 0.3031\ \text{VA}$

27

RLC Circuits: Part 1

Voltage Analysis in Circuits Containing Resistors, Capacitors, and Inductors in Series

27.1 Introduction

Previously in this book, we have examined R_C and R_L series circuits and how capacitors and inductors react when subjected to voltage or current changes.

We saw that capacitors and inductors can store potential energy in the form of electric or magnetic fields.

R_C and R_L circuits are circuits that contain one element that can store energy, and for that reason, all equations we have found are first-order differential equations.

However, RLC circuits contain two elements that can store energy, and at this point, the readers have realized that these circuits will produce second-order differential equations.

Due to the complexity of this theme, we will divide the analysis of RLC series circuits into two parts. In this chapter we will make a voltage analysis and in the next chapter a current analysis.

27.2 A Basic RLC Series Circuit

Figure 27.1 shows a circuit with a voltage source, a switch, a resistor, an inductor, and a capacitor in series.

Figure 27.1 A basic RLC circuit.

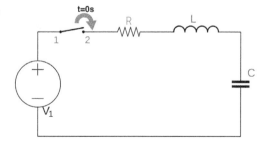

Introductory Electrical Engineering with Math Explained in Accessible Language, First Edition. Magno Urbano.
© 2020 John Wiley & Sons, Inc. Published 2020 by John Wiley & Sons, Inc.

The switch is open, there is no current flowing in the circuit, the inductor is completely de-energized, and the capacitor is completely discharged.

At time t = 0 s, the switch is closed, and current flows and reaches the inductor.

The current that was zero before has now a nonzero value. The inductor hates this current variation and tries to block its flow by generating an opposite electromotive force. This electromotive force manifests itself by making a voltage rise across the inductor with the same value as the power supply but with the inverted polarity.

Therefore, current does not reach the capacitor at this first moment, and the inductor is seen, by the circuit, as an open circuit or a resistor with infinite resistance.

As time passes, current starts to flow through the inductor and reaches the capacitor. Initially the capacitor is completely discharged and will behave like a short circuit, letting all current flow without any resistance. Voltage across the capacitor is 0.

Time passes and the inductor starts to store energy in the form of magnetic field and, at the same time, the capacitor is charging.

As the capacitor reaches its maximum charge, it starts to prevent current from flowing by generating an electromotive force that starts to reduce current gradually.

The inductor sees this current reduction and is trying to prevent that to happen; by using the energy, it has stored in the form of magnetic field to create an electromotive force to push current forward and keep it flowing.

What happens now depends on the values of the capacitor and the inductor, as we will see in the following chapters of this book, but one option is that the inductor will try to force the current to keep flowing against the capacitor's will.

The process will remove charges from the capacitor's plate that was already filled with charges to the other, forcing this component to begin charging with the opposite polarity.

When the capacitor's charge reaches the point where its force is bigger than the inductor's force, the process will reverse: the capacitor will start to discharge through the inductor, forcing current on the opposite direction, and that current will energize the inductor.

The cycle repeats, with one component transferring energy to the other until all energy is gone.

This energy exchange is seen as an oscillation.

The kind of oscillation we just described can be illustrated in terms of voltage variations across the capacitor by the graphics in Figure 27.2. Notice that the voltage oscillates around the power supply voltage V_1, which will be the final capacitor's voltage.

Figure 27.2 A basic RLC circuit.

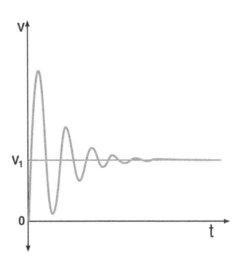

Notice that the voltage rises above the power supply value and oscillates around before stabilizing.

The voltage decays or rises exponentially and has a direct relation to the values of capacitance and inductance, as we will see forward in this book.

The frequency the oscillations happen is called the angular frequency or resonant frequency, also known as the natural resonant frequency, and can be found by the following formula.

ANGULAR FREQUENCY – SERIES RLC CIRCUIT

$$\omega_o = \frac{1}{\sqrt{LC}} \qquad (27.1)$$

ω_o is the natural resonant frequency, measured in radians per second.
L is the inductance, in Henries.
C is the capacitance, in Farads.

Because the angular frequency is related to the frequency by

$$\omega = 2\pi f$$

we can find the circuit's oscillation frequency by using the following formula.

FREQUENCY – SERIES RLC CIRCUIT

$$f = \frac{\omega_o}{2\pi} \qquad (27.2)$$

f is the frequency, in Hertz.
ω_o is the resonant frequency, in radians per second.

As the oscillations happen, they are damped. The much they are dumped are ruled by the dumping factor, calculated by the following formula.

DAMPING COEFFICIENT – SERIES RLC CIRCUIT

$$\alpha = \frac{R}{2L}$$

α is the damping coefficient.
R is the resistance, in Ohms.
L is the inductance, in Henries.

> The results coming from these frequency formulas are approximations, because the damping coefficient interferes in the wave's period. It is like how the frequency reduces slightly as the volume decreases, after hitting a bell with a hammer.

27.2.1 Circuit Analysis

To understand what happens on a series RLC circuit, we must examine the circuit behavior in terms of voltage across the capacitor.

The circuit is the one shown in Figure 27.1 that we repeat again in Figure 27.3 for convenience.

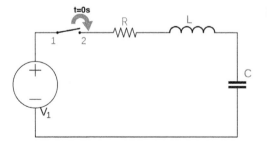

Figure 27.3 A basic RLC circuit.

The switch is open, there is no current flowing in the circuit, the inductor is completely de-energized, and the capacitor is completely discharged.

At time t = 0 s, the switch is closed, and current flows and reaches the inductor.

27.2.2 Voltage Across the Capacitor

If we analyze a series RLC circuit, we get that the power supply voltage is equal to the sum of the voltage drop across the resistor, the voltage across the inductor, and the voltage across the capacitor, or mathematically

$$V_1(t) = V_R(t) + V_L(t) + V_C(t)$$

We know the voltage across the inductor is

$$V_L(t) = L\frac{di}{dt}$$

and the voltage drop across the resistor is

$$V_R(t) = Ri$$

Therefore, the previous equation changes to

$$V_1(t) = Ri + L\frac{di}{dt} + V_C(t) \qquad (27.3)$$

However, we also know that the current across the capacitor is defined as

$$i(t) = C\frac{dV_C(t)}{dt}$$

and that this current, in the example of an RLC circuit in series, is the same current that flows across the circuit.

Therefore, we can substitute i(t) in (27.3) and obtain

$$V_1 = R\left(C\frac{dV_C(t)}{dt}\right) + L\frac{d}{dt}\left(C\frac{dV_C(t)}{dt}\right) + V_C(t)$$

$$V_1 = RC\left(\frac{dV_C(t)}{dt}\right) + LC\left(\frac{d^2V_C(t)}{dt^2}\right) + V_C(t)$$

$$V_1 = LC\frac{d^2V_C(t)}{dt^2} + RC\frac{dV_C(t)}{dt} + V_C(t)$$

We divide everything by LC to write the equation using the standard form as

$$\frac{V_1}{LC} = \frac{d^2V_C(t)}{dt^2} + \frac{R}{L}\frac{dV_C(t)}{dt} + \frac{1}{LC}V_C(t) \tag{27.4}$$

We see, by the squared term in the Eq. (27.4), that this is a second-order differential equation, like we have predicted before.

To solve this equation, we must have a few mathematical concepts in mind.

A Nonhomogeneous Second-Order Differential Equation
That follows the form

$$\frac{d^2f(x)}{dx^2} + a\frac{df(x)}{dx} + bf(x) = g(x)$$

will have a solution formed by two equations. The first one, called homogeneous equation, provides a solution for the transient state, and the second one, called "particular equation," provides a solution for the steady state.

For the case of RLC circuits, we can express the total solution as

$$h(x) = h_{tr}(x) + h_{ss}(x)$$

where $h_{ss}(x)$ is any function that provides a solution for the complete differential equation and $h_{tr}(x)$ is any function that provides a solution for the homogeneous differential equation in the form

$$\frac{d^2f(x)}{dx^2} + a\frac{df(x)}{dx} + bf(x) = 0$$

This later equation has two coefficients, *a* and **b**, that can be found by applying the initial conditions for the circuit.

27.2.3 Back to the Equation

Applying this mathematical concept for the series RLC circuit, we see that the voltage across the capacitor will have two equations: one representing the transient state and another representing the steady state, or

$$V_C(t) = V_{tr}(t) + V_{ss}(t)$$

$V_C(t)$ is the equation for the voltage across the capacitor as a function of time, in Volts.

$V_{tr}(t)$ is the equation for the transient voltage across the capacitor as a function of time, in Volts.

$V_{ss}(t)$ is the equation for the steady-state voltage across the capacitor as a function of time, in Volts.

27.2.3.1 Steady-State Solution

After an infinite time has passed, the voltage across the capacitor will stabilize in a value that we know will be the same as the power supply voltage, but, to confirm this fact, let us work with the equations.

To solve the differential equation in (27.4) for the steady-state solution, consider that

$$V_C(t) = V_{ss}$$

Therefore, the Eq. (27.4) can be written as

$$\frac{V_1}{LC} = \frac{d^2 V_C(t)}{dt^2} + \frac{R}{L}\frac{dV_C(t)}{dt} + \frac{1}{LC}V_C(t)$$

$$\frac{V_1}{LC} = \frac{d^2 V_{ss}}{dt^2} + \frac{R}{L}\frac{dV_{ss}}{dt} + \frac{1}{LC}V_{ss}$$

We know that V_{ss} is a constant value, because this voltage is equal to the constant power supply voltage. Therefore, we can apply the following concept.

> **MATH CONCEPT**   The derivative of a constant is equal to 0.

We can simplify the equation

$$\frac{V_1}{LC} = \frac{d^2 \cancel{V_{ss}}^{\,0}}{\cancel{dt^2}} + \frac{R}{L}\frac{d\cancel{V_{ss}}^{\,0}}{\cancel{dt}} + \frac{1}{LC}V_{ss}$$

$$\frac{V_1}{LC} = \frac{1}{LC}V_{ss}$$

$$V_1 = V_{ss}$$

This result shows that V_{ss} will be equal to V_1, as expected.

27.2.4 Transient Solution

To find the transient solution, we must solve the equation that represents that state, described as follows:

$$\frac{d^2V_{tr}(t)}{dt^2} + \frac{R}{L}\frac{dV_{tr}(t)}{dt} + \frac{1}{LC}V_{tr}(t) = 0$$

This equation can be written in a simplified way:

$$y'' + \frac{R}{L}y' + \frac{1}{LC}y = 0$$

which is a homogeneous differential equation with constant coefficients.

To understand how to solve this kind of equation, we must understand the following mathematical concept.

The first thing to do to solve a second-order differential homogeneous equation in the form

$$Ay'' + By' + Cy = 0$$

is to find the equation represented by y.

If you compare the three terms, you see that the difference between them is only the coefficient, meaning that the equation y never changes.

Therefore, we need an equation that we can take a first and a second derivatives without changing the equation per se.

Mathematically, we know a function that does exactly this and the function is $y = Ke^{\alpha x}$.

The first derivative of this function is $y' = \alpha Ke^{\alpha x}$ and the second derivative is $y'' = \alpha^2 Ke^{\alpha x}$.

See that the base $Ke^{\alpha x}$ continues the same in all cases.

The equation $y = Ke^{-\alpha x}$ also satisfies the requirements.

In practice, the ideal equation will be $y = Ke^{\delta x}$, where $\delta = \pm \alpha$.

By applying this mathematical concept, we see that the equation to be adopted as a solution will have the generic form:

$$y = Ke^{\delta x}$$

By adapting this equation for the given example, we get the transient voltage equation having the form

$$V_{tr} = Ae^{mt}$$

where A is a coefficient, t is the time, and m is the equation's roots.

Therefore, we can convert the equation into

$$\frac{d^2 V_{tr}}{dt^2} + \frac{R}{L}\frac{dV_{tr}}{dt} + \frac{1}{LC}V_{tr} = 0$$

$$\frac{d^2}{dt^2}(Ae^{mt}) + \frac{R}{L}\frac{d}{dt}(Ae^{mt}) + \frac{1}{LC}(Ae^{mt}) = 0$$

We see a lot of terms that are derivatives of exponentials and we can apply the following concept.

MATH CONCEPT The derivative of an exponential

$$\frac{d}{dt}e^{xt} = xe^{xt}$$

Therefore, the terms and their results are

$$\frac{d}{dt}(Ae^{mt}) = m(Ae^{mt})$$

and

$$\frac{d^2}{dt^2}(Ae^{mt}) = m^2(Ae^{mt})$$

Therefore, by substituting these terms, we get

$$\frac{d^2}{dt^2}(Ae^{mt}) + \frac{R}{L}\frac{d}{dt}(Ae^{mt}) + \frac{1}{LC}(Ae^{mt}) = 0$$

$$m^2(Ae^{mt}) + \frac{R}{L}m(Ae^{mt}) + \frac{1}{LC}(Ae^{mt}) = 0$$

$$\left(m^2 + \frac{R}{L}m + \frac{1}{LC}\right)(Ae^{mt}) = 0$$

Looking at this equation, we see that it is formed by two parts,

$$\overbrace{\left(m^2 + \frac{R}{L}m + \frac{1}{LC}\right)}^{1}\overbrace{(Ae^{mt})}^{2} = 0$$

and that the whole equation is equal to 0. For this to be true, one of the parts (1 or 2) must be 0.

For the second term to be 0, or

$$Ae^{mt} = 0$$

the term m will have to be equal to −∞. If this is true, we get

$$Ae^{-\infty} = 0$$

$$Ae^{-\infty^1} = 0$$

$$A = 0$$

We end with A = 0, a situation that does not bring any useful solution for our equations.

Therefore, the first part must be the one equal to 0, or

$$m^2 + \frac{R}{L}m + \frac{1}{LC} = 0$$

27.2.4.1 The Roots of the Equation

The previous equation is a quadratic equation and we know how to solve that and obtain the roots of the equation by using the formula

$$\frac{-b \pm \sqrt{b^2 - 4ac}}{2a}$$

Therefore,

$$\frac{-\dfrac{R}{L} \pm \sqrt{\dfrac{R^2}{L^2} - \dfrac{4}{LC}}}{2}$$

or we can rewrite it as

$$-\frac{R}{2L} \pm \sqrt{\frac{R^2}{4L^2} - \frac{4}{4LC}}$$

$$-\frac{R}{2L} \pm \sqrt{\frac{R^2}{4L^2} - \frac{\cancel{4}}{\cancel{4}LC}}$$

$$-\frac{R}{2L} \pm \sqrt{\frac{R^2}{4L^2} - \frac{1}{LC}}$$

We can simplify this equation by substituting the values inside the radical with those of the angular frequency and damping defined before:

$$\omega_o = \frac{1}{\sqrt{LC}}$$

$$\alpha = \frac{R}{2L}$$

And we get the roots of the equation.

ROOTS – RLC SERIES CIRCUIT

$$-\alpha \pm \sqrt{\alpha^2 - \omega_0^2} \tag{27.5}$$

α is the damping coefficient.
ω_o is the resonant frequency.

> The damping coefficient will modify the circuit behavior. Depending on the case, it will prevent circuit oscillation.

Consequently, the roots defined by (27.5) will be two:

$$m_1 = -\alpha + \sqrt{\alpha^2 - \omega_0^2}$$

$$m_2 = -\alpha - \sqrt{\alpha^2 - \omega_0^2}$$

Looking at the term inside the radical, we see that we can have basically three kinds of results, depending on the term $\alpha^2 - \omega_0^2$, be positive, negative, or 0. In the next sections, we will examine each case.

27.2.4.2 Critically Damped Solution

This is the case where $\alpha = \omega_0$, or the damping factor is equal to the angular frequency and the result of $\alpha^2 - \omega_0^2$, inside the radical, is 0.

This case will give us just one root for the equation, which will be a negative number, as follows:

$$m = -\alpha \pm \overbrace{\sqrt{\alpha^2 - \omega_0^2}}^{0}$$

$$m = -\alpha$$

We know that the complete solution for the equation for voltage across the capacitor will be the sum of the transient solution with the steady-state solution, or

$$V_C(t) = V_{tr}(t) + V_{ss}(t)$$

We also know that the voltage across the capacitor when the circuit reaches the steady state will be the same as the power supply voltage, V_1.

Another thing we know is that the final equation for the transient solution must contain two terms in the form of $y = Ae^{mt}$, something like

$$V_{tr}(t) = Ae^{m_1 t} + Be^{m_2 t}$$

However, we see that the critically damped solution has only one root $m = -\alpha$, which will lead to a transient solution containing only one term, like

$$V_{tr}(t) = Ae^{mt}$$

This equation does not represent the truth. The real equation must contain two terms because the circuit has two initial conditions that must be taken into consideration: the initial voltage across the capacitor and the initial current flowing in the circuit the moment the switch is closed.

We must discard this equation and find a way to discover the correct one.

27.2.4.2.1 Going Back to the Original Equation

$$\frac{d^2V_{tr}(t)}{dt^2} + \frac{R}{L}\frac{dV_{tr}(t)}{dt} + \frac{1}{LC}V_{tr}(t) = 0 \qquad (27.6)$$

We know that

$$\alpha = \frac{R}{2L}$$

Therefore,

$$2\alpha = \frac{R}{L} \qquad (27.7)$$

We also know that

$$\omega_0 = \frac{1}{\sqrt{LC}}$$

Thus,

$$\frac{1}{LC} = \omega_0^2 \qquad (27.8)$$

If we substitute the condition of the critically damped solution, $\omega_0 = \alpha$, in (27.8), we get

$$\frac{1}{LC} = \alpha^2 \qquad (27.9)$$

If we now substitute (27.7) and (27.9) in (27.6), we obtain

$$\frac{d^2V_{tr}(t)}{dt^2} + 2\alpha\frac{dV_{tr}(t)}{dt} + \alpha^2V_{tr}(t) = 0$$

We can rewrite this equation as

$$\frac{d}{dt}\left[\frac{dV_{tr}(t)}{dt} + \alpha V_{tr}(t)\right] + \alpha\left[\frac{dV_{tr}(t)}{dt} + \alpha V_{tr}(t)\right] = 0 \qquad (27.10)$$

To simplify, let us create a new variable y:

$$y = \frac{dV_{tr}(t)}{dt} + \alpha V_{tr}(t) \qquad (27.11)$$

We can use this variable in (27.10) and simplify that equation to

$$\frac{dy}{dt} + \alpha y = 0 \qquad (27.12)$$

Like we have examined before, the solution for the Eq. (27.11) is known to be

$$y = Ae^{mt}$$

Therefore, we can convert Eq. (27.11) into

$$Ae^{mt} = \frac{dV_{tr}(t)}{dt} + \alpha V_{tr}(t)$$

We can manipulate this equation to get A:

$$A = \frac{1}{e^{mt}}\frac{dV_{tr}(t)}{dt} + \frac{1}{e^{mt}}\alpha V_{tr}(t)$$

$$A = e^{-mt}\frac{dV_{tr}(t)}{dt} + e^{-mt}\alpha V_{tr}(t) \qquad (27.13)$$

It may not appear now but the Eq. (27.13) is equal to the following derivative:

$$A = \frac{d}{dt}[e^{-mt}V_{tr}(t)]$$

due to the product rule from calculus.
The following is the mathematical concept.

> **MATH CONCEPT** The derivative of a product of two functions is equal to the derivative of the first function multiplied by the second function added to the first function multiplied by the derivative of the second, or
>
> $$(u.v)' = u'v + uv'$$

By applying this rule, the derivative

$$\frac{d}{dt}[e^{-mt}V_{tr}(t)]$$

will be equal to

$$\frac{d}{dt}[e^{-mt}]V_{tr}(t) + e^{-mt}\frac{d}{dt}[V_{tr}(t)]$$

which is equal to

$$e^{-mt}\frac{dV_{tr}(t)}{dt} + e^{-mt}(-m)V_{tr}(t) \tag{27.14}$$

Compare (27.13) with (27.14). They are the same equation. See how in (27.13) the second term coefficient is equal to α and in (27.14) the same coefficient is equal to $-m$, which is perfectly true because $m = -\alpha$, as we have calculated before.

Therefore, Eq. (27.13) can be converted into

$$A = \frac{d}{dt}[e^{-mt}V_{tr}(t)]$$

If we integrate both sides of this equation with respect to time, we get

$$\int A\,dt = \int \frac{d}{dt}[e^{-mt}V_{tr}(t)]$$

That gives us

$$At + C_1 = e^{-mt} V_{tr}(t) + C_2$$

$$At + C_1 - C_2 = e^{-mt} V_{tr}(t)$$

The integrals give us two constants, C_1 and C_2, because the integral is indefinite.

Constants C_1 and C_2 can be combined and replaced by a single constant B. Consequently, the equation simplifies to

$$At + B = e^{-mt} V_{tr}(t)$$

We can manipulate this equation to find $V_{tr}(t)$:

$$V_{tr}(t) = \frac{At + B}{e^{-mt}}$$

or

$$V_{tr}(t) = (At + B)e^{mt}$$

We can substitute m with $-\alpha$ and get the final transient equation:

$$V_{tr}(t) = (At + B)e^{-\alpha t}$$

Therefore, the final equation for the voltage across the capacitor considering the transient and the steady-state solutions is equal to the following equation.

VOLTAGE ACROSS CAPACITOR – FINAL EQUATION (CRITICALLY DAMPED CASE)

$$V_C(t) = (At + B)e^{-\alpha t} + V_1 \qquad (27.15)$$

$V_C(t)$ is the voltage across capacitor equation as a function of time, in Volts.
A and B are constants, to be determined by the initial conditions.
α is the damping coefficient.
t is the time, in seconds.
V_1 is the voltage supplying the capacitor.

To find the values of A and B, we must use the circuit's initial conditions.

27.2.4.2.2 First Condition ($V_C(0)$)

The first condition to consider in the series RLC circuit is the voltage across the capacitor at time t = 0 s. We call this voltage V_0.

To obtain the first constant, we take Eq. (27.15) and solve it for t = 0 s:

$$V_C(t) = (At + B)e^{-\alpha t} + V_1$$

$$V_C(0) = V_0 = (At + B)e^{-\alpha t} + V_1$$

$$V_0 = \left(A t^0 + B\right)e^{-\alpha t^0} + V_1$$

$$V_0 = (0 + B)e^{-\alpha t^0} + V_1$$

MATH CONCEPT

$e^0 = 1$

$$V_0 = (0 + B)e^{\theta 1} + V_1$$

$$V_0 = B + V_1$$

$$B = V_0 - V_1 \qquad\qquad (27.16)$$

27.2.4.2.3 Second Condition ($dV_C(0)/dt$)

The second condition we must consider is the current flowing in the circuit at time t = 0 s.

Current on a capacitor is defined as the derivative of voltage with respect to time, or

---

$$i = C\frac{dv}{dt}$$

Therefore, all we must do to get the current equation is to find the derivative of the voltage equation (27.15) with respect to time, find the current equation, and solve this other equation for t = 0s.

Thus, we take the equation

$$V_C(t) = (At + B)e^{-\alpha t} + V_1$$

and take the derivative

$$\left.\frac{dV_C(t)}{dt}\right|_{t=0} = A\frac{d}{dt}(te^{-\alpha t}) + B\frac{d}{dt}(e^{-\alpha t}) + \frac{dV_1}{dt} \qquad (27.17)$$

The first term

$$\frac{dV_C(t)}{dt}$$

is equivalent to the current across the capacitor.

To solve this first term, we must consider the following.

We have no means to guarantee that the current across the capacitor will be 0 at time 0 by examining the capacitor alone. Capacitors resist changes in voltages but not in current. However, the capacitor is in series with an inductor and inductors resist current changes. We can, therefore, guarantee that current at time 0 will be 0, and we can simplify (27.17) to

$$\left.\cancel{\frac{dV_C(t)}{dt}}^{\,0}\right|_{t=0} = A\frac{d}{dt}(te^{-\alpha t}) + B\frac{d}{dt}(e^{-\alpha t}) + \frac{dV_1}{dt}$$

$$A\frac{d}{dt}(te^{-\alpha t}) + B\frac{d}{dt}(e^{-\alpha t}) + \frac{dV_1}{dt} = 0$$

The last term is the derivative of a constant.

MATH CONCEPT   The derivative of a constant is equal to 0.

Therefore,

$$A\frac{d}{dt}(te^{-\alpha t}) + B\frac{d}{dt}(e^{-\alpha t}) + \cancel{\frac{dV_1}{dt}}^{\,0} = 0$$

$$A\frac{d}{dt}(te^{-\alpha t}) + B\frac{d}{dt}(e^{-\alpha t}) = 0 \qquad (27.18)$$

To solve the first term

$$A\frac{d}{dt}(te^{-\alpha t})$$

we use the following math concept.

> **MATH CONCEPT** The derivative of a product of two functions is equal to the derivative of the first function multiplied by the second function added to the first function multiplied by the derivative of the second, or
>
> $$(u.v)' = u'v + uv'$$

that gives us this

$$A\left\{\frac{d}{dt}[t] \times (e^{-\alpha t}) + t \times \frac{d}{dt}[e^{-\alpha t}]\right\}$$

producing this solution:

$$A\{e^{-\alpha t} - t\alpha e^{-\alpha t}\} \qquad (27.19)$$

The second term of Eq. (27.18)

$$B\frac{d}{dt}(e^{-\alpha t})$$

is a simple derivative of an exponential, for which we have the following math concept.

> **MATH CONCEPT** The derivative of an exponential
>
> $$\frac{d}{dt}e^{xt} = xe^{xt}$$

that gives the following result:

$$-\alpha Be^{-\alpha t} \qquad (27.20)$$

Putting (27.19) and (27.20) back into (27.18),

$$A\{e^{-\alpha t} - t\alpha e^{-\alpha t}\} - \alpha Be^{-\alpha t} = 0$$

Solving for t = 0

$$-\alpha A t e^{-\alpha t} + A e^{-\alpha t} - \alpha B e^{-\alpha t} = 0$$

$$-\alpha A t e^{-\alpha t} + A e^{-\alpha t} - \alpha B e^{-\alpha t} = 0$$

$$-\alpha A t + A - \alpha B = 0$$

$$-\alpha A t^0 + A - \alpha B = 0$$

which gives us

$$A = \alpha B$$

But B was obtained in (27.16) as

$$B = V_0 - V_1$$

Consequently, A is

$$A = \alpha(V_0 - V_1)$$

27.2.4.2.4 Voltage Curve
The critically damped circuit will not have an oscillatory behavior because the damping coefficient will prevent that from happening.

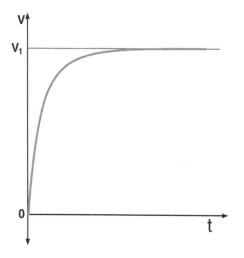

Figure 27.4 Voltage across the capacitor (critically damped solution).

Figure 27.4 shows the curve for the total solution for the voltage across the capacitor for a critically damped series RLC circuit. Notice how the curve starts at 0 and increases exponentially until it stabilizes at a voltage equal to the power supply (V_1).

> This curve was plotted for a critically damped series RLC circuit like the one in Figure 27.3, with a 1H inductor, a 1F capacitor, a 2 Ω resistor, and a power supply V_1 of 10 V.

27.2.4.3 Overdamped Solution

This is the case where $\alpha > \omega_0$, or the damping coefficient is greater than the angular frequency and the result of $\alpha^2 - \omega_0^2$, inside the radical, is positive.

Therefore, the roots m_1 and m_2 will be different, negative, and real numbers as follows:

$$m_1 = -\alpha + \sqrt{\alpha^2 - \omega_0^2}$$

$$m_2 = -\alpha - \sqrt{\alpha^2 - \omega_0^2}$$

Like the critically damped solution, the overdamped solution will not present an oscillatory behavior.

Because the equation has two distinct solutions and we know that the solution for the transient phase respects the form $y = Ae^{mt}$, we conclude that the transient equation for the voltage across the capacitor during the transient phase must have two terms according to the form

$$V_{tr} = Ae^{m_1 t} + Be^{m_2 t}$$

Therefore, the total solution for the voltage across the capacitor must include the voltage across the capacitor during the steady-state phase, which we know to be the same as the power supply V_1.

VOLTAGE ACROSS CAPACITOR – FINAL EQUATION (OVERDAMPED CASE)

$$V_C(t) = Ae^{m_1 t} + Be^{m_2 t} + V_1 \qquad (27.21)$$

$V_C(t)$ is the voltage across capacitor equation as a function of time, in Volts.
A and B are constants, to be determined by the initial conditions.
m_1 and m_2 are the roots of the equation.
t is the time, in seconds.
V_1 is the voltage supplying the capacitor.

To find the values of A and B, we must use the circuit's initial conditions.

27.2.4.3.1 First Condition ($V_C(0)$)

The first condition to consider in the series RLC circuit is the voltage across the capacitor at time t = 0 s. We call this voltage V_0.

To obtain the first constant, we take Eq. (27.21) and solve it for t = 0 s:

$$V_C(t) = Ae^{m_1 t} + Be^{m_2 t} + V_1$$

$$V_C(0) = V_0 = Ae^{m_1 t} + Be^{m_2 t} + V_1$$

$$V_0 = Ae^{m_1 (0)} + Be^{m_2 (0)} + V_1$$

MATH CONCEPT

$e^0 = 1$

$$V_0 = Ae^{0^1} + Be^{0^1} + V_1$$

$$V_0 = A + B + V_1$$

$$A + B = V_0 - V_1 \qquad (27.22)$$

27.2.4.3.2 Second Condition ($dV_C(0)/dt$)

The second condition we must consider is the current flowing in the circuit at time t = 0 s.

Like before, current on a capacitor is defined as the derivative of voltage with respect to time, or

---

$$i = C \frac{dv}{dt}$$

Therefore, all we must do to get the current equation is to find the derivative of the voltage equation (27.21) with respect to time:

$$V_C(t) = Ae^{m_1 t} + Be^{m_2 t} + V_1$$

$$\frac{dV_C(t)}{dt} = \frac{d}{dt}(Ae^{m_1 t}) + \frac{d}{dt}(Be^{m_2 t}) + \frac{d}{dt}V_1$$

$$\frac{dV_C(t)}{dt}^0 = \frac{d}{dt}(Ae^{m_1 t}) + \frac{d}{dt}(Be^{m_2 t}) + \frac{d}{dt}V_1^0$$

The first and the last terms are 0 for the same reason explained on the critically damped solution, because the initial current is 0 and because the derivative of a constant is 0.

Therefore,

$$0 = \frac{d}{dt}(Ae^{m_1 t}) + \frac{d}{dt}(Be^{m_2 t}) + 0$$

$$\frac{d}{dt}(Ae^{m_1 t}) + \frac{d}{dt}(Be^{m_2 t}) = 0$$

At this point we already know how to solve the derivative of an exponential. Thus, the solution for the previous derivatives is

$$Am_1 e^{m_1 t} + Bm_2 e^{m_2 t} = 0$$

Solving this equation for t = 0 s,

$$Am_1 e^{m_1 t^0} + Bm_2 e^{m_2 t^0} = 0$$

$$Am_1 + Bm_2 = 0 \qquad (27.23)$$

We end with two equations, (27.22) and (27.23):

$$\begin{cases} A + B = V_0 - V_1 \\ Am_1 + Bm_2 = 0 \end{cases}$$

From the first equation we get A:

$$A = V_0 - V_1 - B$$

By substituting this value on the second equation, we obtain

$$(V_0 - V_1 - B)m_1 + Bm_2 = 0$$

$$V_0 m_1 - V_1 m_1 - Bm_1 + Bm_2 = 0$$

$$m_1(V_0 - V_1) + B(m_2 - m_1) = 0$$

$$-B(m_2 - m_1) = m_1(V_0 - V_1)$$

$$B(m_1 - m_2) = m_1(V_0 - V_1)$$

CONSTANT B – OVERDAMPED EQUATION

$$B = \frac{V_0 - V_1}{1 - \dfrac{m_2}{m_1}}$$

B is a constant in Eq. (27.21).
V_0 is the capacitor's initial voltage, in Volts.
V_1 is the voltage supplying the capacitor, in Volts.
m_1 and m_1 are the roots of equation.

Upon substituting B in (27.22), we get

$$A + B = V_0 - V_1$$

$$A + \frac{V_0 - V_1}{1 - \frac{m_2}{m_1}} = V_0 - V_1$$

and we will obtain the following formula.

CONSTANT A – OVERDAMPED EQUATION

$$A = \frac{V_0 - V_1}{1 - \frac{m_1}{m_2}}$$

A is a constant in Eq. (27.21).
V_0 is the capacitor's initial voltage, in Volts.
V_1 is the voltage supplying the capacitor, in Volts.
m_1 and m_1 are the roots of equation.

27.2.4.3.3 *Voltage Curve*

The overdamped circuit will not have an oscillatory behavior because the damping coefficient will prevent that from happening after a few brief oscillations.

Figure 27.5 Voltage across the capacitor (overdamped solution).

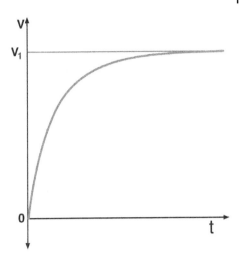

Figure 27.5 shows the curve for the voltage across the capacitor. Notice how the curve grows exponentially and stabilizes in a value greater than the power supply voltage V_1.

> This curve was plotted for an overdamped series RLC circuit like the one in Figure 27.3, with a 1H inductor, a 1F capacitor, a 10 Ω resistor, and a power supply V_1 of 10 V.

27.2.4.4 Underdamped Solution

This is the case where $\alpha < \omega_0$, or the damping coefficient is smaller than the angular frequency and the result of $\alpha^2 - \omega_0^2$, inside the radical, is negative.

This case will give us two roots that are complex numbers as follows:

$$m_1 = -\alpha + j\sqrt{\alpha^2 - \omega_0^2}$$

$$m_1 = -\alpha - j\sqrt{\alpha^2 - \omega_0^2}$$

The equations can be simplified by defining

$$\omega_d = \sqrt{\alpha^2 - \omega_0^2}$$

Therefore, the roots will have the form

$$m_1 = -\alpha + j\omega_d$$
$$m_2 = -\alpha - j\omega_d$$

Like before, we know that the solution for this equation follows the form $y = Ae^{mt}$ and will have the following form for two distinct roots, as defined before in this book:

$$V_C(t) = Ae^{m_1 t} + Be^{m_2 t} + V_1$$

Upon substituting m_1 and m_2 in this equation, we get

$$V_C(t) = Ae^{(-\alpha + j\omega_d)t} + Be^{(-\alpha - j\omega_d)t} + V_1$$

MATH CONCEPT The exponential of a sum of exponents

$$e^{a+b} = e^a \times e^b$$

Therefore,

$$V_C(t) = e^{-\alpha t}\left(Ae^{j\omega_d t} + Be^{-j\omega_d t}\right) + V_1 \tag{27.24}$$

Several terms of Eq. (27.24) are complex exponentials. Fortunately, there is a mathematical definition created by Euler that relates complex exponentials with sines and cosines, which will provide a way to convert equation (27.24) to a friendlier form.

The following is what the definition says.

MATH CONCEPT In mathematics, Euler's identity is the equality

$$e^{j\pi} = 1$$

where e is the Euler number (the base of natural logarithm) and j is the imaginary part.

The π in the Euler's identity formula shows a relation between the complex numbers and the circumference of a circle.

This relation is expressed by the following formula discovered by Euler, where an exponential raised by an imaginary number can be converted into a relation of cosine and sine, as follows:

$$e^{j\theta} = \cos\theta + j\sin\theta$$

or

$$e^{-j\theta} = \cos\theta - j\sin\theta$$

By applying this concept, we can convert equation (27.24) into

$$V_C(t) = e^{-\alpha t}[A(\cos(\omega_d t) + j\sin(\omega_d t)) + B(\cos(\omega_d t) - j\sin(\omega_d t))] + V_1$$

$$V_C(t) = e^{-\alpha t}[(A + B)\cos(\omega_d t) + j(A - B)\sin(\omega_d t)] + V_1$$

and we get the final formula for the voltage across the capacitor for the underdamped solution.

VOLTAGE ACROSS CAPACITOR – FINAL EQUATION (UNDERDAMPED CASE)

$$V_C(t) = e^{-\alpha t}[K_1\cos(\omega_d t) + K_2\sin(\omega_d t)] + V_1 \qquad (27.25)$$

$K_1 = A + B$ is a constant to be discovered by the circuit's initial conditions.
$K_2 = j(A - B)$ is another constant to be discovered by the circuit's initial conditions.
α is the damping coefficient.
t is the given time, in seconds

$$\omega_d = \sqrt{\alpha^2 - \omega_0^2}$$

V_1 is the voltage supplying the capacitor, in Volts.

Again, to find K_1 and K_2, we must apply the circuit's initial conditions, like we did on the previous cases.

27.2.4.4.1 *First Condition ($V_C(0)$)*

The first condition to consider in the series RLC circuit is the voltage across the capacitor at time t = 0 s. We call this voltage V_0.

To obtain the first constant we take Eq. (27.25) and solve it for t = 0 s:

$$V_C(t) = e^{-\alpha t}[K_1\cos(\omega_d t) + K_2\sin(\omega_d t)] + V_1$$

$$V_C(t) = V_0 = e^{-\alpha t^0}\left[K_1\cos\left(\omega_d t^0\right) + K_2\sin\left(\omega_d t^0\right)\right] + V_1$$

$$V_0 = e^0[K_1\cos(0) + K_2\sin(0)] + V_1$$

$$V_0 = e^{\cancel{0}1}\left[K_1 \cos\cancel{(0)}^{\cancel{1}} + K_2 \sin\cancel{(0)}^{\cancel{0}}\right] + V_1$$

$$V_0 = [K_1 + 0] + V_1$$

$$V_0 = K_1 + V_1$$

We get the first constant of Eq. (27.25):

$$K_1 = V_0 - V_1 \tag{27.26}$$

27.2.4.4.2 Second Condition (dV$_C$(0)/dt)

The second condition we must consider is the current flowing in the circuit at time $t = 0$ s.

Current on a capacitor is defined as the derivative of voltage with respect to time, or

$$i = C\frac{dv}{dt}$$

Therefore, all we must do to obtain the current equation is to find the derivative of the voltage equation (27.25) with respect to time.

Therefore, we take the voltage equation

$$V_C(t) = e^{-\alpha t}[K_1\cos(\omega_d t) + K_2\sin(\omega_d t)] + V_1$$

and take the derivative

$$\frac{d}{dt}V_C(t) = \frac{d}{dt}\{e^{-\alpha t}[K_1\cos(\omega_d t) + K_2\sin(\omega_d t)]\} + \frac{d}{dt}(V_1)$$

and rewrite

$$\frac{d}{dt}V_C(t) = \frac{d}{dt}\{K_1 e^{-\alpha t}\cos(\omega_d t) + K_2 e^{-\alpha t}\sin(\omega_d t)\} + \frac{d}{dt}(V_1)$$

To make it easy, let us expand the equation by using the following concept.

> MATH CONCEPT   The derivative of a sum is equal to the sum of derivatives.

Thus, we can expand the equation

$$\frac{d}{dt}V_C(t) = \frac{d}{dt}\{K_1 e^{-\alpha t}\cos(\omega_d t)\} + \frac{d}{dt}\{K_2 e^{-\alpha t}\sin(\omega_d t)\} + \frac{d}{dt}(V_1)$$

MATH CONCEPT The derivative of a constant is equal to 0.

$$\frac{d}{dt}V_C(t)^{0} = \frac{d}{dt}\{K_1 e^{-\alpha t}\cos(\omega_d t)\} + \frac{d}{dt}\{K_2 e^{-\alpha t}\sin(\omega_d t)\} + \frac{d}{dt}(V_1)^{0}$$

$$0 = \frac{d}{dt}\{K_1 e^{-\alpha t}\cos(\omega_d t)\} + \frac{d}{dt}\{K_2 e^{-\alpha t}\sin(\omega_d t)\} + 0$$

Rearranging

$$0 = K_1 \frac{d}{dt}\{e^{-\alpha t}\cos(\omega_d t)\} + K_2 \frac{d}{dt}\{e^{-\alpha t}\sin(\omega_d t)\} \qquad (27.27)$$

Observe Eq. (27.27) in the illustrated form below:

$$0 = K_1 \frac{d}{dt}\left\{\overbrace{e^{-\alpha t}}^{u}\overbrace{\cos(\omega_d t)}^{v}\right\} + K_2 \frac{d}{dt}\left\{\overbrace{e^{-\alpha t}}^{u}\overbrace{\sin(\omega_d t)}^{v}\right\} \qquad (27.28)$$

This equation shows that we are before two derivatives of a product of two functions, u and v, and that we can apply the following math concept to solve both derivatives.

MATH CONCEPT The derivative of a product of two functions is equal to the derivative of the first function multiplied by the second function added to the first function multiplied by the derivative of the second, or

$$(u.v)' = u'v + uv'$$

This will expand Eq. (27.28) into

$$0 = K_1\left\{\overbrace{\frac{d}{dt}[e^{-\alpha t}]\cos(\omega_d t)}^{A} + \overbrace{e^{-\alpha t}\frac{d}{dt}[\cos(\omega_d t)]}^{B}\right\}$$

$$+ K_2\left\{\overbrace{\frac{d}{dt}[e^{-\alpha t}]\sin(\omega_d t)}^{C} + \overbrace{e^{-\alpha t}\frac{d}{dt}[\sin(\omega_d t)]}^{D}\right\} \qquad (27.29)$$

Equation (27.29) can be divided into four parts, namely A, B, C, and D. We will solve these parts independently to make it easy.

27.2.4.4.2.1 Solving Parts A and C Parts A and C are equal. They are both derivatives of an exponential and we already know how to solve them.

Parts A and C will produce the following result:

$$-\alpha e^{-\alpha t} \tag{27.30}$$

27.2.4.4.2.2 Solving Part B To solve part B, we will need a new math concept shown below.

MATH CONCEPT The derivative of
$$\frac{d}{dt}\cos(k\theta) \longrightarrow -k\sin(\theta)$$

By applying this concept, we get the following result for part B:

$$-\omega_d \sin(\omega_d t) \tag{27.31}$$

27.2.4.4.2.3 Solving Part D To solve part D, we will need another new math concept.

MATH CONCEPT The derivative of
$$\frac{d}{dt}\sin(k\theta) \longrightarrow k\cos(\theta)$$

By applying this rule, we get the following result for part D:

$$\omega_d \cos(\omega_d t) \tag{27.32}$$

By putting (27.30), (27.31), and (27.32) back into (27.29), we get

$$0 = K_1 \left\{ -\alpha e^{-\alpha t} \cos(\omega_d t) - e^{-\alpha t} \omega_d \sin(\omega_d t) \right\} + K_2 \left\{ -\alpha e^{-\alpha t} \sin(\omega_d t) \right.$$
$$\left. + e^{-\alpha t} \omega_d \cos(\omega_d t) \right\}$$

Solving for t = 0 s, we get

$$0 = K_1 \left\{ -\alpha e^{-\alpha t^0} \cos\left(\omega_d t^0\right) - e^{-\alpha t^0} \omega_d \sin\left(\omega_d t^0\right) \right\} + K_2 \left\{ -\alpha e^{-\alpha t^0} \sin\left(\omega_d t^0\right) \right.$$
$$\left. + e^{-\alpha t^0} \omega_d \cos\left(\omega_d t^0\right) \right\}$$

$$0 = K_1 \left\{ -\alpha e^0 \cos(0) - e^0 \omega_d \sin(0) \right\} + K_2 \left\{ -\alpha e^0 \sin(0) + e^0 \omega_d \cos(0) \right\}$$

MATH CONCEPT

$e^0 = 1$

$$0 = K_1 \left\{ -\alpha e^{\cancel{0}^1} \cos(0) - e^{\cancel{0}^1} \omega_d \sin(0) \right\} + K_2 \left\{ -\alpha e^{\cancel{0}^1} \sin(0) + e^{\cancel{0}^1} \omega_d \cos(0) \right\}$$

$$0 = K_1 \left\{ -\alpha \cos(0) - \omega_d \sin(0) \right\} + K_2 \left\{ -\alpha \sin(0) + \omega_d \cos(0) \right\}$$

$$0 = K_1 \left\{ -\alpha \cancel{\cos(0)}^1 - \omega_d \cancel{\sin(0)}^0 \right\} + K_2 \left\{ -\alpha \cancel{\sin(0)}^0 + \omega_d \cancel{\cos(0)}^1 \right\}$$

$$0 = -\alpha K_1 + \omega_d K_2$$

$$\alpha K_1 = \omega_d K_2$$

$$K_2 = \frac{\alpha}{\omega_d} K_1$$

But as defined by (27.26),

$$K_1 = V_0 - V_1$$

Therefore, we get the second constant of Eq. (27.25):

$$K_2 = \frac{\alpha}{\omega_d} (V_0 - V_1)$$

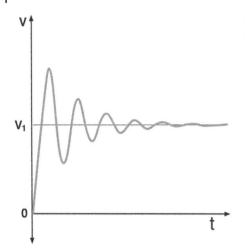

Figure 27.6 Voltage across the capacitor (underdamped solution).

27.2.4.4.3 Voltage Curve

The underdamped circuit will present an oscillatory behavior.

Figure 27.6 shows the curve for the voltage across the capacitor. Notice how the voltage oscillates up and down until it stabilizes at the same value as the power supply V_1.

> This curve was plotted for an overdamped series RLC circuit like the one in Figure 27.3, with a 1H inductor, a 1F capacitor, a 0.31 Ω resistor, and a power supply V_1 of 10 V.

27.3   Examples

27.3.1   Example 1

Figure 27.7 shows a circuit with a direct voltage source, a switch, a resistor, an inductor, and a capacitor in series.

The switch is open, there is no current flowing in the circuit, the inductor is completely de-energized, and the capacitor is completely discharged.

At time t = 0 s, the switch is closed, and current flows and reaches the inductor.

The following is what we want to find about the circuit:

- What is the voltage across the capacitor after the switch is closed?
- What is the voltage across the capacitor 2 sec after the switch is closed?
- What is the circuit's natural resonance frequency?

Figure 27.7 DC powered series RLC
circuit.

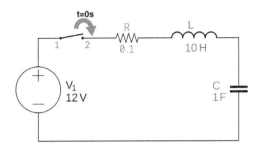

27.3.1.1 What Is the Voltage Across the Capacitor After the Switch Is Closed?

When the switch is closed, current flows and reaches the inductor. The inductor hates sudden variations of current and will prevent current from flowing as soon as the switch is closed.

Therefore, there is no current reaching the capacitor, which is initially discharged, and the voltage across the capacitor is 0.

FINAL RESULT

$$V_C(0^+) = 0$$

27.3.1.2 What Is the Voltage Across the Capacitor Two Seconds After the Switch Is Closed?

We can have three kinds of series RLC circuits, namely, critically damped, overdamped, and underdamped, each one with its own equations for the voltage across the capacitor.

The first thing we need to do is to know which circuit is this one.

The given values are

$$\begin{cases} R = 0.1 \ \Omega \\ L = 10H \\ C = 1F \\ V_1 = 12 \ V \end{cases}$$

We first calculate α and ω_o:

$$\alpha = \frac{R}{2L}$$

$$\alpha = \frac{0.1}{2(10)}$$

$$\alpha = 0.005$$

$$\omega_0 = \frac{1}{\sqrt{LC}}$$

$$\omega_0 = \frac{1}{\sqrt{10(1)}}$$

$$\omega_0 = 0.3162$$

We see that $\alpha < \omega_0$, equivalent to an underdamped circuit.
The equation for the voltage across the capacitor, for this case, is

$$V_C(t) = e^{-\alpha t}[K_1 \cos(\omega_d t) + K_2 \sin(\omega_d t)] + V_1 \qquad (27.33)$$

In the following steps, we will find the several unknowns in the previous equation:

$$\omega_d = \sqrt{\alpha^2 - \omega_0^2}$$

$$\omega_d = \sqrt{0.005^2 - 0.31620^2}$$

$$\omega_d = \sqrt{0.005^2 - 0.31620^2}$$

$$\omega_d = \sqrt{-0.0999}$$

This is the square root of a negative number. We must remember this math concept from the imaginary numbers.

MATH CONCEPT   The imaginary number

$$j = \sqrt{-1}$$

Consequently,

$$j^2 = -1$$

Therefore,

$$\sqrt{-0.0999}$$

can be written as

$$\sqrt{j^2 \times 0.0999}$$

extracting the root of the imaginary part:

$$j\sqrt{0.0999}$$

or

$$j0.3161$$

which results into

$$\omega_d = j0.3161$$

The capacitor is initially discharged; therefore, V_0 is 0. The power supply voltage, V_1, is equal to 12 V.

Therefore, we can find K_1,

$$K_1 = V_0 - V_1$$

$$K_1 = -12$$

and K_2,

$$K_2 = \frac{\alpha}{\omega_d}(V_0 - V_1)$$

$$K_2 = \frac{0.005}{0.3162}(0 - 12)$$

$$K_2 = -0.1897$$

Notice that we just took the number without the imaginary part when we calculate ω_d.

Now that we have all unknowns, the complete formula for the voltage across the capacitor can be assembled like this:

$$V_C(t) = e^{-0.005t}\{-12\cos(0.31618t) - 0.1897\sin(0.31618t)\} + 12$$

Finding the voltage for t = 2 s,

$$V_C(2s) = e^{-0.005(2)}\{-12\cos(0.31618(2)) - 0.1897\sin(0.31618(2))\} + 12$$

FINAL RESULT

$$V_C(2s) = 2.3058 \text{ V}$$

27.3.1.3 What Is the Circuit's Natural Resonance Frequency?

The natural resonance frequency for the series RLC circuit is given by the following formula:

$$f = \frac{\omega_o}{2\pi}$$

$$f = \frac{0.3162}{2\pi}$$

FINAL RESULT

$$f = 0.0503 \text{ Hz}$$

This low value for the frequency was expected due to the high values of capacitance and inductance. Big capacitors and inductors are slow to respond to changes, for example, a transatlantic boat is slow compared with a race boat in the agility to do fast curves and start/stop moving.

27.3.2 Example 2

Consider the same circuit as the previous example. The values of L and C are unknown, but R and V_1 are

$$\begin{cases} R = 8 \ \Omega \\ V1 = 10 \text{ V} \end{cases}$$

Like in the previous circuit, the switch is open.
At time t = 0 s, the switch is closed, and current flows and reaches the inductor.
The circuit is ruled by the following second-order differential equation:

$$\frac{d^2V_C(t)}{dt^2} + 4\frac{dV_C(t)}{dt} + 4V_C(t) = 0 \tag{27.34}$$

The initial conditions for the circuit are as follows: the voltage across the capacitor for t = 0 s is 2 V, and the first derivative of the voltage equation with respect to time, for t = 0 s, is equal to 4 A.
Mathematically speaking,

$$V_C(0) = 2\,V$$
$$V'_C(0) = 4\,A$$

The following is what we want to find about the circuit:

• The complete equation for the voltage across the capacitor.
• The voltage across the capacitor for t = 2 s.

27.3.2.1 The Complete Equation for the Voltage Across the Capacitor

The second-order differential equation for a series RLC circuit follows the form

$$\frac{d^2V_{tr}(t)}{dt^2} + \frac{R}{L}\frac{dV_{tr}(t)}{dt} + \frac{1}{LC}V_{tr}(t) = 0 \tag{27.35}$$

which we repeat as follows:

$$\frac{d^2V_{tr}(t)}{dt^2} + \overbrace{\frac{R}{L}}^{A}\frac{dV_{tr}(t)}{dt} + \overbrace{\frac{1}{LC}}^{B}V_{tr}(t) = 0$$

If we compare A and B of this equation with the same values of Eq. (27.34), which we repeat in the following,

$$\underbrace{\frac{d^2 V_C(t)}{dt^2} + \overbrace{4}^{A} \frac{dV_C(t)}{dt} + \overbrace{4}^{B} V_C(t) = 0}$$

we can extract a lot of elements, for example,

$$\frac{R}{L} = 4 \tag{27.36}$$

In this given example,

$$R = 8\ \Omega$$

By substituting R in (27.36), we get

$$L = 2\,H$$

Another part we can extract by comparing equation (27.34) with (27.35) is

$$\frac{1}{LC} = 4 \tag{27.37}$$

The inductor is

$$L = 2\,H$$

Therefore, by substituting L in (27.37), we get

$$C = \frac{1}{8}F$$

Equation (27.34) can be written in a simplified form as

$$m^2 + 4m + 4 = 0$$

This is a quadratic equation and its roots can be found by using the classical formula:

$$m = \frac{-b \pm \sqrt{b^2 - 4ac}}{2a}$$

$$m = \frac{-4 \pm \sqrt{4^2 - 4(1)(4)}}{2(1)}$$

$$m = \frac{-4 \pm \sqrt{16 - 16}}{2}$$

$$m = -2$$

The equation has only one root, which is typical of a critically damped circuit, governed by an equation for the voltage across the capacitor that follows the form

$$V_C(t) = (At + B)e^{-\alpha t} + V_1 \qquad (27.38)$$

First, we need to find α:

$$\alpha = \frac{R}{2L}$$

$$\alpha = \frac{8}{2(2)}$$

$$\alpha = 2$$

To find the coefficients A and B in Eq. (27.38), we must apply the circuit's initial conditions.

The first condition is the initial voltage across the capacitor for time 0. This is a given condition defined as

$$V_C(0) = 2\,V$$

Therefore, we take the original equation

$$V_C(t) = (At + B)e^{-\alpha t} + V_1$$

and solve it for t = 0 s to find the first coefficient:

$$V_C(0) = 2 = (At + B)e^{-2t} + 10$$

For time t = 0 s,

$$2 = \left(At^0 + B\right)e^{-2t^0} + 10$$
$$2 = (0 + B)e^0 + 10$$
$$2 = Be^{0^1} + 10$$

$$B = -8$$

To find the second coefficient, we must apply the circuit's second condition and solve the equation for t = 0 s. This is also a condition provided by the example definition:

$$V'_C(0) = 4\,A$$

Therefore, we take the original equation (27.38)

$$V_C(t) = (At + B)e^{-\alpha t} + V_1$$

and take the derivative with respect to time:

$$\frac{d}{dt}[V_C(t)] = \frac{d}{dt}[(At + B)e^{-\alpha t}] + \frac{d}{dt}[V_1]$$

The first term, as defined by the problem, is

$$\frac{d}{dt}[V_C(t)] = 4$$

The last term

$$\frac{d}{dt}[V_1]$$

is a derivative of a constant and that is mathematically 0.

The middle term is a derivative of a product of two functions, or

$$\frac{d}{dt}[(At+B)e^{-\alpha t}]$$

which can be solved by applying the following concept.

> **MATH CONCEPT**   The derivative of a product of two functions is equal to the derivative of the first function multiplied by the second function added to the first function multiplied by the derivative of the second, or
>
> $$(u.v)' = u'v + uv'$$

Therefore, the original equation can be expanded into

$$\frac{d}{dt}[(At+B)e^{-\alpha t}] = \frac{d}{dt}[(At+B)]e^{-\alpha t} + (At+B)\frac{d}{dt}[e^{-\alpha t}]$$

and solved as

$$\frac{d}{dt}[V_c(t)]^{\,4} = \frac{d}{dt}[(At+B)]e^{-\alpha t} + (At+B)\frac{d}{dt}[e^{-\alpha t}] + \frac{d}{dt}[V_1]^{\,0}$$

$$4 = \frac{d}{dt}[(At+B)]e^{-\alpha t} + (At+B)\frac{d}{dt}[e^{-\alpha t}]$$

$$4 = Ae^{-\alpha t} - \alpha(At+B)e^{-\alpha t}$$

$$4 = Ae^{-\alpha t} - \alpha Ate^{-\alpha t} - \alpha Be^{-\alpha t}$$

Upon substituting the known values

$$4 = Ae^{-2t} - 2Ate^{-2t} - 2(-8)e^{-2t}$$

$$4 = Ae^{-2t} - 2Ate^{-2t} + 16e^{-2t}$$

and solving for t = 0 s to get the other coefficient,

$$4 = Ae^{-2t^0} - 2At^0e^{-2t^0} + 16e^{-2t^0}$$

$$4 = A - 0 + 16$$

$$A = -12$$

Therefore, the final equation for the voltage across the capacitor is shown as follows.

FINAL RESULT

$$V_C(t) = (-12t - 8)e^{-2t} + 10$$

Calculating the voltage across the capacitor for t = 2 s,

$$V_C(t) = (-12(2) - 8)e^{-2(2)} + 10$$

$$V_C(t) = (-24 - 8)e^{-4} + 10$$

FINAL RESULT

$$V_C(2s) = 9.4139 \text{ V}$$

Exercises

1   Consider the circuit shown in Figure 27.8, composed of a resistor, an inductor, a capacitor, a direct voltage power supply, and a switch.

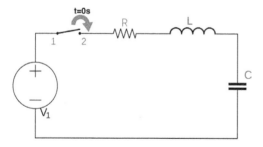

Figure 27.8 DC powered series RLC circuit.

The switch is initially open, the capacitor is completely discharged, and the inductor is completely de-energized.

These are the components:

$$\begin{cases} R = 100 \ \Omega \\ L = 2 \text{ H} \\ C = 1/3 \text{F} \\ V_1 = 12 \text{ V} \end{cases}$$

At time t = 0 s, the switch is closed, and current flows and reaches the inductor.

The following is what we want to find about the circuit:

- What is the voltage across the capacitor as soon as the switch is closed?
- What is the complete equation for the voltage across the capacitor?
- What is the voltage across the capacitor 5 s after the switch is closed?

2 Consider a series RLC circuit like the previous one. The values of all components are the same, except R and V_1, as shown next:

$$\begin{cases} R = 2\ \Omega \\ V_1 = 10\ V \end{cases}$$

The circuit is ruled by the following second-order differential equation:

$$2\frac{d^2V_C}{dt^2} + 8\frac{dV_C}{dt} + 40V_C = 0$$

This circuit's initial conditions are

$$V_C(0) = 0\ V$$
$$V'_C(0) = 6\ A$$

The following is what we want to find about the circuit:

- The complete equation for the voltage across the capacitor.
- The voltage across the capacitor for t = 3 s.

Solutions

1 What is the voltage across the capacitor after the switch is closed?

After the switch is closed, current flows and reaches the inductor. The inductor hates sudden current variations and will not let this new current pass.

Therefore, no current reaches the capacitor, which is continuously discharged.

$$V_C(0^+) = 0$$

What is the complete equation for the voltage across the capacitor?
We can have three kinds of series RLC circuits, namely, critically damped, overdamped, and underdamped, each one with its own equations for the voltage across the capacitor.
The first thing we need to do is to know which circuit is this one.
The given values are

$$\begin{cases} R = 100 \ \Omega \\ L = 2 \ H \\ C = 1/3 \ F \\ V_1 = 12 \ V \end{cases}$$

We first calculate α and ω_o:

$$\alpha = \frac{R}{2L}$$

$$\alpha = \frac{100}{2(2)}$$

$$\alpha = 25$$

$$\omega_o = \frac{1}{\sqrt{LC}}$$

$$\omega_o = \frac{1}{\sqrt{2\left(\frac{1}{3}\right)}}$$

$$\omega_o = 1.2247$$

We see that $\alpha > \omega_o$, equivalent to an overdamped circuit.
The equation for the voltage across the capacitor, for this case, is

$$V_C(t) = Ae^{m_1 t} + Be^{m_2 t} + V_1 \qquad (27.39)$$

The roots of an overdamped solution are

$$m_1 = -\alpha + \sqrt{\alpha^2 - \omega_0^2}$$

$$m_1 = -25 + \sqrt{25^2 - 1.2247^2}$$

$$m_1 = -0.03$$

and

$$m_2 = -25 - \sqrt{25^2 - 1.2247^2}$$

$$m_2 = -49.97$$

The coefficients in Eq. (27.37) are

$$A = \frac{V_0 - V_1}{1 - \dfrac{m_1}{m_2}}$$

$$A = \frac{0 - 12}{1 - \dfrac{(-0.03)}{(-49.97)}}$$

$$A = -12.0072$$

and

$$B = \frac{V_0 - V_1}{1 - \dfrac{m_2}{m_1}}$$

$$B = \frac{0 - 12}{1 - \dfrac{(-49.997)}{(-0.03)}}$$

$$B = 0.0072$$

Having found all unknowns, we can write the final equation for the voltage across the capacitor (27.39) as

$$V_C(t) = Ae^{m_1 t} + Be^{m_2 t} + V_1$$

FINAL RESULT

$$V_C(t) = -12.0072e^{-0.03t} + 0.0072e^{-49.97t} + 12$$

What is the voltage across the capacitor 5s after the switch is closed?
We use the last equation and solve for t = 5 s:

$$V_C(5\,s) = -12.0072e^{-0.03(5)} + 0.0072e^{-49.97(5)} + 12$$

FINAL RESULT

$$V_C(5\,s) = 1.6653 \text{ V}$$

---

2   **The complete equation for the voltage across the capacitor.**
The given circuit is ruled by the following equation:

$$2\frac{d^2V_C}{dt^2} + 8\frac{dV_C}{dt} + 40\,V_C = 0$$

We divide the equation by 2 to write it in the standard form as

$$\frac{d^2V_C}{dt^2} + 4\frac{dV_C}{dt} + 20\,V_C = 0 \tag{27.40}$$

Series RLC circuits are ruled by the following second-order differential equation:

$$\frac{d^2V_C}{dt^2} + \frac{R}{L}\frac{dV_C}{dt} + \frac{1}{LC}V_C = 0$$

which we rewrite as follows:

$$\frac{d^2V_{tr}(t)}{dt^2} + \overbrace{\frac{R}{L}}^{A}\frac{dV_{tr}(t)}{dt} + \overbrace{\frac{1}{LC}}^{B}V_{tr}(t) = 0$$

If we compare A and B of this equation with the same values of Eq. (27.40), which we rewrite in the following,

$$\frac{d^2V_C(t)}{dt^2} + \overbrace{4}^{A}\frac{dV_C(t)}{dt} + \overbrace{20}^{B}V_C(t) = 0 \qquad (27.41)$$

we can extract several values:

$$\frac{R}{L} = 4 \qquad (27.42)$$

In this given example,

$$R = 2\ \Omega$$

By substituting R in (27.42), we get

$$L = 0.5\ H$$

Another part we can extract by comparing the equations is

$$\frac{1}{LC} = 20 \qquad (27.43)$$

Therefore, by substituting L in (27.43), we get

$$C = \frac{1}{20L} = \frac{1}{20(0.5)} = 0.1\ F$$

The next thing we must do is to find if this is a critically damped, overdamped, or underdamped circuit.
We proceed to calculate α

$$\alpha = \frac{R}{2L}$$

$$\alpha = \frac{2}{2(0.5)}$$

$$\alpha = 2$$

and ω_0

$$\omega_0 = \frac{1}{\sqrt{LC}}$$

$$\omega_0 = \frac{1}{\sqrt{0.5(0.1)}}$$

$$\omega_0 = 4.4721$$

We see that $\alpha < \omega_0$. Therefore, this is an underdamped series RLC circuit, ruled by the following equation for the voltage across the capacitor:

$$V_C(t) = e^{-\alpha t}[K_1 \cos(\omega_d t) + K_2 \sin(\omega_d t)] + V_1$$

We proceed finding the unknowns: first K_1

$$K_1 = V_0 - V_1$$

$$K_1 = 0 - 12$$

$$K_1 = -12$$

and then K_2

$$K_2 = \frac{\alpha}{\omega_d}(V_0 - V_1)$$

To find K_2 we need to find ω_d:

$$\omega_d = \sqrt{\alpha^2 - \omega_0^2}$$

$$\omega_d = \sqrt{2^2 - 4.4721^2}$$

$$\omega_d \approx j4$$

Therefore,

$$K_2 = \frac{2}{4}(0 - 12)$$

$$K_2 = -6$$

The complete equation for the voltage across the capacitor can be assembled as follows.

FINAL RESULT

$$V_C(t) = e^{-2t}\left[-12\cos(4t) - 6\sin(4t)\right] + 10$$

The voltage across the capacitor for t = 3 s.
Using the last equation, we can solve for t = 3 s:

$$V_C(3\,s) = e^{-2(3)}\left[-12\cos(4(3)) - 6\sin(4(3))\right] + 10$$

$$V_C(3\,s) = e^{-6}\left[-12\cos(12) - 6\sin(12)\right] + 10$$

FINAL RESULT

$$V_C(3\,s) = 9.9828\,V$$

28

RLC Circuits: Part 2

Current Analysis in Circuits Containing Resistors, Capacitors, and Inductors in Series

28.1   Introduction

In the last chapter, we have performed a voltage analysis in RLC series circuits, finding their voltage equations for various specific cases.

In this chapter we will continue to examine series RLC circuits in terms of current analysis.

28.2   The Circuit

Our analysis will focus on the same circuit we have used on the last chapter: a direct voltage source, a resistor, a capacitor, an inductor, and a switch in series, like shown in Figure 28.1.

Figure 28.1 A basic RLC circuit.

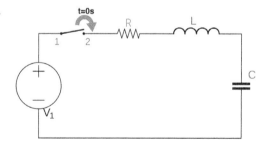

Like before, the switch is initially open, there is no current flowing in the circuit, the inductor is completely de-energized, and the capacitor is completely discharged.

At time t = 0 s, the switch is closed, and current flows and reaches the inductor.

Introductory Electrical Engineering with Math Explained in Accessible Language,
First Edition. Magno Urbano.
© 2020 John Wiley & Sons, Inc. Published 2020 by John Wiley & Sons, Inc.

28.2.1 Current

As soon as the switch is closed, we can apply to the circuit. Kirchhoff's voltage law (KVL) states that the sum of all voltages in a closed loop is 0.

Therefore, the power supply voltage (V_1) is equal to the voltage drop on the resistor added to the voltage across the inductor and capacitor, or

$$V_1 = V_R(t) + V_L(t) + V_C(t) \qquad (28.1)$$

Voltages across a capacitor, an inductor, and a resistor are defined as

$$V_C(t) = \frac{1}{C}\int i\,dt$$

$$V_R(t) = Ri(t)$$

$$V_L(t) = L\frac{di(t)}{dt}$$

By substituting this in (28.1), we get

$$V_1 = Ri(t) + L\frac{di(t)}{dt} + \frac{1}{C}\int i\,dt$$

If we take the derivative of both sides of this equation with respect to time, we obtain

$$\frac{d}{dt}(V_1(t)) = R\frac{d}{dt}i(t) + L\frac{d}{dt}\left(\frac{di(t)}{dt}\right) + \frac{1}{C}\frac{d}{dt}\left(\int i\,dt\right)$$

$V_1(t)$ is the power supply voltage, a constant. We know that the derivative of a constant is 0; therefore,

$$\frac{d}{dt}(\overset{0}{\cancel{V_1(t)}}) = R\frac{d}{dt}i(t) + \frac{d}{dt}\left(\frac{di(t)}{dt}\right) + \frac{1}{C}\frac{d}{dt}\left(\int i\,dt\right)$$

$$0 = R\frac{d}{dt}i(t) + \frac{d}{dt}\left(\frac{di(t)}{dt}\right) + \frac{1}{C}\frac{d}{dt}\left(\int i\,dt\right)$$

The last term is the derivative of an integral, that is, one thing cancels the other, and we have just the function.

Therefore,

$$0 = R\frac{d}{dt}i(t) + \frac{d}{dt}\left(\frac{di(t)}{dt}\right) + \frac{1}{C}i$$

or after rearranging the equation, we have

$$L\frac{d^2i(t)}{dt^2} + R\frac{di(t)}{dt} + \frac{1}{C}i = 0$$

or written in standard form:

$$\frac{d^2i(t)}{dt^2} + \frac{R}{L}\frac{di(t)}{dt} + \frac{1}{LC}i = 0 \tag{28.2}$$

This second-order differential equation for the current (28.2) is, obviously, very similar to the one described on the previous chapter for voltage across a capacitor.

For the same reasons as before, we know that the transient equation for the current will have the form

$$y = Ae^{mt}$$

Consequently, we can convert Eq. (28.2) into

$$\frac{d^2}{dt^2}(Ae^{mt}) + \frac{R}{L}\frac{d}{dt}(Ae^{mt}) + \frac{1}{LC}(Ae^{mt}) = 0 \tag{28.3}$$

The first two terms are derivative of exponentials that we already know how to solve, which will produce the following result for the first term and for the second term:

$$m^2(Ae^{mt})$$

and

$$\frac{R}{L}m(Ae^{mt})$$

By putting these terms back into Eq. (28.3), we get

$$m^2(Ae^{mt}) + \frac{R}{L}m(Ae^{mt}) + \frac{1}{LC}(Ae^{mt}) = 0$$

$$\left(m^2 + \frac{R}{L}m + \frac{1}{LC}\right)(Ae^{mt}) = 0 \qquad (28.4)$$

Equation (28.4) is the same we have found for the voltage across the capacitor in the previous chapter.

Consequently, we know that this equation will have three possible solutions for the current: critically damped, overdamped, and underdamped.

28.3 Current Equations

There is no need to repeat the same calculation we have done in the last chapter. The final equations for the three cases will be the following.

28.3.1 Critically Damped Solution

CURRENT EQUATION – CRITICALLY DAMPED SOLUTION

$$i(t) = (At + B)e^{-\alpha t}$$

i(t) is the equation of current as a function of time, in Amperes.
A and B are coefficients, to be determined by the circuit's initial conditions, i(0) and i′(0).
α is the damping coefficient.
t is the time, in seconds.

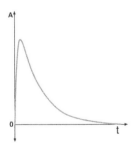

Figure 28.2 Current flowing in the circuit (critically damped solution).

28.3.1.1 Current Curve
Figure 28.2 shows the curve for the current flowing in the critically damped series RLC circuit, as soon as the switch is closed.

Notice how the current grows from 0 and hits a peak on the positive side and then decays slowly to 0.

This curve was plotted for a critically damped series RLC circuit like the one in Figure 28.1, with a 1H inductor, a 1F capacitor, a 2 Ω resistor, and a power supply V_1 of 10 V.

28.3.2 Overdamped Solution

CURRENT EQUATION – OVERDAMPED SOLUTION

$$i(t) = Ae^{m_1 t} + Be^{m_2 t}$$

i(t)is the equation of current as a function of time, in Amperes.
A and B are coefficients, to be determined by the circuit's initial conditions, i(0) and $i'(0)$.
t is the time, in seconds.
α is the damping coefficient.
m_1 and m_2 are the roots equal to

$$m_1 = -\alpha + \sqrt{\alpha^2 - \omega_0^2}$$
$$m_2 = -\alpha - \sqrt{\alpha^2 - \omega_0^2}$$
$$\alpha = \frac{R}{2L}$$
$$\omega_0 = \frac{1}{\sqrt{LC}}$$

28.3.2.1 Current Curve

Figure 28.3 shows the curve for the current flowing in the overdamped series RLC circuit, as soon as the switch is closed.

Notice how the current grows fast from 0 and hits a peak on the positive side and then decays exponentially slowly to 0.

This curve was plotted for an overdamped series RLC circuit like the one in Figure 28.1, with a 1H inductor, a 1F capacitor, a 10 Ω resistor, and a power supply V_1 of 10 V.

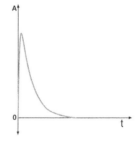

Figure 28.3 Current flowing in the circuit (overdamped solution).

28.3.3 Underdamped Solution

CURRENT EQUATION – UNDERDAMPED SOLUTION

$$i(t) = e^{-\alpha t}[K_1 \cos(\omega_d t) + K_2 \sin(\omega_d t)]$$

i(t) is the equation of current as a function of time, in Amperes.
K_1 and K_2 are coefficients, to be determined by the circuit's initial conditions, i(0) and $i'(0)$.
t is the time, in seconds.
α is the damping coefficient.

$$\alpha = \frac{R}{2L}$$
$$\omega_d = \sqrt{\alpha^2 - \omega_0^2}$$

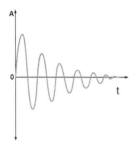

Figure 28.4 Current flowing in the circuit (underdamped solution).

28.3.3.1 Current Curve

Figure 28.4 shows the curve for the current flowing in the underdamped series RLC circuit, as soon as the switch is closed.

Notice how the current oscillates from positive to negative values until it dies.

The amplitude decays according to an exponential pattern.

> This curve was plotted for a critically damped series RLC circuit like the one in Figure 28.1, with a 1H inductor, a 1F capacitor, a 0.3 Ω resistor, and a power supply V_1 of 10 V.

28.4 Examples

28.4.1 Example 1

Figure 28.5 shows a circuit with a direct voltage source, a switch, a resistor, an inductor, and a capacitor in series.

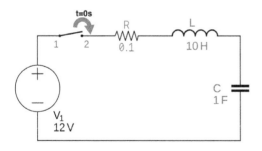

Figure 28.5 A basic RLC circuit.

The switch is open, there is no current flowing in the circuit, the inductor is completely de-energized, and the capacitor is completely discharged.

At time t = 0 s, the switch is closed, and current flows and reaches the inductor.

The following is what we want to find about the circuit:

- What is the current across the circuit after the switch is closed?
- What is the current across the circuit two seconds after the switch is closed?

28.4.1.1 What Is the Current Across the Circuit After the Switch Is Closed?
When the switch is closed, current flows and reaches the inductor. The inductor hates sudden variations of current and will prevent current from flowing as soon as the switch is closed.

FINAL RESULT

$$i(0^+) = 0$$

28.4.1.2 What Is the Current Across the Circuit Two Seconds After the Switch Is Closed?
To find the current equation for this circuit, we must find several parameters and coefficients like α, ω_0, and ω_d. This circuit is the same used in last chapter, and, therefore, the values are already known as

$$\begin{cases} \alpha = 0.005 \\ \omega_0 = 0.3162 \\ \omega_d = j0.3161 \end{cases}$$

This is an underdamped circuit because $\alpha < \omega_0$; therefore, the current equation has the form

$$i(t) = e^{-\alpha t}[K_1 \cos(\omega_d t) + K_2 \sin(\omega_d t)] \tag{28.5}$$

Upon substituting the values, we get

$$i(t) = e^{-0.005t}\{K_1 \cos(0.31618t) + K_2 \sin(0.31618t)\} \tag{28.6}$$

To find K_1 and K_2, we must apply the initial conditions for the circuit.

28.4.1.2.1 *First Condition: Current at Time Zero*
To find the first constant, we must take equation (28.6) and solve for t = 0 s:

$$i(t) = e^{-0.005(0)}\{K_1 \cos(0.31618(0)) + K_2 \sin(0.31618(0))\}$$
$$i(t) = 0 = e^0\{K_1 \cos(0) + K_2 \sin(0)\}$$
$$0 = (1)K_1 + (0)K_2$$

$$K_1 = 0 \qquad\qquad\qquad (28.7)$$

28.4.1.2.2 Second Condition: The Derivative of the Current Equation for Time Zero

To find the other constant, we take the derivative of equation (28.5) with respect to time and solve for t = 0 s.

Therefore, we take the current equation

$$i(t) = e^{-\alpha t}[K_1 \cos(\omega_d t) + K_2 \sin(\omega_d t)]$$

and take the derivative

$$\frac{d}{dt}[i(t)] = \frac{d}{dt}[e^{-\alpha t}[K_1 \cos(\omega_d t) + K_2 \sin(\omega_d t)]]$$

The first term is the derivative of current with respect to time and we know that the voltage across an inductor is equal, exactly, to the derivative of current with respect to time multiplied by the inductance, or

$$V_L = L\frac{di}{dt}$$

Therefore,

$$\frac{di}{dt} = \frac{V_L}{L}$$

We can substitute this relation in the equation

$$\frac{V_L}{L} = \frac{d}{dt}[e^{-\alpha t}[K_1 \cos(\omega_d t) + K_2 \sin(\omega_d t)]]$$

We are before a derivative of a product of functions (u and v) and we can use the product rule:

$$\frac{V_L}{L} = \frac{d}{dt}\left[\overbrace{e^{-\alpha t}}^{u}\overbrace{[K_1 \cos(\omega_d t) + K_2 \sin(\omega_d t)]}^{v}\right]$$

> **MATH CONCEPT** The derivative of a product of two functions is equal to the derivative of the first function multiplied by the second function added to the first function multiplied by the derivative of the second, or
>
> $$(u.v)' = u'v + uv'$$

Therefore, the derivatives will expand to

$$\frac{V_L}{L} = \left\{ \frac{d}{dt}[e^{-\alpha t}] \times [K_1 \cos(\omega_d t) + K_2 \sin(\omega_d t)] \right\}$$
$$+ \left\{ e^{-\alpha t} \times \frac{d}{dt}[K_1 \cos(\omega_d t) + K_2 \sin(\omega_d t)] \right\}$$

that can be expanded to

$$\frac{V_L}{L} = \left\{ \frac{d}{dt}[e^{-\alpha t}] \times [K_1 \cos(\omega_d t) + K_2 \sin(\omega_d t)] \right\}$$
$$+ \left\{ e^{-\alpha t} \times \frac{d}{dt}[K_1 \cos(\omega_d t)] \right\} + \left\{ e^{-\alpha t} \times \frac{d}{dt}[K_2 \sin(\omega_d t)] \right\} \tag{28.8}$$

Equation (28.8) has three derivative terms.

The first term is the derivative of an exponential and will produce the following result:

$$-\alpha e^{-\alpha t}[K_1 \cos(\omega_d t) + K_2 \sin(\omega_d t)] \tag{28.9}$$

The second term is the derivative of a cosine.

> **MATH CONCEPT** The derivative of
>
> $$\frac{d}{dt}\cos(k\theta) \longrightarrow -k\sin(\theta)$$

Therefore, this produces the following result:

$$-e^{-\alpha t}[K_1 \omega_d \sin(\omega_d t)] \tag{28.10}$$

The third term is the derivative of a sine.

> **MATH CONCEPT** The derivative of
> $$\frac{d}{dt}\sin(k\theta)\longrightarrow k\cos(\theta)$$

Therefore, this produces the following result:

$$e^{-\alpha t}[K_2\omega_d\cos(\omega_d t)] \tag{28.11}$$

By putting terms (28.9), (28.10), and (28.11) together in Eq. (28.8), we obtain

$$\frac{V_L}{L} = -\alpha e^{-\alpha t}[K_1\cos(\omega_d t) + K_2\sin(\omega_d t)] - e^{-\alpha t}[K_1\omega_d\sin(\omega_d t)]$$
$$+ e^{-\alpha t}[K_2\omega_d\cos(\omega_d t)]$$

Rewriting

$$\frac{V_L}{L} = -\alpha e^{-\alpha t}[K_1\cos(\omega_d t) + K_2\sin(\omega_d t)] - e^{-\alpha t}[K_1\omega_d\sin(\omega_d t)]$$
$$+ e^{-\alpha t}[K_2\omega_d\cos(\omega_d t)]$$

Substituting the K_1 that we have found in (28.7), we get

$$\frac{V_L}{L} = -\alpha e^{-\alpha t}[(0)\cos(\omega_d t) + K_2\sin(\omega_d t)] - e^{-\alpha t}[(0)\omega_d\sin(\omega_d t)]$$
$$+ e^{-\alpha t}[K_2\omega_d\cos(\omega_d t)]$$

> Because we are taking the derivative at $t = 0$ s, we can discard also the terms containing $\sin(\omega_d t)$, because $\sin(0)$ is zero.
>
> Therefore, we obtain
>
> $$\frac{V_L}{L} = e^{-\alpha t}[K_2\omega_d\cos(\omega_d t)]$$

As soon as the switch is closed, no current crosses the inductor because it generates an electromotive force to block the flow of current. This electromotive force is equal to the power supply voltage. Therefore, the voltage across the inductor V_L, the moment the switch is closed, is equal to V_1.

Therefore, upon substituting the known values on the equation

$$\frac{V_1}{L} = e^{-\alpha t}[K_2\omega_d\cos(\omega_d t)]$$

we get

$$\frac{12}{10} = K_2 e^{-0.005t} [0.3161 \cos(0.3161t)]$$

Solving for t = 0 s

$$\frac{12}{10} = K_2 e^0 [0.3161 \cos(0)]$$

$$\frac{12}{10} = K_2 \, e^{\cancel{01}} [0.3161 \cos(0)]$$

$$\frac{12}{10} = K_2 [0.3161]$$

$$1.2 = 0.3161 K_2$$

$$K_2 = 3.7963$$

We can now assemble the final equation as

$$i(t) = e^{-\alpha t} [K_1 \cos(\omega_d t) + K_2 \sin(\omega_d t)]$$

$$i(t) = e^{-0.005t} [0 + 3.7963 \sin(0.3161t)]$$

FINAL RESULT

$$i(t) = 3.7963 e^{-0.005t} \sin(0.3161t)$$

The current for t = 2 s is

$$i(2\,s) = 3.7963 e^{-0.005(2)} \sin(0.3161(2))$$

FINAL RESULT

$$i(2\,s) = 2.2218\,A$$

28.4.2   Example 2

Consider the same circuit shown in Figure 28.5 in the previous example with two changes: the resistor is an 8 Ω, instead of 0.1 Ω, and the power supply is a 10 V direct voltage source, instead of 12 V.

This new circuit is ruled by the following second-order differential equation:

$$2\frac{di^2(t)}{dt^2} + 2\frac{R}{L}\frac{di(t)}{dt} + \frac{1}{8}i = 0$$

Find the equation for the current that rules this circuit.

28.4.2.1 Solution

We divide the circuit's equation by two to rewrite it using the standard form:

$$\frac{di^2(t)}{dt^2} + \frac{di(t)}{dt} + \frac{1}{16}i = 0$$

Series RLC circuits follow second-order differential equations in the form

$$\frac{di^2(t)}{dt^2} + \frac{R}{L}\frac{di(t)}{dt} + \frac{1}{LC}i = 0$$

By comparing these two equations, we can extract the following information:

$$\frac{R}{L} = 1$$

We get L by substituting R = 8 Ω:

$$L = \frac{8}{1} = 8\,H$$

We can also extract

$$\frac{1}{LC} = \frac{1}{16}$$

Therefore, we obtain C:

$$C = \frac{16}{L} = \frac{16}{8} = 2\,F$$

Using the values of R, L, and C, we can obtain

$$\alpha = \frac{R}{2L}$$

$$\alpha = \frac{8}{2(8)}$$

$$\alpha = 0.5$$

$$\omega_o = \frac{1}{\sqrt{LC}}$$

$$\omega_o = \frac{1}{\sqrt{8(2)}}$$

$$\omega_o = 0.25$$

Because $\alpha > \omega_0$, this is an overdamped series RLC circuit, and the current equation follows the form

$$i(t) = Ae^{m_1 t} + Be^{m_2 t} \qquad (28.12)$$

To find A and B, we apply the initial conditions.

28.4.2.1.1 First Condition: Solve the Current Equation for t = 0 s

We take equation (28.12) and solve for t = 0 s:

$$i(0) = 0 = Ae^{m_1 \cancel{t}^0} + Be^{m_2 \cancel{t}^0}$$

MATH CONCEPT
$$e^0 = 1$$

$$0 = A + B$$
$$0 = A + B$$

$$A = -B$$

28.4.2.1.2 Second Condition: Take the Derivative of the Current Equation and Solve for t = 0 s

If we take the derivative of the current equation (28.12), we get

$$\frac{di(t)}{dt} = \frac{d}{dt}[Ae^{m_1 t}] + \frac{d}{dt}[Be^{m_2 t}] \tag{28.13}$$

The first term is the derivative of current equation with respect to time. But we know that the voltage across the inductor is equal to

$$V_L = L\frac{di(t)}{dt} \tag{28.14}$$

If we rearrange (28.14), we can write it as the derivative of current with respect to time:

$$\frac{di}{dt} = \frac{V_L}{L}$$

which we can substitute in (28.13), making it like

$$\frac{V_L}{L} = \frac{d}{dt}[Ae^{m_1 t}] + \frac{d}{dt}[Be^{m_2 t}]$$

The voltage across the inductor, or V_L, is the voltage across the inductor immediately after the switch is closed, which is known to be the same voltage as the power supply, as explained before.

Therefore, the equation can be converted into

$$\frac{10}{8} = \frac{d}{dt}[Ae^{m_1 t}] + \frac{d}{dt}[Be^{m_2 t}]$$

$$1.25 = \frac{d}{dt}[Ae^{m_1 t}] + \frac{d}{dt}[Be^{m_2 t}]$$

$$1.25 = m_1 Ae^{m_1 t} + m_2 Be^{m_2 t}$$

Solving for t = 0 s,

$$1.25 = m_1 Ae^0 + m_2 Be^0$$

$$1.25 = m_1 A\,e^{\emptyset 1} + m_2 B\,e^{\emptyset 1}$$

$$1.25 = m_1 A + m_2 B$$

We know that A = − B; therefore,

$$1.25 = m_1(-B) + m_2B$$
$$1.25 = B(m_2 - m_1)$$

$$B = \frac{1.25}{m_2 - m_1}$$

To find m_1 and m_2, we must find the equation roots:

$$\frac{d^2V_C}{dt^2} + \frac{dV_C}{dt} + \frac{1}{16}V_C = 0$$

which is equal to

$$m^2 + m + \frac{1}{16} = 0$$

and we can find its roots by using the quadratic equation:

$$m = \frac{-b \pm \sqrt{b^2 - 4ac}}{2a}$$

$$m = \frac{-1 \pm \sqrt{1^2 - 4(1)\left(\frac{1}{16}\right)}}{2(1)}$$

$$m = \frac{-1 \pm \sqrt{0.75}}{2}$$

Therefore, m_1 is

$$m_1 = \frac{-1 + \sqrt{0.75}}{2}$$

$$m_1 = -0.0670$$

and m_2 is

$$m_2 = \frac{-1 - \sqrt{0.75}}{2}$$

$$m_2 = -0.9330$$

Therefore,

$$B = \frac{1.25}{m_2 - m_1}$$

$$B = \frac{1.25}{-0.9330 - (-0.0670)}$$

$$B = -1.4434$$

and A will be

$$A = -B$$

$$A = 1.4434$$

Finally, the equation can be written as follows.

FINAL RESULT

$$i(t) = 1.4434e^{-0.0670t} - 1.4434e^{-0.9330t}$$

Exercises

1   Figure 28.6 shows a circuit like the one we have been analyzing in this chapter, a resistor, an inductor, a capacitor, and a switch in series with a direct voltage power supply.

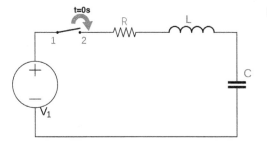

Figure 28.6 A basic RLC circuit.

The switch is initially open, the inductor is completely de-energized, and the capacitor is completely discharged.

The circuit's components have the following values:

$$\begin{cases} R = 100 \ \Omega \\ L = 2 \ H \\ C = 800 \ \mu F \\ V_1 = 30 \ V \end{cases}$$

At time t = 0 s, the switch is closed, and current flows and reaches the inductor.

The following is what we want to find about the circuit:

- What is the current flowing in the circuit as soon as the switch is closed?
- What is the current flowing in the circuit 10 ms after the switch is closed?

2   Consider a series RLC circuit like the previous one. The values of L and C are unknown but the resistor R = 1 Ω and the power supply V_1 = 30 V.

This circuit is ruled by the following equation:

$$400 \frac{d^2 i}{dt^2} + 40 \frac{di}{dt} + i = 0$$

We know that the current is 0 for t = 0 s.
In t = 0 s, the switch is closed.
Find the current equation and its value for t = 1 ms.

Solutions

1   **What is the current flowing in the circuit as soon as the switch is closed?**

After the switch is closed, current flows and reaches the inductor. The inductor hates sudden current variations and will not let this new current pass.

FINAL RESULT

$i(0^+) = 0 \ mA$

What is the current flowing in the circuit 5 ms after the switch is closed?

We can have three kinds of series RLC circuits, namely, critically damped, overdamped, and underdamped, each one with its own equations for the current.

The first thing we need to do is to know which circuit is this one. The given values are

$$\begin{cases} R = 100 \ \Omega \\ L = 2 \ H \\ C = 800 \ \mu F \\ V_1 = 30 \ V \end{cases}$$

We first calculate α and ω_o:

$$\alpha = \frac{R}{2L}$$

$$\alpha = \frac{100}{2(2)}$$

$$\alpha = 25$$

$$\omega_0 = \frac{1}{\sqrt{LC}}$$

$$\omega_0 = \frac{1}{\sqrt{2(800 \times 10^{-6})}}$$

$$\omega_0 = 25$$

We see that $\alpha = \omega_0$, equivalent to a critically damped circuit. The equation for the current, for this case, is

$$i(t) = (At + B)e^{-\alpha t}$$

To find A and B, we apply the circuit's initial conditions.
First Condition: Solve the Current Equation for t = 0 s.
Here we go:

$$i(t) = 0 = \left(A t^0 + B\right)e^{-\alpha t^0}$$

$$B = 0$$

Second Condition: Take the Derivative of the Current Equation and Solve for t = 0 s.
We take the derivative of the current equation

$$i(t) = (At + B)e^{-\alpha t}$$

$$\frac{d}{dt}[i(t)] = \frac{d}{dt}[(At + B)e^{-\alpha t}]$$

Voltage across the inductor is

$$V_L = L\frac{di}{dt}$$

Therefore,

$$\frac{di}{dt} = \frac{V_L}{L}$$

where V_L is the voltage across the inductor as the switch is closed, which is known to be equal to the power supply voltage, as explained before.
Therefore,

$$\frac{d}{dt}[i(t)] = \frac{V_L}{L} = \frac{30}{2} = \frac{d}{dt}[(At + B)e^{-\alpha t}]$$

$$15 = \frac{d}{dt}[(At + B)e^{-\alpha t}]$$

We are before the derivative of a product of functions and we can use the product rule:

> MATH CONCEPT   The derivative of a product of two functions is equal to the derivative of the first function multiplied by the second function added to the first function multiplied by the derivative of the second, or
>
> $(u.v)' = u'v + uv'$

Therefore,

$$15 = \frac{d}{dt}[(At + B)]e^{-\alpha t} + (At + B)\frac{d}{dt}[e^{-\alpha t}]$$

$$15 = Ae^{-\alpha t} + (At + B)\frac{d}{dt}[e^{-\alpha t}]$$

We already know how to solve the derivative of an exponential; therefore,

$$15 = Ae^{-\alpha t} + (At + B)(-\alpha e^{-\alpha t})$$

$$15 = Ae^{-\alpha t} - \alpha(At + B)e^{-\alpha t}$$

$$15 = Ae^{-\alpha t} - \alpha Ate^{-\alpha t} - \alpha Be^{-\alpha t}$$

By substituting B = 0, we get

$$15 = Ae^{-\alpha t} - \alpha Ate^{-\alpha t}$$

Solving for t = 0 s,

$$15 = Ae^{-\alpha t^0} - \alpha A t^0 e^{-\alpha t^0}$$

$$A = 15$$

Thus, we can write the final current equation as follows.

FINAL RESULT

$$i(t) = 15te^{-25t}$$

The current for t = 10 ms is

$$i(10\,ms) = 15(10 \times 10^{-3})e^{-25(10 \times 10^{-3})}$$

FINAL RESULT

$$i(10\,ms) = 0.1168\,A$$

2   We must find the values of L and C.
First, we take the circuit's second-order differential equation and divide it by 400 to write it in the standard form:

$$400\frac{d^2i}{dt^2} + 40\frac{di}{dt} + i = 0$$

The result is

$$\frac{d^2i}{dt^2} + \frac{1}{10}\frac{di}{dt} + \frac{1}{400}i = 0$$

Series RLC circuits are ruled by second-order differential equations in the general form:

$$\frac{di^2(t)}{dt^2} + \frac{R}{L}\frac{di(t)}{dt} + \frac{1}{LC}i = 0$$

By comparing this equation with the original in the standard form, we can extract the following values:

$$\frac{R}{L} = \frac{1}{10}$$

By substituting R = 1 Ω, we get,

$$L = 10\,H$$

Another value we can extract is

$$\frac{1}{LC} = \frac{1}{400}$$

Therefore,

$$C = \frac{400}{L} = \frac{400}{10} = 40\,F$$

Knowing R, L, and C, we can find other parameters:

$$\alpha = \frac{R}{2L}$$

$$\alpha = \frac{1}{2(10)}$$

$$\alpha = 0.05$$

$$\omega_0 = \frac{1}{\sqrt{LC}}$$

$$\omega_0 = \frac{1}{\sqrt{10(40)}}$$

$$\omega_o = 0.05$$

We see that $\alpha = \omega_0$, equivalent to a critically damped circuit. The equation for the current, for this case, is

$$i(t) = (At + B)e^{-\alpha t}$$

To find A and B, we apply the circuit's initial conditions.
First Condition: Solve the Current Equation for t = 0 s.

$$i(0) = 0 = \left(A t^0 + B\right)e^{-\alpha t^0}$$

$$0 = (0 + B)e^0$$

$$B = 0$$

Second Condition: Take the Derivative of the Current Equation and Solve for t = 0 s.
Like we did in the previous example, we take the derivative of the current equation and will get to the following point:

$$\frac{di(t)}{dt} = \frac{V_L}{L} = \frac{d}{dt}[Ate^{m_1 t}] + \frac{d}{dt}[Be^{m_2 t}]$$

V_L is the voltage across the inductor as the switch is closed, known to be the same as the power supply.
Therefore,

$$\frac{di(t)}{dt} = \frac{V_L}{L} = \frac{30}{10} = \frac{d}{dt}[Ate^{m_1 t}] + \frac{d}{dt}[Be^{m_2 t}]$$

$$3 = A\frac{d}{dt}[-te^{-\alpha t}] + B\frac{d}{dt}[e^{-\alpha t}]$$

In the last example, we already explained how to solve this:

$$3 = A\left\{\frac{d}{dt}[t]e^{-\alpha t} + t\frac{d}{dt}[e^{-\alpha t}]\right\} + B\frac{d}{dt}[e^{-\alpha t}]$$

$$3 = A\{1 - \alpha t e^{-\alpha t}\} - \alpha B e^{-\alpha t}$$

Solving for t = 0 s,

$$3 = A - \alpha B$$

By substituting B = 0,

$$A = 3$$

Therefore, the final current for the equation can be written as follows.

FINAL RESULT

$$i(t) = 3te^{-0.05t}$$

Current for t = 1 ms is

$$i(1\,\text{ms}) = 3(1 \times 10^{-3})e^{-0.05(1 \times 10^{-3})}$$

FINAL RESULT

$$i(1\,\text{ms}) = 2.99\,\text{mA}$$

29

Transistor Amplifiers

The Magic Component

29.1   Introduction

In this chapter, we will examine the usage of transistors as amplifiers.

We will show how transistors are able to "amplify"[1] signals and will briefly introduce the reader to how they work.

29.2   Transistor as Amplifiers

Nothing is farther from the truth than thinking transistors are magical devices that can amplify the signals they receive, like a magical portal that receives the signal and amplifies it.

In fact, transistors are like switches that can manipulate and modulate the voltage coming from the power supply to generate a huge copy of the input signal, creating the illusion of amplification.

The output signal is not the input signal amplified but rather the power supply voltage modulated to look like the input signal.

29.3   The Water Storage Tank

Imagine a huge water store tank with a large output pipe. At the bottom of the tank, a tiny water valve is connected electronically to a huge water valve that opens or closes the main pipe. Therefore, if you rotate this tiny valve to one direction, it will make the huge water valve open and let a gigantic amount of water exit from the tank. If you rotate the tiny water valve to the other direction, the huge valve will close.

Thus, small rotations of the tiny valve will produce huge rotations of the largest valve.

---

1 The reader will understand why this word is between quotes later in this book.

Introductory Electrical Engineering with Math Explained in Accessible Language,
First Edition. Magno Urbano.
© 2020 John Wiley & Sons, Inc. Published 2020 by John Wiley & Sons, Inc.

This can be interpreted as an amplification: a small rotation producing a large rotation.

This is what roughly happens with transistors: a tiny current flowing into the input will produce a large current flowing across the output.

29.4   Current Gain

If the reader remembers what we have explained previously about transistors, a transistor, which symbol for an NPN type is shown in Figure 29.1, is a three-terminal device, with a collector (C), a base (B), and an emitter (E).

In an NPN type transistor, a small current flowing from the base to the emitter (base current) will produce a larger current flowing from the collector to the emitter (collector current). This effect is known as "current gain," also known as transistor h_{FE} or β.

Different transistors offer different values for the current gain, which can vary from 10, 20, and 50 to values like 800.

Figure 29.1 NPN transistor.

h_{FE} stands for **h**ybrid parameter, **f**orward transfer, and common **e**mitter.

Therefore, the collector current will be the base current multiplied by h_{FE}, or

$$i_C = h_{FE}\, i_B$$

29.5   Power Supply Rails

All electronic circuits are powered by some kind of voltage source. The voltage source can be a battery or an electrical device of some kind.

Suppose this voltage source is a 9 V battery. This battery has a positive and a negative pole. These two poles or voltage limits are known as rails.

In an amplifier, the output should never exceed the rails or distortion will appear at the output, and damage can happen to the power supply.

29.6   Amplifying

Suppose we want to amplify a sinusoidal wave like the one shown in Figure 29.2. This wave varies from 1 V to −1 V; therefore, this is a 2 V peak-to-peak voltage, or 2 V_{pp}. This means the audio signal varies between positive and negative values.

Figure 29.2 2 Vpp audio signal.

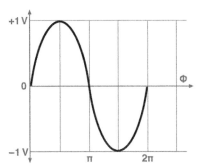

Suppose our amplifier is powered by a 9 V battery. If this is the case, the maximum output signal we can collect at the output, in theory, will be a 9 V_{pp} signal, as shown in Figure 29.3.

Figure 29.3 Input signal and maximum output possible.

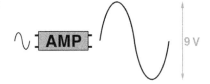

This happens because the amplifier output is limited by the power supply voltage.

Therefore, for an input signal of 2 V_{pp}, the amplifier will only be able to amplify the signal 4.5 times to get a 9 V_{pp} output.

When the output varies the full range possible, in this case from the battery negative pole or 0 V to the maximum battery voltage, or 9 V, equivalent to the battery positive pole, it is said that the output does a full swing or full rail, going from the minimum to the maximum rail, provided by the power supply.

> In the real world, due to project and component limitations, amplifiers will never be able to provide a full swing output.

29.7 Quiescent Operating Point

To work as an amplifier, a transistor must be configured properly to operate within a certain range of parameters. This operating point is called "quiescent operating point" or "Q-point."

Adjusting the operating point of a transistor is called biasing.

Suppose a transistor amplifier will be used to amplify audio signals. An audio signal has positive and negative values. If our amplifier is powered by 9 V battery and the output cannot go beyond this value, we must bias the transistor in a way that the output sits at 4.5 V, the middle of the rail, when there is no input signal. By biasing the transistor like that, the output will be able to swing from 4.5 to 9 V during the output positive cycle and from 4.5 to 0 V during the negative cycle (see Figure 29.4).

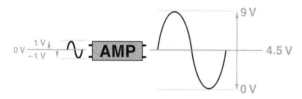

Figure 29.4 Input and output signals.

29.8 Amplifier Classes

Transistors can be used to create amplifiers in very different ways, called "classes." Every class is more or less energy efficient than another.

These classes are named with letters, like A, AB, B, C, D, E, F, G, H, S, and variations of B.

The simplest class of them all is class A and, for that reason, will be the theme of this chapter, specifically in a configuration called common emitter, as we will examine later in this chapter.

29.8.1 Class A

Class A amplifiers are probably the cheapest and the simplest to produce.

Class A amplifiers are good amplifiers with a relatively low output distortion.

The problem with class A is that it consumes too much power. The transistors in a class A configuration are always conducting current, meaning that they are consuming power even if there is no signal at the input.

Therefore, they are not the best amplifiers to use in portable devices, powered by batteries.

To improve the energy efficiency of class A amplifiers, engineers created other classes that improve the energy problem, but these new classes introduce other problems, like signal distortion.

> Like examined previously in this book, there are two kinds of transistors: NPN and PNP. All explanations in this chapter will be for amplifiers using NPN transistors because all principles can be applied for amplifiers using PNP transistors.

29.8.1.1 Common Emitter

To make a transistor work as an amplifier, it must be properly biased.

Biasing a transistor means adjusting its voltages and currents, so it can operate as an amplifier.

Common-emitter mode is the most popular way to use a transistor as an amplifier.

It is called "common-emitter" amplifier since the emitter is common to both the input and the output circuit.

Common-emitter amplifiers have the following characteristics:

• Signal enters by the base and exits by the collector.
• Medium input and output impedances.
• Medium current and voltage gains.
• High power gain.
• Input and output have a phase relationship of 180°.

29.8.1.1.1 Biasing a Transistor

Biasing a transistor to work as a common-emitter amplifier can be done in three basic ways: fixed base biasing, collector feedback biasing, and voltage divider biasing.

We will examine these configurations in the next sections.

Fixed Base Biasing This method for biasing a common-emitter class A amplifier is seen in Figure 29.5.

In Figure 29.5, V_{CC} represents the power supply's positive pole, and the triangle pointing down, known as the ground point, represents the power supply's negative pole. R_C is the collector resistor and R_B is the base resistor.

This fixed base biasing configuration uses a resistor connected between V_{CC} and the collector and another one from the same point to the transistor base. The purpose of these resistors is to provide voltage levels and currents for the transistor operation. The emitter is connected to the ground, as shown in Figure 29.5.

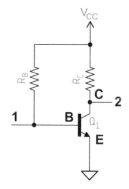

Figure 29.5 Fixed base biasing common-emitter amplifier.

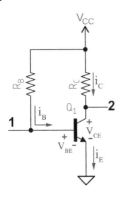

Figure 29.6 Currents and voltages in a fixed base biasing common-emitter amplifier.

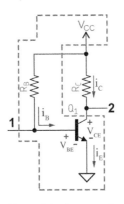

Figure 29.7 Input mesh of a fixed base biasing common-emitter amplifier.

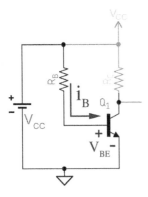

Figure 29.8 Equivalent input mesh of a fixed base biasing common-emitter amplifier.

Transistors are generally represented in diagram by the uppercase letter Q or T.

In the common-emitter configuration, the input signal enters by the transistor base, point 1 in Figure 29.5, and exits by the collector (C), or point 2. Therefore, the input signal will be connected between point 1 and ground, and the output will be collected between point 2 and ground.

To set the operating point for the fixed base biasing configuration, we start by specifying the currents. Figure 29.6 shows the several currents of this kind of configuration: the collector current (i_C), the base current (i_B), and the emitter current (i_E).

Figure 29.6 also shows the voltage across collector–emitter (V_{CE}) and the voltage base–emitter (V_{BE}).

Figure 29.6 shows that the base is positive compared with the emitter, meaning that the base must have a greater electrical potential. In fact, to work properly, the voltage at the base must be at least 0.7 V greater than the emitter. If so, current will flow from collector to emitter.

The signals of V_{CE} in Figure 29.6 show that the voltage at the collector will be greater than the emitter during normal operation.

Currents We start by applying Kirchhoff's voltage law (KVL) to the circuit's input mesh shown in Figure 29.7, composed of the base resistor, the power supply, and the base–emitter junction.

This mesh is equivalent to the circuit shown in Figure 29.8.

If we apply KVL to this mesh, we get

$$V_{CC} - R_B i_B - V_{BE} = 0$$

Therefore, we can find the base current by rearranging the equation

$$R_B i_B = V_{CC} - V_{BE}$$

$$i_B = \frac{V_{CC} - V_{BE}}{R_B} \tag{29.1}$$

We know that the collector current is

$$i_C = h_{FE}\, i_B \qquad (29.2)$$

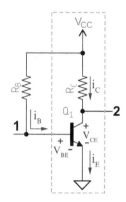

Figure 29.9 Output mesh of a fixed base biasing common-emitter amplifier.

In the next step, we find the voltage across collector–emitter by applying KVL to the output mesh, shown in Figure 29.9.

This output mesh is composed of the collector resistor, the collector–emitter block, and the power supply, as shown in Figure 29.9, and can be redrawn as shown in Figure 29.10.

If we apply KVL to this mesh, we get

$$V_{CC} - R_C i_C - V_{CE} = 0$$

Therefore,

$$V_{CE} = V_{CC} - R_C i_C$$

and

$$i_C = \frac{V_{CC} - V_{CE}}{R_C} \qquad (29.3)$$

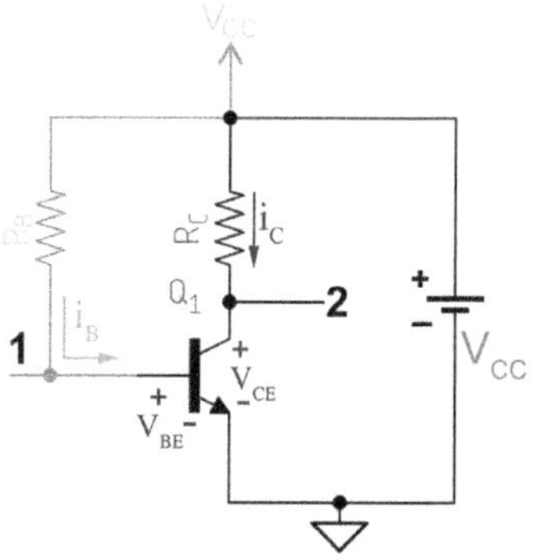

Figure 29.10 Output mesh equivalent of a fixed base biasing common-emitter amplifier.

A transistor can be one of three states, namely, cutoff, saturated, or conducting, this later being equal to the normal operating mode as an amplifier.

If a transistor is saturated, the collector–emitter voltage, V_{CE}, will be 0. Therefore, current will be flowing at maximum intensity, just limited by the collector resistor.

In this case, the saturation collector current will be

$$i_{C_{(sat)}} = \frac{V_{CC} - \cancel{V_{CE}}^{0}}{R_C}$$

$$i_{C_{(sat)}} = \frac{V_{CC}}{R_C}$$

On the other operation extreme, or when the transistor is cutoff or not conducting, V_{CE} will be equal to the power supply voltage, V_{CC}, because there is no current flowing across the collector; therefore,

$$V_{CE_{(off)}} = V_{CC}$$

If we want the transistor to work correctly as an amplifier, we must make sure that it operates in the middle point between saturation and cutoff in what is called the "active region." We also have to be sure that the input signal does not drive the circuit out of this region.

Setting It Up   We will set this project to use the low power general-purpose silicon-based transistor model 2N2222A and to be powered by a 9 V battery.

Transistor manufacturers publish datasheets about their products where the electric characteristics of their devices are specified at detail. The datasheet for transistor 2N2222A is available at MIT.

You can access their document using our shortcut at http://transistor.katkay.com

The datasheet shows the collector–emitter saturation voltage curve, shown in Figure 29.11.

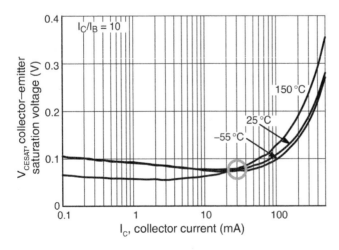

Figure 29.11 Collector–emitter saturation voltage for transistor 2N2222A.

This graph shows that the collector–emitter voltage keeps its value relatively stable until the collector current reaches about 30 mA. After that, its values increase exponentially. Therefore, we must choose a collector current that is below 30 mA to keep the transistor operating in a stable region. We choose 10 mA for this project.

We read the datasheet again and this time we look for the current gain curve (h_{FE}), shown in Figure 29.12.

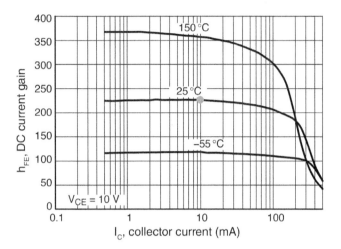

Figure 29.12 Current gain curve for transistor 2N2222A.

Figure 29.12 shows a value for the h_{FE} parameter equal to 225 for a current of 10 mA at room temperature.

Thermal Runaway   Figure 29.12 shows three curves for h_{FE}, one for each different temperature. The higher the temperature, the bigger the h_{FE} value, making it clear that h_{FE} increases when the temperature rises.

A transistor will normally heat during normal operation due to the current flowing between collector and emitter. This temperature rise will increase the h_{FE} value. The collector–emitter current is product of the base current and the h_{FE}. If h_{FE} rises, the collector–emitter current will increase and so on, which can lead to the transistor destruction.

This effect is called "thermal runaway," and the best way to prevent this situation is to add a resistor between the emitter and the ground, something that this circuit lacks.

The lack of an emitter resistor in this circuit is condition enough not to use this kind of project, but we will show other situations that makes this, in this fixed base biasing method not good as an amplifier.

The Design   We have established four values for our amplifier design so far: the power supply voltage, the collector current, the current gain, and the base–emitter voltage:

$$\begin{cases} V_{CC} = 9 \text{ V} \\ i_C = 10 \text{ mA} \\ h_{FE} = 225 \\ V_{BE} = 0.7 \text{ V} \end{cases}$$

The ideal amplifier would be the one that can provide a full swing output. In this given example, it would be an amplifier that could provide an output varying from 0 to 9 V.

But like we have explained before, a few conditions will prevent an amplifier from using the full rail. One of these conditions is the collector–emitter voltage.

Reading the graph in Figure 29.11, we see that the transistor 2N2222A has a collector–emitter saturation voltage of, approximately, 0.1 V for 10 mA. As a rule of thumb, we think that it is too dangerous, in terms of transistor stabilization, to have a collector–emitter voltage below 0.7 V, and we recommend always having it bigger than 1 V.

To make sure we will have a stable amplifier, we will choose to have this minimum voltage set to 2 V through this chapter. Therefore, the minimum collector–emitter we choose is

$$V_{CE_{MIN}} = 2 \text{ V}$$

The emitter is connected to the ground, as we can see in Figure 29.9. Consequently, 2 V is also the minimum collector voltage, or in other words, the collector voltage when the output is at its minimum value.

The collector is where the signal output exits, meaning that the output is collected between collector and ground.

We have just selected 2 V as the minimum voltage for the collector, meaning that no collector voltage must be below that point. That is the same as saying that the amplifier will only be able to vary from 2 to 9 V. Therefore, the larger output signal may have only the difference between these two values, that is, 7 V peak to peak.

Consequently, the output wave will only be able to vary by 3.5 V up and the same amount down.

If the maximum allowable voltage for the collector is 9 V, then it must be adjusted to be 3.5 V below that point when at rest, equivalent to 5.5 V = 9 − 3.5, a voltage we call "middle point collector voltage," or "collector at rest voltage," which can also be found by the following formula:

$$V_{C_{middle}} = \frac{V_{CC} - V_{C_{MIN}}}{2} + V_{C_{MIN}}$$

or

$$V_{C_{middle}} = \frac{V_{CC} + V_{C_{MIN}}}{2}$$

V_{CC} is the power supply voltage.
$V_{C_{MIN}}$ is the minimum collector voltage.
$V_{C_{middle}}$ is the collector voltage at rest.

Therefore,

$$V_{C_{middle}} = \frac{9 + 2}{2}$$

$$V_{C_{middle}} = 5.5 \, V$$

V_C is the collector voltage, measured between the collector and the ground. V_{CE} is the collector–emitter voltage, measured between these two elements.

Now the output can swing 3.5 V up, from 5.5 to 9 V during the positive cycle and 3.5 V down, from 5.5 to 2 V during the negative cycle.

Resistors We must calculate, now, the resistors that will make every chosen voltage and current possible.

This is what we have chosen so far:

$$\begin{cases} V_{CC} = 9\,V \\ i_C = 10 \text{ mA} = 10 \times 10^{-3}\,A \\ V_{CE} = 5.5 \text{ V} \end{cases}$$

We can start by finding the collector resistor by using Eq. (29.3):

$$R_C = \frac{V_{CC} - V_{CE}}{i_C}$$

$$R_C = \frac{9 - 5.5}{10 \times 10^{-3}}$$

$$R_C = 350 \ \Omega$$

We can use Eq. (29.2) to find the base current:

$$i_C = h_{FE} i_b$$

$$i_B = \frac{i_C}{h_{FE}}$$

$$i_B = \frac{10 \times 10^{-3}}{225}$$

$$i_B = 44.44 \ \mu A$$

But we also know by Eq. (29.1) that

$$i_B = \frac{V_{CC} - V_{BE}}{R_B}$$

and we can use this equation to find the base resistor:

$$R_B = \frac{V_{CC} - V_{BE}}{i_B}$$

$$R_B = \frac{9 - 0.7}{44.44 \ \mu A}$$

$$R_B = 186.75 \ k\Omega$$

AC Analysis   What happens to the circuit when an audio signal is applied to the input?

First, we must remember that the way we have designed this amplifier was in such a way to establish the currents flowing in the transistor, the base voltage as 0.7 V and the collector voltage at rest, to be at 5.5 V.

Suppose we now inject a sinusoidal audio signal at the transistor base, like the one shown in Figure 29.13. This signal can be from a microphone, for example, and varies from +40 to −40 μA.

Figure 29.13 Audio signal.

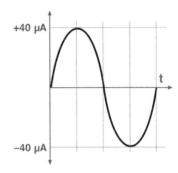

The Positive Cycle At the beginning, the audio signal sits at 0 μA (see Figure 29.13). Therefore, there is no current injected at the transistor base, and all other currents or voltages in the circuit keep their calculated values.

As time passes, the audio signal increases from 0 μA and reaches its peak at +40 μA.

We have calculated the base current to be 44.44 μA, but now, there is this new additional current of +40 μA coming from the audio signal, making the base current increase to 84.44 μA.

This new base current will make the collector current increase to

$$i_C = h_{FE}\, i_B$$

$$i_C = (225)\left(84.44 \times 10^{-6}\right)$$

$$i_C = 19 \text{ mA}$$

In other words, the collector current increased from 10 to 19 mA.

If we go back to the output mesh's KVL equation and solve for V_{CE}, using this new collector current, we get the new collector–emitter voltage:

$$V_{CE} = V_{CC} - R_C i_C$$

$$V_{CE} = 9 - (350)(19 \times 10^{-3})$$

$$V_{CE} = 2.35\,V$$

In resume, during the positive cycle of the audio signal, it varied from 0 to +40 µA, making the base current increase from 44.44 to 84.44 µA, forcing the collector voltage to vary from 5.5 to 2.35 V.

In other words, an increase in the input current produced a decrease in the output voltage, showing that this amplifier is an inverter, meaning that all signals injected at the input will be collected at the output inverted, with a 180° phase offset.

The Negative Cycle   In this section we will see what happens during the negative cycle of the audio input.

The time passes and the audio signal decreases to −40 µA. This new negative current is added to the base current, making it drop from 44.44 to only 4.44 µA. This new current will produce the following collector current:

$$i_C = h_{FE}\,i_B$$

$$i_C = (225)(4.44 \times 10^{-6})$$

$$i_C \approx 1\,mA$$

If we go back again to the output mesh's KVL equation and solve for V_{CE} using this new current, we get

$$V_{CE} = V_{CC} - R_C i_C$$

$$V_{CE} = 9 - (350)(1 \times 10^{-3})$$

$$V_{CE} = 8.65\,V$$

In resume, the audio signal varied from 0 to −40 μA, making the base current drop from 44.44 to 4.44 μA, forcing the collector voltage to vary from 5.5 to 8.65 V.

In other words, a decrease in the input current produced an increase in the output voltage, confirming this amplifier as an inverter.

Conclusion This fixed biased transistor amplifier works as an amplifier within certain parameters, but it is rarely used in real life because it has a lot of problems.

The first problem is lacking an emitter resistor, which can lead the transistor to self-destruction, for the reasons we have explained.

The other problem is that the design is highly dependable on the h_{FE} value. In real life, even the same type of transistor can have different values for h_{FE}. The slightest difference in h_{FE} will result in different voltages and currents across the circuit.

To see the dependence of h_{FE} of this kind of circuit, consider the voltage gain of this amplifier.

The voltage gain will be the output voltage divided by the input voltage.

Looking at Figure 29.6, we can see that the output voltage is

$$V_{OUT} = V_{CC} - R_C i_C$$

and that the input voltage is

$$V_{IN} = V_{CC} - R_B i_B$$

Thus, the voltage gain is

$$V_G = \frac{V_{OUT}}{V_{IN}} = \frac{V_{CC} - R_C i_C}{V_{CC} - R_B I_B}$$

We have i_C and i_B in the formula and we know that

$$i_C = h_{FE} i_B$$

Therefore, the voltage gain is

$$V_G = \frac{V_{CC} - R_C h_{FE} i_B}{V_{CC} - R_B i_B}$$

You can see that h_{FE} is in the voltage gain formula and that i_B is in both nominator and denominator, making the voltage gain dependent on these values.

Even the same kind of transistor hardly has the same h_{FE}. Imagine designing a product to be sold to thousands of customers that behave differently for every person.

With this kind of design, voltages and currents do not remain stable during transistor operation and can vary enormously, and even if the transistor does not self-destruct, due to the thermal runaway problem, its temperature will affect and change h_{FE}, increasing the output current and disturbing all other voltages and currents across the circuit.

For all these problems, this circuit design is not recommended.

In the following section, we will examine other types of transistor biasing and see if we can solve this problem.

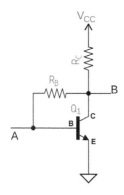

Figure 29.14 Collector feedback biasing.

Collector Feedback Biasing Figure 29.14 shows a collector feedback biasing circuit where R_C is the collector resistor and R_B is the base resistor or the feedback resistor. The base resistor is connected to the collector, instead of being connected to the power supply.

This situation creates what is called collector feedback. The word feedback is used because part of the collector current is injected into the base. The collector feedback current ensures that the transistor is always biased in the active region.

Like we have seen previously, when the collector current increases, the voltage at the collector drops, because the voltage drop across the collector resistor increases. This reduces the base current that in turn reduces the collector current and so on until it stabilizes at a certain value.

We will see in the next paragraphs that this kind of amplifier design is also an inverter amplifier.

This feedback resistor is, in fact, creating a negative feedback into the input, because part of the inverted output signal is being injected into the input.

There are several benefits of negative feedback in any kind of amplifier like improving system stabilization, reducing distortion and noise, and improving the system bandwidth. However, a reduction in gain is present for all amplifiers using feedback.

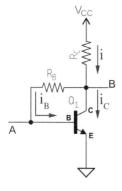

Figure 29.15 Collector feedback biasing (currents).

Currents The first thing is to establish the currents flowing in the circuit, like shown in Figure 29.15.

By the diagram in Figure 29.15, we can establish that

$$i = i_B + i_C$$

We know that

$$i_C = h_{FE}\, i_B$$

Therefore, current across R_C will be

$$i = i_B + h_{FE} i_B$$

$$i = (h_{FE} + 1)i_B \qquad\qquad (29.4)$$

Consequently, voltage drop across R_C is

$$V_{R_C} = (h_{FE} + 1)i_B R_C$$

If we apply KVL to the input mesh (Figure 29.16), formed by the collector and the base resistor and the base–emitter junction, we get

$$V_{CC} - V_{R_C} - V_{R_B} - V_{BE} = 0$$

which means that the power supply voltage subtracted from the voltage drops across the collector and the base resistors subtracted from the base–emitter voltage is equal to 0.
This can be expanded into

$$V_{CC} - iR_C - R_B i_B - V_{BE} = 0$$

Upon substituting (29.4),

$$V_{CC} - (h_{FE} + 1)i_B R_C - R_B i_B - V_{BE} = 0$$

Therefore, the base current will be

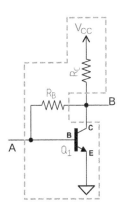

Figure 29.16 Input mesh of a collector feedback biasing circuit.

$$i_B = \frac{V_{CC} - V_{BE}}{R_B + (h_{FE} + 1)R_C} \quad (29.5)$$

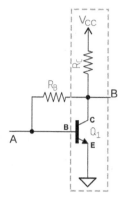

If we apply KVL to the output mesh (Figure 29.17), formed by the collector resistor and the collector–emitter block, we get

$$V_{CC} - V_{R_C} - V_{CE} = 0$$

This can be expanded into

Figure 29.17 Output mesh of a collector feedback biasing circuit.

$$V_{CC} - R_C i - V_{CE} = 0$$

Upon substituting (29.4),

$$V_{CC} - (h_{FE} + 1)i_B R_C - V_{CE} = 0$$

Therefore,

$$i_C = \frac{h_{FE}(V_{CC} - V_{BE})}{R_B + (h_{FE} + 1)R_C} \quad (29.6)$$

By comparing the base current (29.5) and the collector current (29.6), it is clear that both the collector and the base currents are highly dependable upon h_{FE}.

To see the dependence of h_{FE} of this kind of circuit, consider the voltage gain of this amplifier.

The voltage gain will be the output voltage divided by the input voltage. Looking at Figure 29.6, we can see that the output voltage is

$$V_{OUT} = V_{CC} - R_C i_C$$

and that the input voltage is

$$V_{IN} = V_{CC} - R_C i_C - R_B i_B$$

Thus, the voltage gain is

$$V_G = \frac{V_{OUT}}{V_{IN}} = \frac{V_{CC} - R_C i_C}{V_{CC} - R_C i_C - R_B i_B}$$

We have i_C and i_B in the formula and we know that

$$i_C = h_{FE} i_B$$

Therefore, the voltage gain is

$$V_G = \frac{V_{CC} - R_C h_{FE} i_B}{V_{CC} - R_C h_{FE} i_B - R_B i_B}$$

You can see that h_{FE} is in the voltage gain formula and that i_B is in both nominator and denominator, making the voltage gain dependent on these values.

This design shows the same problem as the fixed base biasing one: currents and voltages will vary wildly if h_{FE} changes. An additional issue with this design is the lack of an emitter resistor to prevent the thermal runaway problem.

We must continue our epic journey to find another biasing method that can solve these problems.

29.8.1.1.2 *Voltage Divider Biasing*

This method of transistor biasing is shown in Figure 29.18, known as voltage divider biasing, and is the most popular method of biasing a transistor.

In this kind of biasing, R_C is the collector resistor and R_B is the base resistor.

R_1 and R_2 are called the bias resistors, configured as a voltage divider that runs a large current across, compared with the base current, and helps to establish a stable base voltage and minimize any variations in the base current.

The thermal runaway problem is finally solved by adding an emitter resistor.

The emitter resistor R_E is, in fact, the sum of two resistors: the emitter resistor R_E itself and a resistor we will call "gain resistor," proper from the transistor, that controls the voltage gain or the amplification factor. We will disregard this resistor in this book because it falls outside the book's objective.

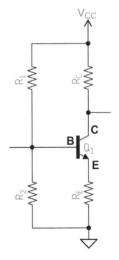

Figure 29.18 Voltage divider biasing.

Parameters The first thing we must do is to establish the transistor operating point.

We have added an emitter resistor to solve the thermal runaway. A current will flow across this resistor and cause a voltage drop. Therefore, the emitter (Figure 29.18) will have a voltage that is not 0, like the previous designs we have examined, because the emitter is not connected to the ground anymore.

If our intention is to make an amplifier with the maximum output voltage swing possible, we must choose a value for the emitter voltage that is the lowest we can get, because this voltage will reduce the output swing by the same amount.

Remember when we first examined this that the possible output voltage range is VCC subtracted from the collector–emitter voltage. Now with another voltage drop on the output mesh, the maximum output range possible will be VCC subtracted from the collector–emitter voltage and the emitter resistor voltage drop.

It is recommended to have the voltage drop across the emitter resistor around 1 V.

Like the last time, we chose a collector current of 10 mA, which will result in a value for h_{FE} equal to 225 for transistor 2N2222A.

This transistor is silicon based; therefore, its required base–emitter voltage is 0.7 V.

We will use a 12 V power supply for this project, instead of a 9 V battery. These are the parameters we have so far:

$$\begin{cases} V_{CC} = 12 \text{ V} \\ i_C = 10 \text{ mA} \\ h_{FE} = 225 \\ V_{BE} = 0.7 \text{ V} \\ V_E = V_{R_E} = 1 \text{ V} \end{cases}$$

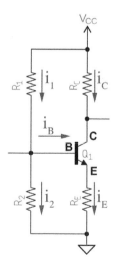

Figure 29.19 Voltage divider biasing (currents).

Currents   Figure 29.19 shows the currents flowing in the voltage divider biasing circuit.

To find the current equations, we start by defining the circuit's input and output meshes, shown in Figure 29.20.

By applying KVL to the input mesh, we get

$$V_{CC} - V_{R_1} - V_{R_2} = 0$$

$$V_{CC} - R_1 I_1 - R_2 I_2 = 0 \qquad (29.7)$$

Doing the same for the output mesh, we obtain

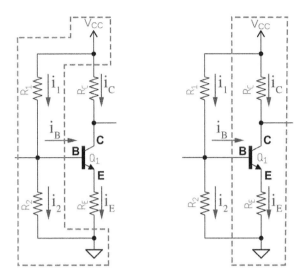

Figure 29.20 Voltage divider biasing (input and output meshes).

$$V_{CC} - V_{R_C} - V_{CE} - V_{R_E} = 0$$

$$V_{CC} - R_C I_C - V_{CE} - R_E I_E = 0 \qquad (29.8)$$

Design   Like the previous project, the collector needs a minimum voltage that is recommended not to be less than 2 V.

Therefore,

$$V_{CE_{MIN}} = 2 \text{ V}$$

If the emitter voltage was chosen to be 1 V and $V_{CE_{MIN}}$ is 2 V, then the collector minimum voltage will be the sum of these two values, or

$$V_{C_{MIN}} = V_{CE\,MIN} + V_{R_E}$$
$$V_{C_{MIN}} = 2 + 1$$

$$V_{C_{MIN}} = 3 \text{ V}$$

This means that the output voltage will not be able not to vary from 0 to 12 V but rather from 3 to 12 V.

Therefore, to have maximum output amplitude, we must set the collector voltage to be at the middle of this range, or in the middle value between 3 and 12 V, or

$$V_C = \frac{V_{CC} + V_{C_{MIN}}}{2}$$

$$V_C = \frac{12 + 3}{2}$$

$$V_C = 7.5 \text{ V}$$

Consequently, the output voltage will be able to vary from 7.5 to 12 V during the positive cycle and from 7.5 to 3 V during the negative cycle, a 4.5 V variation for both sides.

We have the collector current and h_{FE}; therefore we can find the base current:

$$i_B = \frac{i_C}{h_{FE}}$$

$$i_B = \frac{10 \times 10^{-3}}{225}$$

$$i_B = 44.4444 \ \mu A$$

The emitter current is equal to the sum of collector and base currents, but because the base current is infinitely less than the collector current, we can ignore it and say that, for most purposes, the emitter current is equal to the collector current.

However, in this book, we will always consider the base current.

Therefore, the emitter current is

$$i_E = i_C + i_B$$

$$i_E = \left(10 \times 10^{-3}\right) + \left(44.4444 \times 10^{-6}\right)$$

$$i_E = 10.0444 \text{ mA}$$

Knowing these values, we can find the emitter resistor:

$$V_{R_E} = R_E i_E$$

$$1 = \left(10.0444 \times 10^{-3}\right) R_E$$

$$R_E \approx 100 \ \Omega$$

Similarly, the voltage drop across the collector resistor is

$$V_{R_C} = R_C i_C \qquad (29.9)$$

The collector voltage is also equal to the sum of the voltage across the collector–emitter and the voltage across the emitter resistor, or

$$V_C = V_{CE} + V_{R_E}$$

If we substitute this in the output mesh equation in equation (29.8), we get another way to represent the voltage across the collector resistor:

$$V_{CC} - V_{R_C} - V_{CE} - V_{R_E} = 0$$
$$V_{CC} - V_{R_C} - V_C = 0$$

$$V_{R_C} = V_{CC} - V_C \qquad (29.10)$$

which is the power supply voltage subtracted from the collector voltage we have already established before.

Therefore, because Eqs. (29.9) and (29.10) represent the same thing, we can equate them and find the collector resistor:

$$V_{CC} - V_C = R_C i_C$$

$$12 - 7.5 = \left(10 \times 10^{-3}\right) R_C$$

$$R_C = 450 \ \Omega$$

The bias resistors R_1 and R_2 are configured as a voltage divider.
Current i_1 flows across R_1, and a current i_2 flows across R_2 (Figure 29.19).
Current i_1 will be the sum of the base current i_B with i_2, or

$$i_1 = i_B + i_2$$

What this means is that the base current will be a fraction of the current flowing across R_1. In other words, the base current is drained from the current flowing across R_1.

We must consider that the currents across R_1 and R_2 will establish a voltage divider that will establish the base voltage. Therefore, if we want this base voltage to be as invariable as possible, knowing the base current will be drained from the current i_1 across R_1, we must choose a value for i_1 that is much larger than the base current. This will make sure that the base current being drained from i_1 will not disturb the voltage set by the divider.

As a rule of thumb, we generally choose i_1 to be 10 times smaller than the collector current i_C, or

$$i_1 = 0.1\,i_C$$

We already know the collector current. Therefore, we can calculate i_1:

$$i_1 = 0.1(10 \text{ mA})$$

$$i_1 = 1 \text{ mA}$$

We know that the base voltage will be equal to the sum of the emitter voltage and the base–emitter voltage, which, for silicon-based transistors, is approximately equal to 0.7 V.

We also know the emitter voltage; therefore, we can find the base voltage:

$$V_B = V_{BE} + V_E$$

$$V_B = 0.7 + 1$$

$$V_B = 1.7 \text{ V}$$

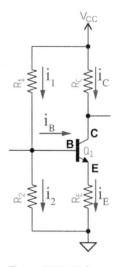

Figure 29.21 Voltage divider biasing – The base.

In Figure 29.19, we see that the base is connected between resistors R_1 and R_2, as shown in Figure 29.21.

Resistor R_2 is connected between the base and ground. The voltage across this resistor is equal to the product of its resistance and current, which is also the base voltage in relation to ground, or

$$V_{R_2} = R_2 i_2 = V_B$$

We have two equations for V_A. Consequently we can equate them:

$$R_2 I_2 = V_B$$

We already know that

$$\begin{cases} i_1 = 1 \text{ mA} \\ i_B = 44.4444 \text{ μA} \end{cases}$$

and

$$i_1 = i_2 + i_B$$

Therefore,

$$i_2 = i_1 - i_B$$

$$i_2 = 1 \text{ mA} - 44.4444 \text{ μA}$$

$$i_2 = 10^{-3} - 44.4444 \times 10^{-6}$$

$$i_2 = 1 \text{ mA} - 44.4444 \text{ μA}$$

$$i_2 = 955.5555 \text{ μA}$$

We can now find R_2:

$$V_{R_2} = R_2 i_2$$

$$R_2 = \frac{V_{R_2}}{i_2} = \frac{V_B}{i_2}$$

$$R_2 = \frac{V_B}{i_2}$$

Substituting

$$\begin{cases} V_{R_2} = V_B = 1.7 \text{ V} \\ i_2 = 955.5555 \text{ μA} \end{cases}$$

$$R_2 = \frac{1.7}{955.5555 \times 10^{-6}}$$

$$R_2 = 1779.06 \ \Omega$$

To find R_1, we go back to the input mesh equation:

$$V_{CC} - V_{R_1} - V_{R_2} = 0$$

We already know that

$$\begin{cases} V_{R_2} = V_B = 1.7 \, V \\ V_{CC} = 12 \, V \end{cases}$$

Therefore,

$$V_{R_1} = V_{CC} - V_{R_2}$$

$$V_{R_1} = 12 - 1.7$$

$$V_{R_1} = 10.3 \ V$$

We already know i_1, and therefore we can find R_1:

$$V_{R_1} = R_1 i_1$$

$$R_1 = \frac{V_{R_1}}{i_1}$$

$$R_1 = \frac{10.3}{1 \times 10^{-3}}$$

$$R_1 = 10.3 \ k\Omega$$

In resume, these are the values we have designed so far:

$$\begin{cases} R_1 = 10.3 \ k\Omega \\ R_2 = 1779.06 \ \Omega \\ R_C = 450 \ \Omega \\ R_E = 100 \ \Omega \end{cases}$$

However, when choosing components like resistors, capacitors, inductors, and so on, we must rely on values that are available for sale on the market and some of the calculated values are not.

Checking Appendix G, we choose the final values for 20% tolerance resistors:

$$\begin{cases} R_1 = 10 \text{ k}\Omega \\ R_2 = 1.5 \text{ k}\Omega \\ R_C = 470 \ \Omega \\ R_E = 100 \ \Omega \end{cases}$$

These new values will change voltages and currents across the circuit. By the end of this chapter, at the exercises section, we will examine the effect of such changes.

29.8.1.1.3 Conclusion

One thing that should be observed about the voltage divider method is that the base current and the base voltage depend basically on the currents circulating through R_1 and R_2 and the voltages established by these resistors are independent from h_{FE}.

To see the dependence of h_{FE} of this kind of circuit, consider the voltage gain of this amplifier.

The voltage gain will be the output voltage divided by the input voltage. Looking at Figure 29.19, we can see that the output voltage is

$$V_{OUT} = V_{CC} - R_C i_C$$

and the input voltage is

$$V_{IN} = V_{CC} \times \frac{R_2}{R_1 + R_2}$$

Therefore, the voltage gain is

$$V_G = \frac{V_{OUT}}{V_{IN}} = \frac{(V_{CC} - R_C i_C) \times (R_1 + R_2)}{V_{CC} R_2}$$

We do not see h_{FE} on the voltage gain formula because there is no dependence on the base current too, showing how the gain is independent from h_{FE}.

Exercises

1   Consider the voltage divider biasing common-emitter class A amplifier shown in Figure 29.22.
   The circuit uses the NPN silicon-based low power general-purpose transistor type 2N2222A.

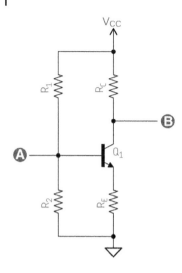

Figure 29.22 Voltage divider biasing.

Figure 29.23 shows the h_{FE} curve for this transistor.

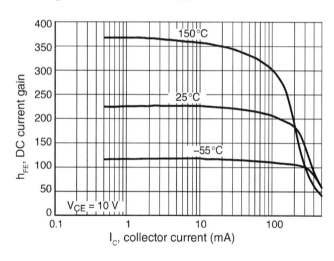

Figure 29.23 h_{FE} DC current gain (transistor 2N2222A).

The power supply voltage and the collector current should be

$$\begin{cases} V_{CC} = 18 \text{ V} \\ i_C = 8 \text{ mA} \end{cases}$$

Design this amplifier for maximum output swing possible.

Solutions

1 The first thing we do is to define a voltage for the emitter. This will be the voltage across the emitter resistor and as a rule of thumb should be around 1 V to keep the amplifier stable and prevent the thermal runaway.

The current collector was chosen to be 8 mA. This value gives us a value of h_{FE} equal to 225 for this 2N2222A transistor.

This is a silicon transistor; therefore, the base–emitter voltage is approximately 0.7 V.

These are the values we have so far:

$$\begin{cases} V_{CC} = 18 \text{ V} \\ i_C = 8 \text{ mA} \\ h_{FE} = 225 \\ V_{BE} = 0.7 \text{ V} \end{cases}$$

The base current can be calculated:

$$i_B = \frac{i_C}{h_{FE}}$$

$$i_B = \frac{8 \times 10^{-3}}{225}$$

$$i_B = 35.5555 \text{ μA}$$

We have chosen the emitter voltage to be 1 V. As a rule of thumb, we know that the collector–emitter voltage should be at least 2 V to keep the transistor out of the saturation zone:

$$V_{CE_{MIN}} = 2 \text{ V}$$

Therefore, the minimum value for the collector voltage during the negative cycle of the output will be

$$V_{C_{MIN}} = V_{CE_{MIN}} + V_E$$

$$V_{C_{MIN}} = 2 + 1$$

$$V_{C_{MIN}} = 3 \text{ V}$$

This means that the collector will be able to swing from 3 to 18 V and consequently will have an amplitude of 15 V peak to peak, equivalent to the subtraction of 18 and 3 V.

Therefore, for maximum swing, we must position the collector voltage at a middle point between 3 and 18 V, or

$$V_C = \frac{V_{CC} + V_{C_{MIN}}}{2}$$

$$V_C = \frac{18 + 3}{2}$$

$$V_C = 10.5 \text{ V}$$

The collector sitting at 10.5 V will be able to swing from 10.5 to 18 V during the positive cycle and from 10.5 to 3 V during the negative cycle of the output signal.

The emitter current is the sum of the collector current and base current, or

$$i_E = i_C + i_B$$

$$i_E = \left(8 \times 10^{-3}\right) + \left(35.5555 \times 10^{-6}\right)$$

$$i_E = 8.0355 \text{ mA}$$

Now we can find the emitter resistor:

$$V_{R_E} = i_E R_E$$

$$1 = \left(8.0355 \times 10^{-3}\right) R_E$$

$$R_E = 124.4447 \text{ }\Omega$$

Similarly, the voltage drop across the collector resistor is

$$V_{R_C} = i_C R_C$$

and is also

$$V_{R_C} = V_{CC} - V_C$$

Therefore, we can equate these two equations and find the collector resistor:

$$V_{CC} - V_C = R_C i_C$$

$$18 - 10.5 = R_C(8 \times 10^{-3})$$

$$R_C = 937.5 \ \Omega$$

R_1 and R_2 are a voltage divider. As a rule of thumb, the current across R_1 should be 10 times less than the collector current.
We can use this to get the current across R_1:

$$i_1 = \frac{i_C}{10}$$

$$i_1 = \frac{8 \times 10^{-3}}{10}$$

$$i_1 = 800 \ \mu A$$

We have defined the voltage across the emitter resistor to be 1 V.
We know that the voltage across base–emitter, for silicon-based transistors, is around 0.7 V.
Therefore, we can find the base voltage:

$$V_B = V_{BE} + V_E$$

$$V_B = 0.7 + 1$$

$$V_B = 1.7 \ V$$

R_2 is connected between the base and ground. Therefore, the voltage across this resistor is the same as the base voltage, or

$$V_{R_2} = V_B$$

Therefore,

$$R_2 i_2 = V_B$$

$$R_2 = \frac{V_B}{i_2}$$

But we know that

$$i_2 = i_1 - i_B$$

Therefore, we can find R_2:

$$R_2 = \frac{V_B}{i_1 - i_B}$$

$$R_2 = \frac{1.7}{(800 \times 10^{-6}) - (35.5555 \times 10^{-6})}$$

$$R_2 \approx 2224\ \Omega$$

If R_1 and R_2 are a voltage divider and we know the voltage between them and the current flowing, we can use this to find R_1:

$$R_1 = \frac{V_{CC} - V_B}{i_1}$$

$$R_1 = \frac{18 - 1.7}{800 \times 10^{-6}}$$

$$R_1 = 20375\ \Omega$$

We have finished the project.
These are the values we have so far:

$$\begin{cases} R_1 = 20375\ \Omega \\ R_2 = 2224\ \Omega \\ R_C = 937.5\ \Omega \\ R_E = 124.4447\ \Omega \end{cases}$$

However, most of these values do not exist in the market. We must replace them with the available values listed on Appendix G.

The next list contains the values we have chosen for 20% tolerance resistors:

$$\begin{cases} R_1 = 18\ k\Omega \text{ or } 22\ k\Omega? \\ R_2 = 2.2\ k\Omega \\ R_C = 1\ k\Omega \\ R_E = 120\ \Omega \end{cases}$$

The calculated value for R_1 = 20375 Ω is approximately equidistant of the available values of 18 and 22 kΩ. For this reason, we are not exactly sure which one we must choose.

To know which one is the best one for our purposes, let us test both options using a software simulator.[2]

The results provided by the simulator were the following:

Choosing R_1 equal to 18 kΩ will increase the current across R_1 from 800 μA to about 900 μA and an increase on the current across R_2 from 764 μA to around 852 μA. These two changes modify the base voltage from 1.69 to 1.87 V.

Collector voltage drops from 9.81 to 7.52 V and the emitter voltage rises from 1.08 to 1.3 V. In technical terms, the output swing is now limited to vary from 3.3 to 18 V. This is not the middle point between 3.3 and 18 V.

If the collector voltage is now sitting at 7.52 V, it means that it can vary 10.48 V up, from 7.52 to 18 V, and just 4.22 V down, from 7.52 to 3.3 V. Because the output amplification must always be equally up and down, we must limit the amplification to swing 4.22 V up and down. So, the maximum output voltage we may have without clipping is a wave with 8.44 V peak to peak.

However, if we choose R_1 to be 22 kΩ, the simulator tells us that the current across R_1 will drop from 800 to 746 μA and the current across R_2 will drop from 764 to 714 μA. These changes will modify the base voltage from 1.7 to 1.57 V.

The collector voltage will rise from 9.81 to 10.01 V, and the emitter voltage will drop from 1 V to 961 mV.

The reader can already realize the changes now are less dramatic, compared with what we have designed.

The collector voltage sitting at 10.01 V provides output variations of 8 V up to 18 V and 7.049 V down to 2.961 V.

For the same reasons as before, the output must be limited to 7.049 V up and down, making the total output's peak-to-peak amplitude equal to 14.098 V, almost twice the previous case.

Therefore, choosing R_1 equal to 22 kΩ makes a much better amplifier, closer to what we have designed.

---

2 www.systemvision.com, for example, or any other the reader prefers.

30

Operational Amplifiers

A Brief Introduction

30.1 Introduction

In this chapter, we will do a brief explanation about operational amplifiers.

30.2 Operational Amplifiers

An operational amplifier or an op-amp is a high-gain voltage amplifier with a differential input and, generally, a single output. In this configuration, an op-amp produces an output voltage that is commonly hundreds of thousands of times larger than the difference voltage between its input terminals.

Figure 30.1 shows the symbol for the operational amplifier.

Figure 30.1 Operational amplifier.

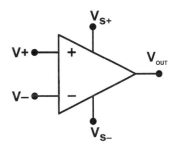

V_+ is the non-inverting input.
V_- is the inverting input.
V_{S+} is to be connected to the positive power rail from the power supply.
V_{S-} is to be connected to the negative power rail from the power supply.
V_{OUT} is the output.

Introductory Electrical Engineering with Math Explained in Accessible Language,
First Edition. Magno Urbano.
© 2020 John Wiley & Sons, Inc. Published 2020 by John Wiley & Sons, Inc.

30.3 How Op-Amp Works

An op-amp is basically a voltage amplifier to be used with external feedback components like resistors and capacitors between its inputs and output terminals. These components will determine the operation to be performed by the amplifier, for example, sum, subtraction, integration, differentiation, etc., which is the reason why it is called operational amplifier.

Op-amps have two inputs, a non-inverting input and an inverting input. These inputs are called differential inputs.

Internally they are as shown in Figure 30.2.

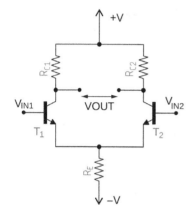

Figure 30.2 Operational amplifier (internal circuit).

Signals applied to the non-inverting input will appear at the output amplified but with the same phase as the input, like shown in Figure 30.3.

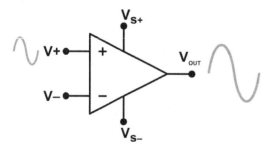

Figure 30.3 Non-inverting input.

However, if the same signal is applied to the inverting input, it will appear 180° inverted in phase at the output, as shown in Figure 30.4.

Figure 30.4 Inverting input.

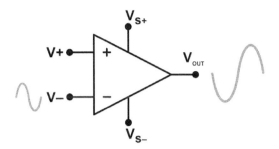

As mentioned earlier, op-amps have differential inputs, meaning that if the same signal is applied at both of its inputs, the output will be the difference, or the subtraction of both signals, as shown in Figure 30.5.

Figure 30.5 Differential inputs.

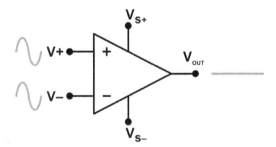

In Figure 30.5, the line at the output can be interpreted as 0 or as no change, meaning that the output will be constant if both inputs are the same. This is known as "common mode rejection ratio."

V_{S+} and V_{S-} terminals should be connected to the positive and negative power rails. Op-amps generally like to operate with symmetrical power supplies, meaning that the power supply should provide a positive voltage, a negative voltage, and a ground reference.

If the reader remembers the mountain analogy, negative and positive voltages are relative and, therefore, can be simulated by, for example, establishing a voltage divider using two resistors, like shown in Figure 30.6.

Figure 30.5 shows two resistors in series connected to a 9-V battery to create a voltage divider and simulate a symmetrical power supply with +4.5 V, −4.5 V, and ground (GND) in a configuration known as virtual ground. This method is very limited, and any current gain above certain levels will make the ground level to change, unbalancing the voltages.

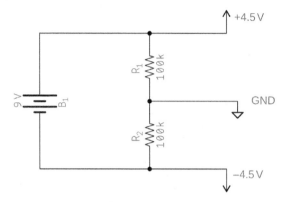

Figure 30.6 Virtual ground.

30.4 Op-Amp Characteristics

- The voltage gain is very high, something that can go up to 10000 or more.
- Very high input impedance, generally near infinity, resulting in basically no current flowing into either of its two inputs and zero offset between the same inputs.
- Very low output impedance, generally near 0.
- The output of an op-amp will be the difference between the voltage signals applied to its inputs multiplied by a certain gain.
- The gain of an op-amp will be infinite if there is no element connecting the input and the output. In this case, the gain is known as "open-loop gain."
- If an element, like a resistor, is connected between the input and the output, the total gain will be reduced and therefore can be adjusted to a predetermined level.

30.5 Typical Configurations

An op-amp can be basically configured as an inverting op-amp and a non-inverting op-amp.

30.5.1 Inverting Op-Amp

Figure 30.7 shows a typical inverting op-amp circuit.

Figure 30.7 Inverting op-amp.

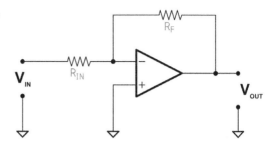

To find the output voltage and consequently the gain, let us establish some points and currents, as shown in Figure 30.8.

Figure 30.8 Inverting amplifier (currents).

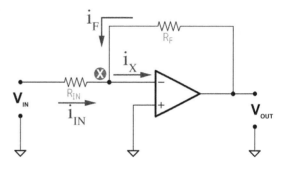

By using Kirchhoff's current law, we know that

$$i_{IN} + i_F = i_X$$

i_{IN} is the input current.
i_F is the feedback current.
i_X is the inverting input current.

and that the current flowing across an op-amp input is extremely low, is negligible, and can be considered 0.

Therefore,

$$i_{IN} + i_F = 0$$

$$i_{IN} = -i_F \tag{30.1}$$

We get the feedback current by applying nodal analysis to the output node:

$$i_F = \frac{V_{OUT} - V_X}{R_F}$$

The same nodal analysis can be applied to node X to get i_{IN}:

$$i_{IN} = \frac{V_{IN} - V_X}{R_{IN}}$$

We can substitute both equations in (30.1) and obtain

$$\frac{V_{IN} - V_X}{R_{IN}} = -\frac{V_{OUT} - V_X}{R_F} \tag{30.2}$$

We know that the difference between the op-amps inputs is 0, and we notice that the non-inverting input is connected to ground. Therefore, the voltage across the non-inverting input is 0, and consequently the voltage of the inverting input will also be 0, or

$$V_X = 0$$

Substituting this in (30.2),

$$\frac{V_{IN} - 0}{R_{IN}} = -\frac{V_{OUT} - 0}{R_F}$$

$$\frac{V_{IN}}{R_{IN}} = -\frac{V_{OUT}}{R_F}$$

We get the final output voltage.

INVERTING OP-AMP OUTPUT VOLTAGE

$$V_{OUT} = -\frac{R_F}{R_{IN}} V_{IN}$$

V_{IN} is the input voltage.
V_{OUT} is the output voltage.
R_F is the feedback resistor.
R_{IN} is the input resistor.

Consequently, the gain is

INVERTING OP-AMP VOLTAGE GAIN

$$A_V = \frac{V_{OUT}}{V_{IN}} = -\frac{R_F}{R_{IN}}$$

A_V is the voltage gain.
V_{IN} is the input voltage.
V_{OUT} is the output voltage.
R_F is the feedback resistor.
R_{IN} is the input resistor.

30.5.2 Non-inverting Op-Amp

Figure 30.9 shows a typical non-inverting op-amp circuit.

Figure 30.9 Non-inverting op-amp.

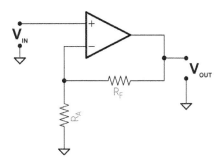

This kind of amplifier will amplify the signal and keep the signal with the same phase as the input.

To find the output voltage and consequently the gain, let us establish some points and currents, as shown in Figure 30.10.

Figure 30.10 Non-inverting amplifier (currents).

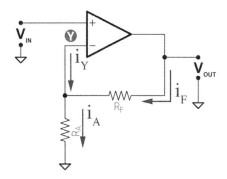

By applying Kirchhoff's current law in node Y, we get

$$i_A = i_F + i_Y$$

i_{IN} is the input current.
i_F is the feedback current.
i_X is the inverting input current.

Like in the last example, we know that the current flowing across an op-amp input is extremely low, is negligible, and can be considered 0.
Therefore,

$$i_A = i_F + 0$$

$$i_A = i_F \qquad\qquad (30.3)$$

We get the feedback current by applying nodal analysis to the output node:

$$i_F = \frac{V_{OUT} - V_Y}{R_F}$$

The same nodal analysis can be applied to node Y to get i_A:

$$i_A = \frac{V_Y}{R_A}$$

We can substitute both equations in (30.3):

$$\frac{V_{OUT} - V_Y}{R_F} = \frac{V_Y}{R_A}$$

$$V_Y = \frac{(V_{OUT} - V_Y)R_A}{R_F}$$

$$V_Y R_F = (V_{OUT} - V_Y)R_A$$

$$V_Y R_F = V_{OUT}R_A - V_Y R_A$$

$$V_{OUT}R_A = V_Y R_F + V_Y R_A$$

$$V_{OUT} = \frac{V_Y R_F + V_Y R_A}{R_A}$$

$$V_{OUT} = V_Y \left(\frac{R_F + R_A}{R_A} \right)$$

$$V_{OUT} = V_Y \left(\frac{R_F}{R_A} + 1 \right)$$

But we know that for op-amps the input voltages are the same; therefore,

$$V_Y = V_{IN}$$

Consequently, we get the final formula for the output voltage,

INVERTING OP-AMP OUTPUT VOLTAGE

$$V_{OUT} = V_{IN} \left(\frac{R_F}{R_A} + 1 \right)$$

A_V is the voltage gain.
V_{IN} is the input.
V_{OUT} is the output.
R_F is the feedback resistor.
R_A is the voltage divider resistor.

and for the gain,

INVERTING OP-AMP GAIN

$$A_V = \frac{V_{OUT}}{V_{IN}} = 1 + \frac{R_F}{R_A}$$

A_V is the voltage gain.
V_{IN} is the input.
V_{OUT} is the output.
R_F is the feedback resistor.
R_A is the voltage divider resistor.

30.5.3 Voltage Follower

Figure 30.11 shows an amplifier where the feedback resistor, R_F, was replaced with a wire.

This kind of configuration is called "voltage follower" or "buffer," also known as "unity-gain amplifier" or "isolation amplifier," because this amplifier has a gain of 1, as follows:

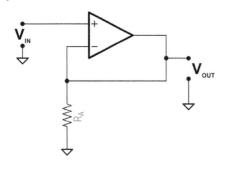

Figure 30.11 Voltage follower.

$$A_V = \frac{V_{OUT}}{V_{IN}} = 1 + \frac{R_F}{R_A}$$

But

$$R_F = 0$$

therefore,

$$A_V = 1 + \frac{0}{R_A}$$

$$A_V = 1 + 0$$

VOLTAGE FOLLOWER OP-AMP GAIN

$$A_V = 1$$

A voltage follower does not provide any amplification to the signal. The reason it is called a voltage follower is because the output voltage directly follows the input voltage, meaning the output voltage is the same as the input voltage. A voltage follower acts as an isolation, a buffer, providing no amplification or attenuation to the signal and can be used, for example, as an interface to interconnect two circuits, providing isolation between the two.

One example of the advantage of having isolation can be explained using the voltage divider circuit we have mentioned in Figure 30.6.

As explained earlier, this virtual ground works by the principle that the 9 V provided by the battery will be split into +4.5 V, −4.5 V, and ground. This will only work and keep the symmetry if the ground value remains stable at 0 V.

The big problem of this circuit is that the negative and the positive part of the power supply must supply circuits that drain the same current from both sides. If this is not true, the virtual ground will move from zero and the power supply will not be symmetrical anymore.

Let us see what happens if we connect a load of 10 kΩ to the negative part, as shown in Figure 30.12.

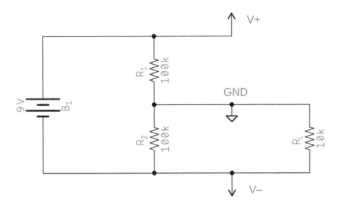

Figure 30.12 Virtual ground plus load.

Resistors R_2 and R_L are now in parallel. According to the first Ohm law, they can be replaced by an equivalent:

$$\frac{1}{R_{EQ}} = \frac{1}{R_2} + \frac{1}{R_L}$$

$$\frac{1}{R_{EQ}} = \frac{1}{100000} + \frac{1}{10000}$$

$$R_{EQ} = 9090.90 = 90.9090 \text{ k}\Omega$$

The voltage divider will now be equal to what is shown in Figure 30.13.

Figure 30.13 Voltage divider as virtual ground.

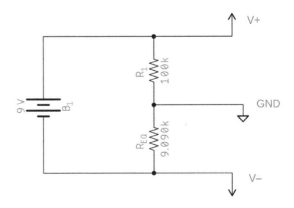

The voltage divider that was composed of two 100 kΩ resistors is now composed of a resistor R_1 of 100 kΩ and another resistor, R_{EQ}, of approximately 9 kΩ. This means that the voltage between ground and V_ that was −4.5 V before is now just −0.75 V or six times less.

If the voltage dropped in the negative part, it raised on the positive side. Now the voltage between ground and V_+ that was +4.5 V is now +8.25 V.

The conclusion is that the ground voltage is not in the middle between V_ and V_+, anymore.

Thus, a voltage divider using resistors is not a good solution to build a symmetrical power supply from a nonsymmetrical voltage source. We need something that can isolate the load from the divider. Any ideas how we can accomplish this?

The solution to our problem is the one shown in Figure 30.14.

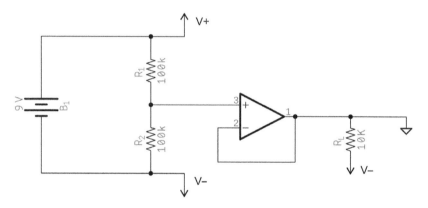

Figure 30.14 Voltage divider improved with a voltage follower.

Now, the voltage follower input is connected between the load R_L and the voltage divider, isolating one from the other. The current drained by the load will not affect the voltage divider that much and will keep its value stable. This solution is not completely perfect but is better than having the voltage divider alone.

This solution works because the op-amp offers a very high input impedance and, for that reason, will drain very little current from the voltage divider. With a very high input impedance, something around 100 MΩ or more, it is now that impedance in parallel with R_2 that gives an equivalent resistor still equal to R_2, keeping the relationship between R_1 and R_2 the same and consequently the voltage balance stable.

The following are simple examples of circuits using op-amps.

30.5.4 Non-inverting Summing Amplifier

Figure 30.15 shows a non-inverting summing amplifier that can be used to add multiple voltages.

Figure 30.15 Non-inverting summing amplifier.

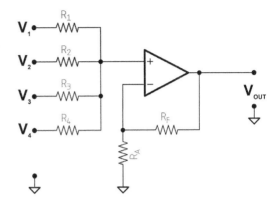

To find the output voltage, let us establish some points and currents, as shown in Figure 30.16.

Figure 30.16 Non-inverting summing amplifier (currents).

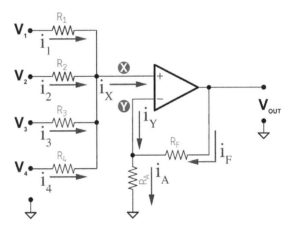

We have already calculated previously in this chapter that the output voltage for the non-inverting amplifier is

$$V_{OUT} = \left(1 + \frac{R_F}{R_A}\right) V_{IN}$$

or in this given example,

$$V_{OUT} = \left(1 + \frac{R_F}{R_A}\right) V_X \qquad (30.4)$$

We know that both op-amps have the same voltage, or

$$V_X \approx V_Y$$

Therefore, the following formula is also true:

$$V_{OUT} = \left(1 + \frac{R_F}{R_A}\right) V_Y$$

We also know that the sum of the input currents is equal to the current flowing into the non-inverting input, or

$$i_1 + i_2 + i_3 + i_4 + \cdots + i_N = i_X$$

However, the current flowing across an op-amp input is extremely low and negligible, and we can consider it to be 0.
Therefore,

$$i_A = i_F + 0$$

$$i_1 + i_2 + i_3 + i_4 + \cdots + i_N = 0 \qquad (30.5)$$

We get each current by applying nodal analysis to node X, for each input. Therefore,

$$i_1 = \frac{V_X - V_1}{R_1}$$

$$i_2 = \frac{V_X - V_2}{R_2}$$

$$i_3 = \frac{V_X - V_3}{R_3}$$

$$i_4 = \frac{V_X - V_4}{R_4}$$

and so on for the multiple inputs.

By substituting these equations into (30.5), we obtain

$$\frac{V_X - V_1}{R_1} + \frac{V_X - V_2}{R_2} + \frac{V_X - V_3}{R_3} + \frac{V_X - V_4}{R_4} + \cdots + \frac{V_X - V_N}{R_N} = 0$$

We can expand this equation into

$$\frac{V_X}{R_1} - \frac{V_1}{R_1} + \frac{V_X}{R_2} - \frac{V_2}{R_2} + \frac{V_X}{R_3} - \frac{V_3}{R_3} + \frac{V_X}{R_4} - \frac{V_4}{R_4} + \cdots + \frac{V_X}{R_N} - \frac{V_N}{R_N} = 0$$

Isolating all V_X on one side,

$$\frac{V_X}{R_1} + \frac{V_X}{R_2} + \frac{V_X}{R_3} + \frac{V_X}{R_4} + \cdots + \frac{V_X}{R_N} = \frac{V_1}{R_1} + \frac{V_2}{R_2} + \frac{V_3}{R_3} + \frac{V_4}{R_4} + \cdots + \frac{V_N}{R_N}$$

$$V_X \left(\frac{1}{R_1} + \frac{1}{R_2} + \frac{1}{R_3} + \frac{1}{R_4} + \cdots + \frac{1}{R_N} \right) = \frac{V_1}{R_1} + \frac{V_2}{R_2} + \frac{V_3}{R_3} + \frac{V_4}{R_4} + \cdots + \frac{V_N}{R_N} \tag{30.6}$$

We notice that the block

$$\frac{1}{R_1} + \frac{1}{R_2} + \frac{1}{R_3} + \frac{1}{R_4} + \cdots + \frac{1}{R_N}$$

is equal to the equivalent resistance of resistors in parallel and we can, consequently, replace the whole block with that resistance, or

$$\frac{1}{R_1} + \frac{1}{R_2} + \frac{1}{R_3} + \frac{1}{R_4} + \cdots + \frac{1}{R_N} = \frac{1}{R_{EQ}}$$

Therefore, by substituting this into (30.6), we get

$$V_X \left(\frac{1}{R_{EQ}} \right) = \frac{V_1}{R_1} + \frac{V_2}{R_2} + \frac{V_3}{R_3} + \frac{V_4}{R_4} + \cdots + \frac{V_N}{R_N}$$

$$V_X = R_{EQ} \left(\frac{V_1}{R_1} + \frac{V_2}{R_2} + \frac{V_3}{R_3} + \frac{V_4}{R_4} + \cdots + \frac{V_N}{R_N} \right) \tag{30.7}$$

Upon substituting (30.7) in (30.4), we get the final equation for the output voltage.

NON-INVERTING SUMMING AMPLIFIER – OUTPUT VOLTAGE

$$V_{OUT} = \left(1 + \frac{R_F}{R_A}\right) R_{EQ} \left(\frac{V_1}{R_1} + \frac{V_2}{R_2} + \frac{V_3}{R_3} + \frac{V_4}{R_4} + \cdots + \frac{V_N}{R_N}\right) \qquad (30.8)$$

V_{OUT} is the output voltage.
R_F is the feedback resistor.
R_A is the grounding resistor.
R_{EQ} is the equivalent resistance to all input resistors in parallel.
$V_1, V_2, \ldots$ are the individual input voltages.
$R_1, R_2, \ldots$ are the individual input resistors.

30.5.4.1 Summing Two Inputs

If we are summing just two signals and the input resistors are the same, a very typical case in the real world, the output voltage will be

$$V_{OUT} = \left(1 + \frac{R_F}{R_A}\right) R_{EQ} \left(\frac{V_1}{R} + \frac{V_2}{R}\right)$$

$$V_{OUT} = \left(1 + \frac{R_F}{R_A}\right) R_{EQ} \left(\frac{V_1 + V_2}{R}\right) \qquad (30.9)$$

We know that the equivalent resistor of two resistors in parallel is

$$R_{EQ} = \frac{R_1 \times R_2}{R_1 + R_2}$$

For two equal resistors, we have

$$R_{EQ} = \frac{R \times R}{R + R}$$

$$R_{EQ} = \frac{R^2}{2R}$$

$$R_{EQ} = \frac{R}{2}$$

By substituting into (30.9), we get the final formula for the output voltage for two inputs,

$$V_{OUT} = \left(1 + \frac{R_F}{R_A}\right) \left(\frac{R}{2}\right) \left(\frac{V_1 + V_2}{R}\right)$$

$$V_{OUT} = \left(1 + \frac{R_F}{R_A}\right)\left(\frac{\cancel{R}}{2}\right)\left(\frac{V_1 + V_2}{\cancel{R}}\right)$$

NON-INVERTING SUMMING AMPLIFIER – OUTPUT VOLTAGE FOR TWO INPUTS

$$V_{OUT} = \left(1 + \frac{R_F}{R_A}\right)\left(\frac{V_1 + V_2}{2}\right)$$

V_{OUT} is the output voltage.
R_F is the feedback resistor.
R_A is the grounding resistor.
$V_1, V_2, \ldots$ are the individual input voltages.

30.5.4.2 Summing Three Inputs

If we have three inputs, the result is slightly different.

$$V_{OUT} = \left(1 + \frac{R_F}{R_A}\right) R_{EQ}\left(\frac{V_1}{R} + \frac{V_2}{R} + \frac{V_3}{R}\right) \tag{30.10}$$

We know that the equivalent resistor of three resistors in parallel is

$$\frac{1}{R_{EQ}} = \frac{1}{R_1} + \frac{1}{R_2} + \frac{1}{R_3}$$

If we solve that for equal resistors, we have

$$\frac{1}{R_{EQ}} = \frac{1}{R} + \frac{1}{R} + \frac{1}{R}$$

$$\frac{1}{R_{EQ}} = \frac{3}{R}$$

$$R_{EQ} = \frac{R}{3}$$

By substituting into (30.10), we get the final formula for the output voltage for three inputs,

$$V_{OUT} = \left(1 + \frac{R_F}{R_A}\right)\left(\frac{R}{3}\right)\left(\frac{V_1 + V_2 + V_3}{R}\right)$$

$$V_{OUT} = \left(1 + \frac{R_F}{R_A}\right)\left(\frac{\cancel{R}}{3}\right)\left(\frac{V_1 + V_2 + V_3}{\cancel{R}}\right)$$

NON-INVERTING SUMMING AMPLIFIER – OUTPUT VOLTAGE FOR THREEE INPUTS

$$V_{OUT} = \left(1 + \frac{R_F}{R_A}\right)\left(\frac{V_1 + V_2 + V_3}{3}\right)$$

$V_1, V_2, \dots$ are the individual input voltages.
V_{OUT} is the output voltage.
R_F is the feedback resistor.
R is the input resistor.

30.5.5   Inverting Summing Amplifier

This case is like the previous one, where we want to sum several input signals. However, this is an inverter amplifier, where the output is 180° out of phase, compared to the input.

To sum more than two voltages, the inverting summing amplifier circuit, as shown in Figure 30.17, is the best one to choose.

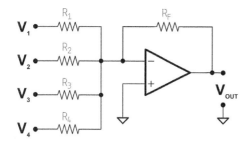

Figure 30.17 Inverting summing amplifier.

To find the output voltage, let us establish some points and currents, as shown in Figure 30.18.

Figure 30.18 Inverting summing amplifier (currents).

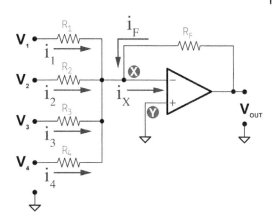

We have already calculated previously in this chapter that the output voltage for the inverting amplifier is

$$V_{OUT} = \left(\frac{R_F}{R_A}\right) V_{IN}$$

or in this given example,

$$V_{OUT} = \left(\frac{R_F}{R_A}\right) V_X \tag{30.11}$$

We know that both inputs have the same voltage. Because the non-inverting input is connected to ground, we have

$$V_X \approx V_Y = 0 \tag{30.12}$$

We know that the sum of the input currents is equal to the current flowing into the inverting input, or

$$\left(i_1 + i_2 + i_3 + i_4 + \cdots + i_N\right) + i_F = i_X$$

However, the current flowing across an op-amp input is extremely low and negligible, and we can consider it to be 0.

Therefore,

$$(i_1 + i_2 + i_3 + i_4 + \cdots + i_N) + i_F = 0$$
$$(i_1 + i_2 + i_3 + i_4 + \cdots + i_N) = -i_F$$

$$i_F = -(i_1 + i_2 + i_3 + i_4 + \cdots + i_N) \qquad (30.13)$$

We get each current by applying nodal analysis to node X, for each input. Therefore,

$$i_1 = \frac{V_1 - V_X}{R_1}$$

$$i_2 = \frac{V_2 - V_X}{R_2}$$

$$i_3 = \frac{V_3 - V_X}{R_3}$$

$$i_4 = \frac{V_4 - V_X}{R_4}$$

$$i_F = \frac{V_{OUT} - V_X}{R_F}$$

and so on for the multiple inputs.
By substituting (30.12) into these equations, we get

$$i_1 = \frac{V_1}{R_1}$$

$$i_2 = \frac{V_2}{R_2}$$

$$i_3 = \frac{V_3}{R_3}$$

$$i_4 = \frac{V_4}{R_4}$$

$$i_F = \frac{V_{OUT}}{R_F}$$

Therefore, Equation (30.13) is now

$$\frac{V_{OUT}}{R_F} = -\left(\frac{V_1}{R_1} + \frac{V_2}{R_2} + \frac{V_3}{R_3} + \frac{V_4}{R_4} + \cdots + \frac{V_N}{R_N}\right)$$

and we get the final equation for the output voltage.

INVERTING SUMMING AMPLIFIER – OUTPUT VOLTAGE

$$V_{OUT} = -R_F\left(\frac{V_1}{R_1} + \frac{V_2}{R_2} + \frac{V_3}{R_3} + \frac{V_4}{R_4} + \cdots + \frac{V_N}{R_N}\right)$$

$V_1, V_2, \ldots$ are the individual input voltages.
V_{OUT} is the output voltage.
R_F is the feedback resistor.
$R_1, R_2, \ldots$ are the individual input resistors.

30.5.6 Integrator

Like explained at the beginning of this chapter, operational amplifiers have this name because they can perform mathematical operations. Performing the integral of the input signal is just one of these operations.

The output of this circuit is the integral of the input (see Figure 30.19).

Figure 30.19 Op-amp integrator.

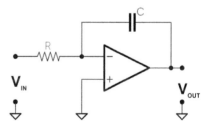

To find the output voltage equation, we start by specifying some currents and points, like shown in Figure 30.20.

By applying Kirchhoff's current law to node X, we get

$$i_R + i_C = i_X$$

However, we know that the current flowing across any of the op-amp inputs is negligible and can be considered to be 0.

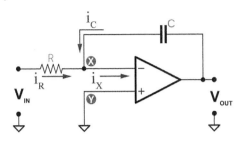

Figure 30.20 Op-amp integrator (currents).

Therefore,

$$i_R + i_C = 0$$

$$i_R = -i_C \qquad (30.14)$$

By applying nodal analysis to node X, we find the current across R:

$$i_R = \frac{V_{IN} - V_X}{R}$$

We know that both op-amp inputs have the same voltage, which in this case is 0, because the non-inverting input is connected to ground.

Thus,

$$V_X = 0$$

and consequently,

$$i_R = \frac{V_{IN}}{R}$$

For the capacitor case, we know that the current flowing across this element is

$$i = C\frac{dV}{dt}$$

The voltage across the capacitor is

$$V_{OUT} - V_X$$

Therefore, the current is

$$i_C = C\frac{d(V_{OUT} - V_X)}{dt}$$

By substituting

$$V_X = 0$$

we get

$$i_C = C\frac{dV_{OUT}}{dt}$$

We can substitute i_R and i_F in (30.14):

$$i_R = -i_C$$

$$\frac{V_{IN}}{R} = -C\frac{dV_{OUT}}{dt}$$

or,

$$\frac{dV_{OUT}}{dt} = -\frac{V_{IN}}{RC}$$

If we integrate both sides, we get the formula for the voltage output.

INTEGRATOR – OUTPUT VOLTAGE

$$V_{OUT} = -\frac{1}{RC}\int V_{IN}dt$$

V_{IN} is the input voltage.
V_{OUT} is the output voltage.
R is the input resistor.
C is the feedback capacitor.

30.5.7 Differentiator

The differentiator circuit, able to differentiate the input signal, is shown in Figure 30.21.

Figure 30.21 Differentiator.

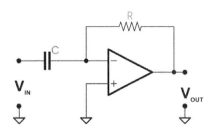

To get the output voltage equation, we do the same thing we did in the previous example to get

$$i_C = -i_R$$

where

$$i_C = C\frac{dV_{IN}}{dt}$$

$$i_R = \frac{V_{OUT}}{R}$$

Therefore,

$$C\frac{dV_{IN}}{dt} = -\frac{V_{OUT}}{R}$$

We get the output voltage equation as follows.

DIFFERENTIATOR – OUTPUT VOLTAGE

$$V_{OUT} = -RC\frac{dV_{IN}}{dt}$$

V_{IN} is the input voltage.
V_{OUT} is the output voltage.
R is the feedback resistor.
C is the input capacitor.

31

Instrumentation and Bench

A Brief Introduction

31.1 Introduction

In this chapter, we introduce the reader to instrumentation, a bench equipment, essential to those willing to project and build electronic circuits.

31.2 Multimeter

A multimeter is a test tool used to measure multiple electrical values, principally voltage (Volts), current (Amperes), and resistance (Ohms).

Modern multimeters generally group a lot more functions, and some models can measure frequency (Hertz), capacitance (Farads), and inductance (Henries) and perform tests on diodes and transistors.

Modern multimeters are digital devices, like the one shown in Figure 31.1, but, in the past, they were analog needle-based mechanical devices.

Figure 31.1 Digital multimeter.

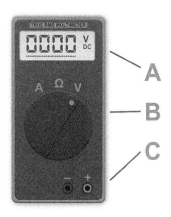

Introductory Electrical Engineering with Math Explained in Accessible Language,
First Edition. Magno Urbano.
© 2020 John Wiley & Sons, Inc. Published 2020 by John Wiley & Sons, Inc.

Multimeters are formed by three basic parts:

- Display (A), where measurement readouts can be viewed.
- Dial or rotary switch (B) for selecting primary measurement values (voltage, current, frequency, etc.).
- Input jacks (C), where the test leads are inserted.

31.2.1   True RMS Multimeter

Measuring AC currents and voltages can be a problem if a multimeter is not prepared to do that.

An average responding multimeter uses average mathematical formulas to measure pure sinusoidal waves with a certain accuracy. If the wave being measured is not sinusoidal, like a square or triangular wave, the measurement accuracy decreases immensely. In many cases, a regular multimeter will offer an accuracy that can be 40% off the real value.

The usage of non-sinusoidal waves increased greatly in recent years in fields like robotics, computers, control of variable speed motors, and more.

Therefore, it is essential to have a multimeter that can measure such wave forms with precision, like true root mean square (RMS) multimeters can do.

31.3   Voltmeter

Voltmeters are tools that measure the difference in potential (voltage) between two points.

Voltmeters are part of multimeters. To use the voltmeter part of a multimeter, its rotary switch must be rotated to the correspondent position, like illustrated with the V letter in Figure 31.1.

> Some multimeters may have two or more positions for voltages: VDC and VAC for direct and alternating voltages.

To measure voltage, the voltmeter test leads must be connected in parallel with the circuit being measured.

Figure 31.2 shows a multimeter measuring the voltage of a battery.

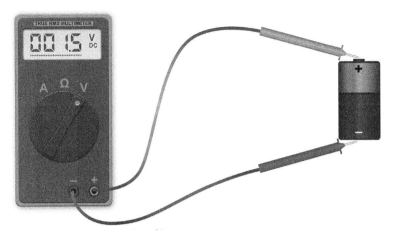

Figure 31.2 Voltmeter measuring a battery.

31.4 Ammeter

Ammeters are tools that measure the current flowing across two points in a circuit.

Ammeters are also part of multimeters. To use the ammeter part of a multimeter, its rotary switch must be rotated to the correspondent position, like illustrated with the A letter in Figure 31.3.

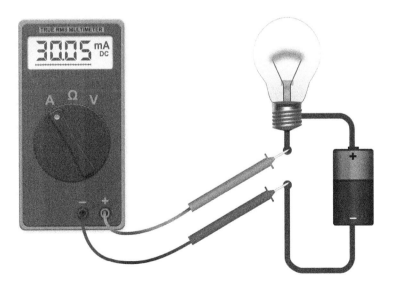

Figure 31.3 Ammeter measuring current.

> Some multimeters may have two or more positions for current: DC and AC, for direct and alternating currents.

To measure current, the voltmeter's test leads must be connected in series with the circuit being measured.

Figure 31.3 shows a multimeter measuring the current flowing in a circuit with a battery and a light bulb.

31.5 Ohmmeter

Ohmmeters are tools that measure the electrical resistance between two points in a circuit.

Ohmmeters are also part of multimeters. To use the ohmmeter part of a multimeter, its rotary switch must be rotated to the correspondent position, like illustrated with the Ω letter in Figure 31.4.

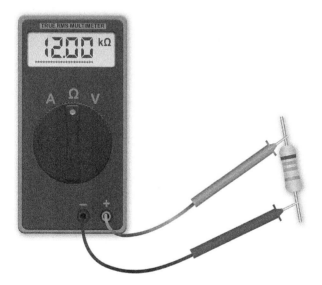

Figure 31.4 Ohmmeter measuring a resistor.

> Some multimeters may have two or more positions for resistance, one for every range of resistances.

To measure resistance, the multimeter test leads must be connected between two points in a circuit, and the circuit must be turned off, without any power supply or stored charge.

> A multimeter can be easily damaged if connected to a circuit that is being powered by a power supply or has charge stored in capacitors.

Figure 31.4 shows a multimeter measuring the resistance of a resistor.

31.6 Oscilloscope

Oscilloscopes are an important tool in any electrical engineer lab, because they allow you to see electric signals as they vary over time.

Figure 31.5 shows an illustration representing the main parts of an oscilloscope:

Figure 31.5 Oscilloscope.

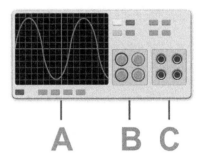

- Display (A), where measurement readouts and the waveforms can be viewed.
- Buttons where the amplitude or width of the waveform can be adjusted for every channel.
- Input jacks (C), where the test leads for every channel are inserted.

Oscilloscopes are a kind of voltmeter where you can see the waveform being measured. However, oscilloscopes have channels that allow multiple waveforms to be viewed and measured at the same time in the same display.

Oscilloscopes are used in the same way you would use a voltmeter.

31.7 Breadboards

Electrical circuits in their physical form are usually formed by components soldered to a fiberglass board containing faces covered with copper tracks, which interconnect the several points of a circuit.

A breadboard, on the other hand, is a solderless reusable device for temporary prototype with electronics and test circuit designs. This makes it easy to use for creating temporary prototypes and experimenting with circuit design.

Electronic components can be interconnected by inserting their leads or terminals into the holes of a breadboard.

The board has strips of metal underneath the board, connecting rows of pins.

Figure 31.6 shows an illustration representing the parts of a typical breadboard:

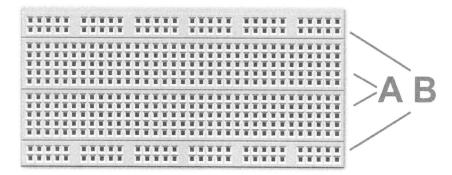

Figure 31.6 Breadboard.

- Component area (A), where components are inserted.
- Power and other buses (B), where ground, power supply, and other elements are established.

The rows and columns inside areas A and B are internally connected as shown in Figure 31.7.

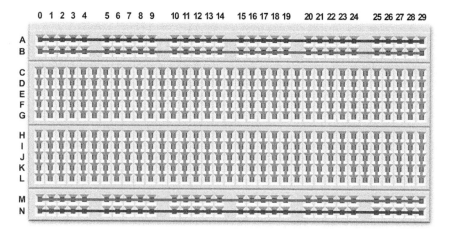

Figure 31.7 Breadboard (internal connections).

Examining Figure 31.7, we see that all holes in row A are electrically connected, from column 0 through 29. Consequently, a component inserted at hole A0 and another one at hole A29 will be electrically connected. The same is true for rows B, M, and N.

However, every column from 0 to 20 is electrically connected from C to G and from H to L. In column 0, for example, holes C0, D0, E0, F0, and G0 are connected to each other. Following that, holes H0, I0, J0, K0, and L0, are also connected to each other. However, these two groups of five holes are not connected, being insulated from each other.

If a component is inserted at C0 and another one at G0, they will be electrically connected. However, if we insert a component at G0 and another one at H0, they will not be electrically connected.

The same is true for all holes in lines from C to L in Figure 31.7.

Figure 31.8 shows a picture of a typical breadboard in use.

Figure 31.8 Breadboard in use.

31.8   Wire Diameter

Most wires are, in general, cylindrical tubes covered with insulating plastic.

Due to electrical resistance, current flowing through a wire makes the wire heat.

According to the second Ohm's law, the resistance in a wire is inversely proportional to the wire's cross-sectional area, or

$$R = \frac{\rho L}{A}$$

For that reason, the smaller the wire's cross section, the lesser the current a wire can handle without overheating.

Appendix F shows the American standard for wire gauges or American Wire Gauges (AWG) and the international millimetric standard.

In electrical engineering one must calculate the right wire's cross section based on the current this wire must handle. As a rule of thumb, we generally use a wire that can handle twice the current we want it to handle.

31.8.1 Types of Wires

There are basically two types of wires or cables in the market: stranded wire and solid core wire. Figure 31.9 shows a picture of both.

Figure 31.9 Stranded and solid wire.

Solid core cables use one solid copper wire per conductor. It has a lower attenuation and is less costly than stranded cable; however it is designed for situations where they stay put without flexing once installed. The reason for that is that solid core wires are more rigid and flexing them constantly will make them break.

Stranded cables consist of several strands of wires wrapped around each other in each conductor. They are much more flexible and consequently suited to applications that demand flexibility. Stranded cables have higher attenuation; therefore, they are better used over shorter distances.

31.9 Power Supply

Power supplies are another of the bench equipment.

A bench power supply is a practical tool and allows circuits to be powered and consequently tested before they are finished.

This way, you do not need to have a specific power supply for each project you want to test, providing you with a reliable and stabilized source of power at different voltages and currents.

Symmetrical power supplies are formed by three basic parts:

- Display (A), where voltage and current levels are viewed.
- Rotary buttons (B) for adjusting voltage and current levels.
- Output jacks (C), where the selected voltage and current are delivered.

Notice that the output jacks provide symmetrical voltage. In the given example in Figure 31.10, the power supply is adjusted to provide 24 V and 2.3 A. At the output jacks, the power supply will provide –24 V, ground or 0 V, and +24 V, at the black (minus), green (ground), and red (positive) terminals, respectively.

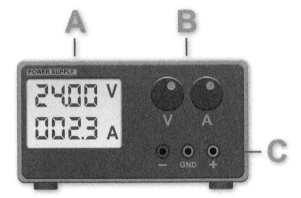

Figure 31.10 Typical symmetrical power supply.

31.10   Soldering Station

A soldering station, mostly used in electronics and electrical engineering, consists of one or more soldering tools connected to the main unit, which includes temperature adjustment, holders and stands, soldering tip cleaners, etc.

Soldering stations are widely used in electronics repair workshops or electronic laboratories or for household applications and hobbies.

31.11   Soldering Fume Extractors

Soldering produce fumes. Soldering fume extractors or absorbers are equipment that can extract or absorb these fumes generated by any soldering iron.

31.12   Lead-Free Solder

Solder generally contain lead that is a known neurotoxin and can pose other significant chronic health effects. Therefore, solder that contains lead is considered to be toxic.

The electronics industry has shifted to lead-free solder, but most hobbyists are still using lead in their solder.

Traditional solder generally contains around 63% of tin and 27% of lead and melts at 183 °C (361.4 °F). Lead-free solder, on the other hand, contains around 96.5% of tin, 3% of silver, and 0.5% of copper and melts at 228 °C (442.4 °F), making it much harder to melt, requiring solder stations that can go that hot.

Solder is sold in reels and can be either flux cored, with flux in the core of the wire, or solid, with no flux in the core. Cored solder wire is hollow solder wire with flux in the core.

Solder flux is a kind of chemical used to remove any oxide, oil, grease, and other unwanted materials from the surfaces prior to soldering. Solder flux helps to deoxidize metals and helps better soldering and wetting. These impurities can affect the performance of circuits and cause future circuit failures. Therefore, using flux-cored solder saves both time and money.

31.13   A Few Images of Real Products

In the following images, nicely provided by Fluke and Weller, we show a few real equipment.

Figure 31.11 Fluke 87V (true RMS) multimeter.

Figure 31.12 Fluke 190-204 scopemeter (portable oscilloscope).

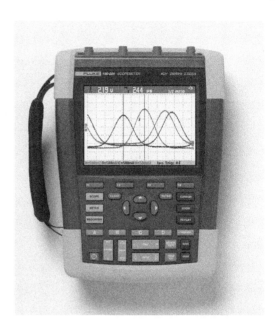

From back to front in Figure 31.13, we see the temperature module, the iron support and cleaner, and the soldering iron.

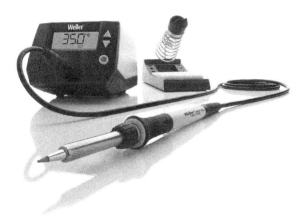

Figure 31.13 Weller WE1010 soldering station.

The two elements with tubes in the background in Figure 31.14 are the fume extractor and the hot air soldering station.

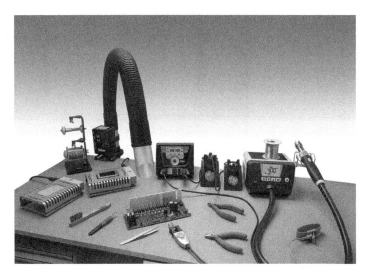

Figure 31.14 Weller working bench.

Thanks Weller and Fluke for the images.

Appendix A

International System of Units (SI)

Prefix	Symbol	Multiplier
Yotta	Y	10^{24} = 1000000000000000000000000
Zetta	Z	10^{21} = 1000000000000000000000
Exa	E	10^{18} = 1000000000000000000
Peta	P	10^{15} = 1000000000000000
Tera	T	10^{12} = 1000000000000
Giga	G	10^{9} = 1000000000
Mega	M	10^{6} = 1000000
Kilo	k	10^{3} = 1000
Hecto	h	10^{2} = 100
Deca	da	10^{1} = 10
Deci	d	10^{-1} = 0.1
Centi	c	10^{-2} = 0.01
Milli	m	10^{-3} = 0.001
Micro	μ	10^{-6} = 0.000001
Nano	n	10^{-9} = 0.000000001

(Continued)

Introductory Electrical Engineering with Math Explained in Accessible Language,
First Edition. Magno Urbano.
© 2020 John Wiley & Sons, Inc. Published 2020 by John Wiley & Sons, Inc.

Prefix	Symbol	Multiplier
Pico	p	$10^{-12} = 0.000000000001$
Femto	f	$10^{-15} = 0.000000000000001$
Atto	a	$10^{-18} = 0.000000000000000001$
Zepto	z	$10^{-21} = 0.000000000000000000001$
Yocto	y	$10^{-24} = 0.000000000000000000000001$

Appendix B

Color Code: Resistors

First stripe	Second stripe	Third stripe							
		Gold	Black	Brown	Red	Orange	Yellow	Green	Blue
Brown	Black	1	10	100	1k	10k	100k	1M	10M
Brown	Brown	1.1	11	110	1.1k	11k	110k	1.1M	11M
Brown	Red	1.2	12	120	1.2k	12k	120k	1.2M	12M
Brown	Orange	1.3	13	130	1.3k	13k	130k	1.3M	13M
Brown	Green	1.5	15	150	1.5k	15k	150k	1.5M	15M
Brown	Blue	1.6	16	160	1.6k	16k	160k	1.6M	16M
Brown	Gray	1.8	18	180	1.8k	18k	180k	1.8M	18M
Red	Black	2	20	200	2k	20k	200k	2M	20M
Red	Red	2.2	22	220	2.2k	22k	220k	2.2M	22M
Red	Yellow	2.4	24	240	2.4k	24k	240k	2.4M	24M
Red	Violet	2.7	27	270	2.7k	27k	270k	2.7M	27M
Orange	Black	3	30	300	3k	30k	300k	3M	30M
Orange	Orange	3.3	33	330	3.3k	33k	330k	3.3M	33M
Orange	Blue	3.6	36	360	3.6k	36k	360k	3.6M	36M

(Continued)

Introductory Electrical Engineering with Math Explained in Accessible Language,
First Edition. Magno Urbano.
© 2020 John Wiley & Sons, Inc. Published 2020 by John Wiley & Sons, Inc.

First stripe	Second stripe	Third stripe							
		Gold	Black	Brown	Red	Orange	Yellow	Green	Blue
Orange	White	3.9	39	390	3.9k	39k	390k	3.9M	39M
Yellow	Orange	4.3	43	430	4.3k	43k	430k	4.3M	43M
Yellow	Violet	4.7	47	470	4.7k	47k	470k	4.7M	47M
Green	Brown	5.1	51	510	5.1k	51k	510k	5.1M	51M
Green	Blue	5.6	56	560	5.6k	56k	560k	5.6M	56M
Blue	Red	6.2	62	620	6.2k	62k	620k	6.2M	62M
Blue	Gray	6.8	68	680	6.8k	68k	680k	6.8M	68M
Violet	Green	7.5	75	750	7.5k	75k	750k	7.5M	75M
Gray	Red	8.2	82	820	8.2k	82k	820k	8.2M	82M
White	Brown	9.1	91	910	9.1k	91k	910k	9.1M	91M

Appendix C

Root Mean Square (RMS) Value

C.1 Introduction

In this appendix, we will demonstrate how the root mean square (RMS) value is derived in terms of infinitesimal calculus for the sinusoidal function.

C.2 RMS Value

A sinusoidal function is defined by the following generic equation.

$$f(t) = A \sin (\omega t + \theta) \tag{C.1}$$

ω is the angular frequency, in radians per second.
t is the time, in seconds.
θ is the initial phase angle, in radians.[1]
A is the amplitude.

where

$$\omega = \frac{2\pi}{T} \tag{C.2}$$

ω is the angular frequency, in radians per second.
T is the period, in seconds.

---

1 In other words, the angle where the wave started.

Introductory Electrical Engineering with Math Explained in Accessible Language,
First Edition. Magno Urbano.
© 2020 John Wiley & Sons, Inc. Published 2020 by John Wiley & Sons, Inc.

In statistics, the RMS value of any function is given by the following equation:

$$V_{RMS} = \sqrt{\frac{1}{T}\int_0^T f(t)^2.dt}$$

We can simplify this equation by squaring both sides:

$$V_{RMS}^2 = \frac{1}{T}\int_0^T f(t)^2.dt$$

If we substitute f(t) by the sinusoidal function (C.1), we get

$$V_{RMS}^2 = \frac{1}{T}\int_0^T \{A\sin(\omega t + \theta)\}^2.dt$$

Solving...

$$V_{RMS}^2 = \frac{1}{T}\int_0^T A^2\sin^2(\omega t + \theta).dt$$

$$V_{RMS}^2 = \frac{1}{T}A^2\int_0^T \sin^2(\omega t + \theta).dt$$

For the purposes of this calculation, we can disregard θ by making it 0:

$$V_{RMS}^2 = \frac{1}{T}A^2\int_0^T \sin^2(\omega t + 0).dt$$

$$V_{RMS}^2 = \frac{1}{T}A^2\int_0^T \sin^2(\omega t).dt$$

MATH CONCEPT

$\int \sin^2(\theta) = \frac{\theta}{2} - \frac{1}{4}\sin(2\theta)$

If we apply this concept, we obtain

$$V_{RMS}^2 = \frac{1}{T}A^2\left[\frac{t}{2} - \frac{1}{4}\sin(2\omega t)\right]\Big|_0^T$$

Solving for the interval,

$$V_{RMS}^2 = \frac{1}{T}A^2\left\{\left[\frac{T}{2} - \frac{1}{4}\sin(2\omega T)\right] - \left[\frac{\omega(0)}{2} - \frac{1}{4}\sin(2\omega(0))\right]\right\}$$

$$V_{RMS}^2 = \frac{1}{T}A^2 \left\{ \left[\frac{T}{2} - \frac{1}{4}\sin(2\omega T) \right] - \left[0 - \frac{1}{4}\sin(0) \right] \right\}$$

MATH CONCEPT

$\sin(0) = \sin(x\pi) = 0$

$$V_{RMS}^2 = \frac{1}{T}A^2 \left\{ \left[\frac{T}{2} - \frac{1}{4}(0) \right] - \left[0 - \frac{1}{4}(0) \right] \right\}$$

$$V_{RMS}^2 = \frac{1}{T}A^2 \left\{ \frac{T}{2} \right\}$$

Simplifying T,

$$V_{RMS}^2 = \frac{A^2}{2}$$

$$V_{RMS} = \sqrt{\frac{A^2}{2}}$$

RMS VALUE

$$V_{RMS} = \frac{A}{\sqrt{2}}$$

V_{RMS} is the RMS value.
A is the amplitude value.

Appendix D

Complex Numbers

D.1 Real Numbers

Real numbers can be seen as numbers that exist in a horizontal line. Zero is at the center. Positive numbers extend to the right and negative numbers to the left, like shown in Figure D.1.

Figure D.1 Real numbers.

$$-\infty \; \cdots \; -3 \; -2 \; -1 \quad 0 \quad 1 \quad 2 \quad 3 \; \cdots \; +\infty$$

D.2 Complex Numbers

A complex number is represented by two axes, a real (horizontal) and an imaginary (vertical) one, like shown in Figure D.2.

This characteristic of complex numbers provides these numbers with the ability to express three quantities: two numerical values and an angle.

An imaginary number has the following form, called a rectangular form.

> a + bi

where a is the real part and b the imaginary part.
The imaginary part is represented with a letter, generally i or j.

Introductory Electrical Engineering with Math Explained in Accessible Language,
First Edition. Magno Urbano.
© 2020 John Wiley & Sons, Inc. Published 2020 by John Wiley & Sons, Inc.

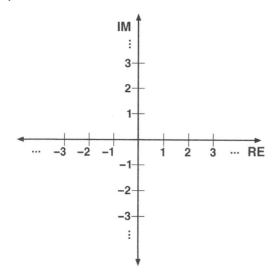

Figure D.2 Imaginary number axes.

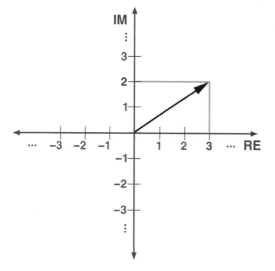

Figure D.3 Imaginary number example.

3 + 2i is an imaginary number that can be represented on the complex number plane as seen in Figure D.3.

D.2.1 Operations with Complex Numbers

D.2.1.1 Addition
To add two complex numbers, the real and the imaginary parts must be added alone.

Therefore,

$$+\ \frac{\begin{array}{r} 3+2i \\ 5-10i \end{array}}{8-8i}$$

D.2.1.2 Subtraction

To subtract two complex numbers, the real and the imaginary parts must be subtracted alone.

Therefore,

$$-\ \frac{\begin{array}{r} 13+12i \\ 15-100i \end{array}}{-2-88i}$$

D.2.1.3 Multiplication

The product of two complex numbers follows the distributive law.

The multiplication of a + bi and c + di will be

$$ac + a(di) + (bi)c + (bi)(di)$$

Therefore,

$$(7+10i) \times (4-9i)$$

will be

$$(7 \times 4) + (7 \times (-9i)) + (10i \times 4) + (10i \times (-9i))$$

$$28 - 63i + 40i - 90i^2$$

MATH CONCEPT

$i^2 = -1$

Therefore,

$$28 - 23i - 90(-1)$$

$$28 - 23i + 90$$

FINAL RESULT

$$118 - 23i$$

D.2.1.4 Division

The division of two complex numbers is made by multiplying both numbers by the conjugate of the second number.

The conjugate of a complex number is the same number with the imaginary part taken with the inverse sign. The conjugate of a + bi is a – bi.

Thus,

$$\frac{4+2i}{2-i}$$

$$\frac{(4+2i)(2+i)}{(2-i)(2+i)}$$

$$\frac{(4\times2)+(4\times i)+(2i\times2)+(2i\times i)}{(2\times2)+(2\times i)+(-i\times2)+(-i\times i)}$$

$$\frac{8+4i+4i+2i^2}{4+2i-2i-i^2}$$

> **MATH CONCEPT**
>
> $i^2 = -1$

$$\frac{8+4i+4i+2(-1)}{4+2i-2i-(-1)}$$

$$\frac{8+4i+4i-2}{4+2i-2i+1}$$

FINAL RESULT

$$\frac{6+8i}{5}$$

D.3 Polar Numbers

By observing Figure D.3, it is clear that number 3 + 2i can be seen as a vector with a certain magnitude R and an angle θ with the real axis, as shown in Figure D.4.

Figure D.4 Imaginary number represented as a vector.

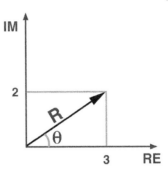

Figure D.5 Imaginary number – real and imaginary parts.

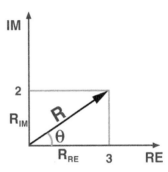

By trigonometry, the vector can be decomposed into real R_{RE} and imaginary R_{IM} parts as illustrated in Figure D.5.

REAL COMPONENT

$$R_{RE} = R\cos(\theta)$$

R_{RE} is the real component.
θ is the angle of R with the real axis.

IMAGINARY COMPONENT

$$R_{IM} = R\sin(\theta)$$

R_{IM} is the imaginary component.
θ is the angle of R with the real axis.

Therefore, the magnitude can be found by the following formula.

MAGNITUDE

$$R = \sqrt{R_{RE}^2 + R_{IM}^2}$$

R is the magnitude.
R_{RE} is the real component.
R_{IM} is the imaginary component.

where

$$\theta = A\tan\left(\frac{R_{IM}}{R_{RE}}\right)$$

This formula will always be true for angles, one positive and one negative, meaning two angles separated by 180°, or π. The best way to always find the correct value for θ is to use one of the two following formulas.

USE THIS FORMULA IF THE REAL PART IS NEGATIVE

$$\theta = A\tan\left(\frac{R_{IM}}{R_{RE}}\right) + \pi$$

θ is the angle of R with the real axis.
R_{RE} is the real component.
R_{IM} is the imaginary component.

USE THIS FORMULA IF THE REAL PART IS POSITIVE

$$\theta = A\tan\left(\frac{R_{IM}}{R_{RE}}\right)$$

θ is the angle of R with the real axis.
R_{RE} is the real component.
R_{IM} is the imaginary component.

Modern calculators have the function ATAN2 that calculates the angle correctly, because they take into consideration the real part sign.

Therefore, complex numbers can also be expressed in phasor form.

FORMULA

$$Z = \sqrt{R_{RE}^2 + R_{IM}^2} \quad \angle\theta°$$

Z is the number in polar form.
θ is the angle of R with the real axis.
R_{RE} is the real component.
R_{IM} is the imaginary component.

D.4 Trigonometric Numbers

Complex numbers can also be expressed as trigonometric numbers in the following form.

FÓRMULA

$$Z = R(\cos(\theta) + i\sin(\theta))$$

Z is the trigonometric number.
R is the magnitude.
θ is the angle of R with the real axis.

Appendix E

Table of Integrals

Integral	Result
$\int x^n dx$	$\dfrac{1}{n+1} x^{n+1}$
$\int \frac{1}{x} dx$	$\ln(x) \rightarrow x > 0$ $-\ln(x) \rightarrow x < 0$
$\int \dfrac{c}{ax+b} dx$	$\dfrac{c}{a} \ln|ax+b|$
$\int \dfrac{1}{(x+a)^2} dx$	$-\dfrac{1}{x+a}$
$\int \dfrac{1}{\sqrt{a-x}} dx$	$2\sqrt{a-x}$
$\int (x+a)^n dx$	$(x+a)^n \left(\dfrac{a}{1+x} + \dfrac{x}{1+n} \right)$ if $x \neq -1$
$\int \ln(x) dx$	$x \ln(x) - x$
$\int \ln(ax) dx$	$x \ln(ax) - x$
$\int \dfrac{\ln(ax)}{x} dx$	$\dfrac{1}{2}[\ln(ax)]^2$
$\int e^{ax} dx$	$\dfrac{1}{a} e^{ax}$
$\int x e^x dx$	$(x-1)e^x$

(Continued)

Introductory Electrical Engineering with Math Explained in Accessible Language, First Edition. Magno Urbano.
© 2020 John Wiley & Sons, Inc. Published 2020 by John Wiley & Sons, Inc.

Integral	Result
$\int xe^{ax}dx$	$\left(\dfrac{x}{a} - \dfrac{1}{a^2}\right)e^{ax}$
$\int xe^{-ax^2}dx$	$-\dfrac{1}{2a}e^{-ax^2}$
$\int e^x \sin(x)dx$	$\dfrac{1}{2}e^x[\sin(x) - \cos(x))$
$\int \sin(x)dx$	$-\cos(x)$
$\int \cos(x)dx$	$\sin(x)$
$\int \tan(x)dx$	$\ln|\sec(x)|$
$\int \tan(x)dx$	$-\ln[\cos(x)]$
$\int x\sin(x)dx$	$-x\cos(x) + \sin(x)$
$\int x\cos(x)dx$	$\cos(x) + x\sin(x)$
$\int \sin(x)\cos(x)dx$	$-\dfrac{1}{2}[\cos(x)]^2$
$\int \sin^2(x)dx$	$\dfrac{x}{2} - \dfrac{1}{4}\sin(2x)$
$\int \cos^2(x)dx$	$\dfrac{x}{2} + \dfrac{1}{4}\sin(2x)$
$\int \sinh(x)dx$	$\cosh(x)$
$\int \cosh(x)dx$	$\sinh(x)$

Appendix F

AWG Versus Metric System: Wire Cross Sections

Metric system (mm²)	Diameter (inches)	Diameter (mm)	Cross section (mm²)
0000 (4/0)	0.460	11.7	107
000 (3/0)	0.410	10.4	85.0
00 (2/0)	0.365	9.27	67.4
0 (1/0)	0.325	8.25	53.5
1	0.289	7.35	42.4
2	0.258	6.54	33.6
3	0.229	5.83	26.7
4	0.204	5.19	21.1
5	0.182	4.62	16.8
6	0.162	4.11	13.3
7	0.144	3.67	10.6
8	0.129	3.26	8.36
9	0.114	2.91	6.63
10	0.102	2.59	5.26
11	0.0907	2.30	4.17

(Continued)

Introductory Electrical Engineering with Math Explained in Accessible Language,
First Edition. Magno Urbano.
© 2020 John Wiley & Sons, Inc. Published 2020 by John Wiley & Sons, Inc.

Metric system (mm²)	Diameter (inches)	Diameter (mm)	Cross section (mm²)
12	0.0808	2.05	3.31
13	0.0720	1.83	2.63
14	0.0641	1.63	2.08
15	0.0571	1.45	1.65
16	0.0508	1.29	1.31
17	0.0453	1.15	1.04
18	0.0403	1.02	0.82
19	0.0359	0.91	0.65
20	0.0320	0.81	0.52
21	0.0285	0.72	0.41
22	0.0254	0.65	0.33
23	0.0226	0.57	0.26
24	0.0201	0.51	0.20
25	0.0179	0.45	0.16
26	0.0159	0.40	0.13

X/0 is used in the American Wire Gauge (AWG) system for wires with diameters greater than 1. The number X represents the number of zeros and pronounced like "aught." 3/0, for example, pronounced "three aught" means "three zeros," or AWG 000.

Appendix G

Resistors: Commercial Values

0.1%, 0.25%, and 0.5% resistors											
10	10.1	10.2	10.4	10.5	10.6	10.7	10.9	11	11.1	11.3	11.4
11.5	11.7	11.8	12	12.1	12.3	12.4	12.6	12.7	12.9	13	13.2
13.3	13.5	13.7	13.8	14	14.2	14.3	14.5	14.7	14.9	15	15.2
15.4	15.6	15.8	16	16.2	16.4	16.5	16.7	16.9	17.2	17.4	17.6
17.8	18	18.2	18.4	18.7	18.9	19.1	19.3	19.6	19.8	20	20.3
20.5	20.8	21	21.3	21.5	21.8	22.1	22.3	22.6	22.9	23.2	24.4
23.7	24	24.3	24.6	24.9	25.2	25.5	25.8	26.1	26.4	26.7	27.1
27.4	27.7	28	28.4	28.7	29.1	29.4	29.8	30.1	30.5	30.9	31.2
31.6	32	32.4	32.8	33.2	33.6	34	34.4	34.8	35.2	35.7	36.1
36.5	37	37.4	37.9	38.3	38.8	39.2	39.7	40.2	40.7	41.2	41.7
42.2	42.7	43.2	43.7	44.2	44.8	45.3	45.9	46.4	47	47.5	48.1
48.7	49.3	49.9	50.5	51.1	51.7	52.3	53	53.6	54.2	54.9	55.6
56.2	56.9	57.6	58.3	59	59.7	60.4	61.2	61.9	62.9	63.4	64.2
64.9	65.7	65.5	67.3	68.1	69	69.8	70.6	71.5	72.3	73.2	74.1
75	75.9	76.8	77.7	78.7	79.6	80.6	81.6	82.5	83.5	84.5	85.6
86.6	87.6	88.7	89.8	90.9	92	93.1	94.2	95.3	96.5	97.6	98.8

Introductory Electrical Engineering with Math Explained in Accessible Language,
First Edition. Magno Urbano.
© 2020 John Wiley & Sons, Inc. Published 2020 by John Wiley & Sons, Inc.

1% resistors											
10	10.2	10.5	10.7	11	11.3	11.5	11.8	12.1	12.4	12.7	13
13.3	13.7	14	14.3	14.7	15	15.4	15.8	16.2	16.5	16.9	17.4
17.8	18.2	18.7	19.1	19.6	20	20.5	21	21.5	22.1	22.6	23.2
23.7	24.3	24.9	25.5	26.1	26.7	27.4	28	28.7	29.4	30.1	30.9
31.6	32.4	33.2	34	34.8	35.7	36.5	37.4	38.3	39.2	40.2	41.2
42.2	43.2	44.2	45.3	46.4	47.5	48.7	49.9	51.1	52.3	53.6	54.9
56.2	57.6	59	60.4	61.9	63.4	64.9	66.5	68.1	69.8	71.5	73.2
75	76.8	78.7	80.6	82.5	84.5	86.6	88.7	90.9	93.1	95.3	97.6

5% resistors											
10	11	12	13	15	16	18	20	22	24	27	30
33	36	39	43	47	51	56	62	68	75	82	91

10% resistors											
10	12	15	18	22	27	33	39	47	56	68	82
100	120	150	180	220	270	330	390	470	560	680	820
1000	1200	1500	1800	2200	2700	3300	3900	4700	5600	6800	8200
10000	12000	15000	18000	22000	27000	33000	39000	47000	56000	68000	82000
100000	120000	150000	180000	220000	270000	330000	390000	470000	560000	680000	820000

20% resistors					
10	15	22	33	47	68
100	150	220	330	470	680
1000	1500	2200	3300	4700	6800
10000	15000	22000	33000	47000	68000
100000	150000	220000	330000	470000	680000

Appendix H

Capacitors: Commercial Values

Electrolytic capacitors									
0.1 µF	12 µF	60 µF	220 µF	460 µF	1000 µF	3500 µF	7400 µF	25000 µF	62000 µF
0.15 µF	15 µF	68 µF	230 µF	470 µF	1100 µF	3600 µF	7600 µF	26000 µF	66000 µF
0.22 µF	16 µF	72 µF	233 µF	480 µF	1200 µF	3700 µF	7800 µF	27000 µF	68000 µF
0.33 µF	18 µF	75 µF	240 µF	500 µF	1300 µF	3900 µF	8200 µF	28000 µF	76000 µF
0.47 µF	20 µF	82 µF	243 µF	510 µF	1400 µF	4000 µF	8300 µF	30000 µF	0.1 F
0.68 µF	21 µF	88 µF	250 µF	520 µF	1500 µF	4100 µF	8400 µF	31000 µF	0.11 F
1 µF	22 µF	100 µF	270 µF	540 µF	1600 µF	4200 µF	8700 µF	32000 µF	0.12 F
1.5 µF	24 µF	108 µF	300 µF	550 µF	1700 µF	4300 µF	9000 µF	33000 µF	0.15 F
2 µF	25 µF	120 µF	320 µF	560 µF	1800 µF	4600 µF	9600 µF	34000 µF	0.22 F
2.2 µF	27 µF	124 µF	324 µF	590 µF	2000 µF	4700 µF	10000 µF	36000 µF	0.33 F

(Continued)

Introductory Electrical Engineering with Math Explained in Accessible Language,
First Edition. Magno Urbano.
© 2020 John Wiley & Sons, Inc. Published 2020 by John Wiley & Sons, Inc.

Electrolytic capacitors									
3 µF	30 µF	130 µF	330 µF	620 µF	2100 µF	4800 µF	11000 µF	37000 µF	0.47 F
3.3 µF	33 µF	140 µF	340 µF	645 µF	2200 µF	5000 µF	12000 µF	38000 µF	0.666 F
4 µF	35 µF	145 µF	350 µF	650 µF	2500 µF	5100 µF	13000 µF	39000 µF	
4.7 µF	36 µF	150 µF	370 µF	680 µF	2600 µF	5400 µF	15000 µF	40000 µF	
5 µF	39 µF	161 µF	378 µF	700 µF	2700 µF	5500 µF	16000 µF	41000 µF	
5.6 µF	40 µF	170 µF	380 µF	708 µF	2800 µF	5600 µF	17000 µF	47000 µF	
6.8 µF	43 µF	180 µF	390 µF	730 µF	2900 µF	5800 µF	18000 µF	48000 µF	
7 µF	47 µF	189 µF	400 µF	800 µF	3000 µF	6000 µF	20000 µF	50000 µF	
8 µF	50 µF	200 µF	420 µF	820 µF	3100 µF	6500 µF	22000 µF	55000 µF	
8.2 µF	53 µF	210 µF	430 µF	850 µF	3300 µF	6800 µF	23000 µF	56000 µF	
10 µF	56 µF	216 µF	450 µF	860 µF	3400 µF	7200 µF	24000 µF	60000 µF	

Ceramic and other types of capacitors							
Code	µF	nF	pF	Code	µF	nF	pF
100K	0.00001	0.01	10	339K	0.0000033	0.0033	3.3
101K	0.0001	0.1	100	390K	0.000039	39	39
102K	1	1	1000	391K	0.00039	0.39	390
102K	1	1	1000	392K	0.0039	3.9	3900
103K	0.01	10	10000	393K	39	39	39000
104K	0.1	100	100000	394K	0.39	390	390000

Ceramic and other types of capacitors							
Code	μF	nF	pF	Code	μF	nF	pF
105K	1	1000	1000000	399K	0.0000039	0.0039	3.9
109K	0.000001	1	1	400K	0.00004	0.04	40
120K	0.000012	12	12	401K	0.0004	0.4	400
121K	0.00012	0.12	120	402K	4	4	4000
122K	0.0012	1.2	1200	403K	0.04	40	40000
123K	12	12	12000	404K	0.4	400	400000
124K	0.12	120	120000	409K	0.000004	4	4
129K	0.0000012	0.0012	1.2	470K	0.000047	47	47
150K	0.000015	15	15	471K	0.00047	0.47	470
151K	0.00015	0.15	150	472K	0.0047	4.7	4700
152K	0.0015	1.5	1500	473K	47	47	47000
153K	15	15	15000	474K	0.47	470	470000
154K	0.15	150	150000	479K	0.0000047	0.0047	4.7
159K	0.0000015	0.0015	1.5	500K	0.00005	0.05	50
180K	0.000018	18	18	501K	0.0005	0.5	500
181K	0.00018	0.18	180	502K	5	5	5000
182K	0.0018	1.8	1800	503K	0.05	50	50000
183K	18	18	18000	504K	0.5	500	500000
184K	0.18	180	180000	509K	0.000005	5	5
189K	0.0000018	0.0018	1.8	560K	0.000056	56	56
200K	0.00002	0.02	20	561K	0.00056	0.56	560
201K	0.0002	0.2	200	562K	0.0056	5.6	5600

(Continued)

Ceramic and other types of capacitors							
Code	μF	nF	pF	Code	μF	nF	pF
202K	2	2	2000	563K	56	56	56000
203K	0.02	20	20000	564K	0.56	560	560000
204K	0.2	200	200000	569K	0.0000056	0.0056	5.6
209K	0.000002	2	2	600K	0.00006	0.06	60
220K	0.000022	22	22	601K	0.0006	0.6	600
221K	0.00022	0.22	220	602K	6	6	6000
222K	0.0022	2.2	2200	603K	0.06	60	60000
223K	22	22	22000	604K	0.6	600	600000
224K	0.22	220	220000	609K	0.000006	6	6
229K	0.0000022	0.0022	2.2	680K	0.000068	68	68
250K	0.000025	25	25	681K	0.00068	0.68	680
251K	0.00025	0.25	250	682K	0.0068	6.8	6800
252K	0.0025	2.5	2500	683K	68	68	68000
253K	25	25	25000	684K	0.68	680	680000
254K	0.25	250	250000	689K	0.0000068	0.0068	6.8
259K	0.0000025	0.0025	2.5	700K	0.00007	0.07	70
270K	0.000027	27	27	701K	0.0007	0.7	700
271K	0.00027	0.27	270	702K	7	7	7000
272K	0.0027	2.7	2700	703K	0.07	70	70000
273K	27	27	27000	704K	0.7	700	700000
274K	0.27	270	270000	709K	0.000007	7	7
279K	0.0000027	0.0027	2.7	800K	0.00008	0.08	80
300K	0.00003	0.03	30	801K	0.0008	0.8	800

Ceramic and other types of capacitors							
Code	μF	nF	pF	Code	μF	nF	pF
301K	0.0003	0.3	300	802K	8	8	8000
302K	3	3	3000	803K	0.08	80	80000
303K	0.03	30	30000	804K	0.8	800	800000
304K	0.3	300	300000	809K	0.000008	8	8
309K	0.000003	3	3	820K	0.000082	82	82
330K	0.000033	33	33	821K	0.00082	0.82	820
331K	0.00033	0.33	330	822K	0.0082	8.2	8200
332K	0.0033	3.3	3300	823K	82	82	82000
333K	33	33	33000	824K	0.82	820	820000
334K	0.33	330	330000	829K	0.0000082	0.0082	8.2

Appendix I

Inductors: Commercial Values

Inductors (pH)	
100	500
150	560
180	600
200	670
220	680
270	700
300	780
330	800
390	820
400	900
470	

Inductors (nH)									
1	4.7	8.7	20	50	93	157	247	380	590
1.1	4.8	8.8	20.8	51	94	160	250	386	592
1.15	4.9	8.9	21	52	95	163	252	387	600

(Continued)

Introductory Electrical Engineering with Math Explained in Accessible Language,
First Edition. Magno Urbano.
© 2020 John Wiley & Sons, Inc. Published 2020 by John Wiley & Sons, Inc.

Inductors (nH)									
1.2	5	9	21.5	53	96	164	256	390	604
1.3	5.1	9.1	21.7	54	97	165	257	397	610
1.4	5.2	9.2	22	55	98	166	260	400	620
1.5	5.3	9.3	23	55.5	98.5	169	262	405	624
1.6	5.4	9.4	23.5	56	98.7	170	264	410	630
1.65	5.45	9.5	24	58	99	175	269	416	640
1.7	5.5	9.6	25	59	100	178	270	420	650
1.8	5.6	9.7	26	59.5	101	179.6	272	422	660
1.9	5.7	9.8	26.5	60	102	180	276	430	670
2	5.8	9.85	27	61	105	182	280	434	674
2.1	5.9	9.9	27.3	61.5	106	185	282	435	680
2.2	6	10	28	62	108	187	285	440	690
2.3	6.1	10.2	29	64	110	188	288	442	700
2.4	6.2	10.4	30	65	111	190	290	450	708
2.5	6.3	11	31	66	114	191	292	460	709
2.55	6.4	11.2	32	67	115	192	298	463	720
2.6	6.5	11.9	32.75	67.8	117	194	300	465	730
2.7	6.6	12	33	68	119.7	195	307	468	740
2.8	6.68	12.1	34	69	120	196	310	470	750
2.9	6.7	12.3	35	70	122	197	311	477	760
3	6.8	12.5	35.5	71	123	200	315	480	764
3.1	6.9	12.55	36	72	125	202	317	490	770
3.2	7	13	37	73	126	205	320	491	780
3.3	7.1	13.7	37.5	74	130	206	322	500	783
3.32	7.15	13.8	38.5	75	132	207	325	505	790

Inductors (nH)									
3.4	7.2	14	39	76	134	210	330	509	800
3.5	7.3	14.7	40	77	135	215	331	510	814
3.6	7.4	14.9	41	78	139	216	333	520	820
3.7	7.5	15	42	79	140	217	337	523	845
3.8	7.6	15.7	43	80	141	220	340	530	846
3.85	7.7	16	43.5	82	144	222	342	538	850
3.9	7.8	16.6	44	82.7	145	223	344	540	870
4	7.9	17	44.5	83	146	225	350	545	880
4.1	8	17.5	45	84	149	226	360	548	896
4.2	8.1	17.8	46	85	149.4	230	363	550	900
4.3	8.2	17.95	46.5	86	150	240	365	556	908
4.4	8.3	18	46.6	88	152	241	369	560	909.5
4.5	8.4	18.5	47	90	154	242	370	570	910
4.55	8.5	19	48	91	155	245	375	571	950
4.6	8.6	19.4	49	92	156	246	378	580	978

Inductors (µH)									
1	3.3	7.8	12.2	22	38.07	69.7	125	252	521
1.01	3.32	7.8	12.3	22.3	38.3	69.77	128	253	530
1.02	3.35	7.9	12.5	22.39	39	70	130	256	546
1.06	3.4	7.94	12.5	22.5	39.5	70.05	131	258	550
1.08	3.5	8	12.6	22.6	39.9	70.5	131.8	260	560
1.09	3.6	8	12.7	22.7	40	71.1	132	262	561
1.1	3.7	8.06	12.8	22.8	41	73	133.2	264.7	562

(*Continued*)

Inductors (µH)									
1.15	3.76	8.1	12.9	23	41.1	73.7	134.2	270	570
1.18	3.8	8.17	13	23.3	41.5	75	137.5	272	575
1.2	3.9	8.2	13.1	23.7	42	76.8	140	275	579.4
1.23	4	8.23	13.2	24	43	77	143	285	580
1.25	4.1	8.3	13.3	24.	43.3	77.5	143.5	290	590
1.26	4.2	8.32	13.5	24.2	44	78	145	296.7	595.7
1.27	4.23	8.4	13.9	24.31	45	78.4	146	300	600
1.3	4.3	8.45	14	24.4	46	79	148	303.3	602.5
1.32	4.4	8.5	14.2	25	46.3	80	148.5	305	620
1.33	4.5	8.6	14.3	25.7	46.7	80.75	150	307	620
1.4	4.6	8.7	14.6	26	46.8	82	151	316	630
1.45	4.7	8.8	14.67	26.2	47	82.2	156	320	640
1.47	4.74	8.8	14.7	27	47.1	83	159.65	324	645
1.5	4.76	8.83	14.75	27.1	47.4	84	160	325	648
1.58	4.8	8.84	14.8	27.16	47.7	85	160.6	325	650
1.6	4.9	8.86	14.9	27.4	49.7	86	165	325.1	656.2
1.65	4.95	8.9	15	27.9	50	87	166	330	670
1.68	5	9	15.14	28	51	88	167	332	680
1.7	5.1	9.1	15.4	28.2	51.2	88.2	167.2	335	682.8
1.75	5.2	9.2	15.5	28.4	51.9	89.5	168	350	696
1.7929	5.3	9.3	15.8	29	52	89.6	170	357	700
1.8	5.4	9.4	15.9	29.1	52.5	89.7	173	360	749
1.83	5.5	9.5	16	29.5	52.6	90	175	370	750
1.84	5.6	9.6	16.1	29.7	53	91	176	372	760

Inductors (µH)									
1.87	5.7	9.62	16.2	29.8	54	92	176.1	374	770
1.875	5.78	9.65	16.3	30	54.2	92.5	176.4	375	780
1.9	5.8	9.67	16.5	30.6	54.4	93	178	377	781
1.98	5.9	9.7	16.6	31	54.81	95	180	380	786
2	6	9.8	16.7	31.8	55	96	181	383	790
2.02	6.1	9.9	16.76	31.84	55.2	97	187.4	390	800
2.03	6.2	10	16.8	32	55.23	97.7	190	399	807
2.1	6.23	10	17	32.5	55.4	98	192	400	810
2.15	6.3	10.1	17.6	32.6	55.6	98.8	197	402	820
2.16	6.4	10.2	18	32.8	56	99.3	200	417.4	830
2.17	6.4	10.27	18.2	32.9	57	99.5	201.6	420	832
2.2	6.5	10.4	18.3	33	57.8	100	203.7	426.3	836
2.3	6.6	10.5	18.5	33.15	58	101	205	429.6	849
2.32	6.63	10.6	18.64	33.2	58.1	101.7	207.2	430	850
2.4	6.7	10.68	18.85	33.3	60	102	209.6	439	860
2.42	6.8	10.7	19	33.33	61	105	210	440	880
2.45	6.9	10.79	19.4	33.7	61.9	107.5	211	450	900
2.5	7	10.9	19.5	35	62	110	216.7	450	910
2.54	7.03	11	19.7	35.2	63.2	111.6	217	458.7	915
2.6	7.1	11.1	20	35.3	64	112	218.8	460	940
2.7	7.2	11.2	20.5	35.48	64.2	113.2	220	470	950
2.76	7.22	11.3	20.8	35.5	65	114	223	471	955
2.8	7.28	11.3	20.9	35.6	65.04	115	224	472	960

(Continued)

Inductors (µH)									
2.84	7.3	11.4	21	36	66.5	116	225	476	980
2.9	7.4	11.5	21.47	36.6	66.89	118	230	480	
3	7.5	11.6	21.5	37	67	120	231	497	
3.05	7.5	11.8	21.6	37.5	67.6	121	240	500	
3.1	7.6	11.9	21.65	37.6	68	122.4	245.8	501	
3.2	7.63	12	21.8	37.9	68.22	122.5	248	510	
3.22	7.7	12	21.9	38	68.37	123	250	518	

Inductors (mH)									
1	1.8	3.1	5.3	8.3	13.2	21.5	37.8	61	142
1	1.82	3.2	5.4	8.4	13.4	22	38	62	150
1.0017	1.9	3.3	5.5	8.8	13.6	23	39	65.6	158
1.008	1.95	3.4	5.6	9	14	23.7	40	66	180
1.03	1.97	3.5	5.7	9.1	14.5	24	40.5	67.5	190
1.05	2	3.6	5.85	9.2	14.6	24.5	40.8	68	200
1.07	2.1	3.7	5.9	9.3	14.8	25	41	70	208
1.08	2.2	3.8	6	9.35	15	25.5	41.5	70.3	220
1.1	2.2	3.9	6.1	9.4	15.1	26	42	75	250
1.134	2.204	4	6.2	9.5	15.4	26.2	42.5	76	265
1.17	2.25	4.1	6.4	9.6	15.5	26.5	43.6	77	270
1.186	2.3	4.15	6.5	9.8	16	26.7	44	78	300
1.2	2.35	4.2	6.6	9.9	16.2	27	45	82	320
1.203	2.36	4.25	6.7	10	16.4	27.7	46.7	82.5	330
1.25	2.37	4.3	6.8	10.2	16.9	28	47	90	370
1.3	2.38	4.4	6.9	10.5	17	28.4	48	91	500

Inductors (mH)									
1.31	2.4	4.5	7	10.8	17.1	28.6	49	92	560
1.32	2.48	4.6	7.1	10.9	17.7	29	50	92.5	600
1.34	2.5	4.66	7.2	11	17.8	29.6	50.7	93	680
1.4	2.55	4.7	7.3	11.2	18	30	51	100	900
1.47	2.6	4.75	7.4	11.4	18.4	32	52	105	
1.48	2.61	4.8	7.5	11.5	18.5	32.4	53.5	110	
1.499	2.7	4.81	7.6	12	18.52	33	54	115.7	
1.5	2.8	4.9	7.7	12.2	19	34	55	120	
1.58	2.88	4.91	7.8	12.3	19.4	35	56	125	
1.6	2.89	5	8	12.5	19.6	36	57	135	
1.62	2.9	5.1	8.1	12.8	20	37	59.6	138	
1.7	3	5.2	8.2	13	21.2	37.7	60	140	

Inductors (H)			
1	3	8	20
1.5	3.5	9	30
2	4	10	40
2.2	5	12	60
2.5	6	14	150
2.6	7	15	

Appendix J

Simulation Tools

The lists that follow show the best simulation tools you will find around, in our opinion, that can help you create, test, and simulate electronic devices. The list is in alphabetic order. The bold ones are the ones we prefer:

Online simulators
www.circuitlab.com
www.everycircuit.com
www.partsim.com/simulator
www.systemvision.com

Software (Windows and Mac)
www.analog.com/en/design-center/design-tools-and-calculators/ltspice-simulator.html
www.autodesk.com/products/eagle/free-download
www.circuitmaker.com

Introductory Electrical Engineering with Math Explained in Accessible Language,
First Edition. Magno Urbano.
© 2020 John Wiley & Sons, Inc. Published 2020 by John Wiley & Sons, Inc.

Appendix K

Glossary

Glossary	
Ω	Omega is the symbol used to express resistance, in Ohms.
A	Ampere is the unit used to express current.
AC	Alternating current is a kind of current that has its value changing cyclically over time according to a sinusoidal function.
Admittance	In electrical engineering, admittance is a measure of how easily a circuit or device will allow a current to flow. It is defined as the reciprocal of impedance.
Anode	An anode is the positively charged electrode of a device.
C	Coulomb is the unit used to express electric charge.
Capacitance	Capacitance is the ability of a body to store potential energy in the form of electrical charge, electrical field.
Capacitor	Is a passive device that can store potential energy in the form or electric field. See capacitance.
Cathode	A cathode is the negatively charged electrode of a device.
DC	Direct current is a kind of current that keeps its value constant over time.
DC bias	See DC offset.
DC offset	Is a mean amplitude displacement in AC voltage by the existence or a spurious DC component added to it.

(Continued)

Introductory Electrical Engineering with Math Explained in Accessible Language,
First Edition. Magno Urbano.
© 2020 John Wiley & Sons, Inc. Published 2020 by John Wiley & Sons, Inc.

Glossary	
Dielectric	A dielectric is an electrical insulator that can be polarized by an applied electric field.
Diode	Is a semiconductor device that can conduct current in one direction only.
Electrolytic capacitor	Is a polarized capacitor that can, among other things, have high capacitance values without increasing too much in size. See capacitor.
F	Farad is the unit used to express capacitance. See capacitor.
g	Gram is the unit used to express mass.
H	Henry is the unit used to express inductance. See inductor.
Hz	Hertz is the unit used to express frequency.
Impedance	The effective resistance of an electric circuit or component to alternating current, arising from the combined effects of ohmic resistance and reactance.
Inductance	Inductance is the ability of a body to store potential energy in the form of magnetic field. See inductor.
Inductor	Is a passive device that can store potential energy in the form or magnetic field. See inductance.
J	Joule is the unit used to express energy.
m	Meter is the unit used to express distance.
N	Newton is the unit used to express force.
Permittivity	Permittivity is the measure of a material's ability to store an electric field in the polarization of the medium.
Phase	Is a difference in angle between two alternating signals.
Potentiometer	Is a three terminal variable resistor. See resistor.
Resistance	Is the property of a body that opposes the flow of current.
Resistor	Is a passive device that resists current flow.
S	Siemens is the unit used to express admittance.
Transformer	Is a device that can be used to reduce or to increase the voltage of an alternating current signal.

Glossary
Transistor	Is an active semiconductor device that can be used to perform many functions like amplification, rectification, switching, etc.
Trimmer	Is a miniature adjustable component that can be a variable potentiometer, a variable capacitor, or a variable inductor. See resistor, capacitor, and inductor.
Trimpot	Is a miniature adjustable potentiometer. See potentiometer.
V	Volt is the unit used to express electric potential.
VA	Volt-ampere is the unit used to express apparent power.
VAR	Volt-ampere reactive is the unit used to express reactive power.
W	Watt is the unit used to express real or active power.

Index

Introductory Electrical Engineering with Math Explained in Accessible Language,
First Edition. Magno Urbano.
© 2020 John Wiley & Sons, Inc. Published 2020 by John Wiley & Sons, Inc.